Wine Regions of the World

By the same authors

The New Wine Companion

Wine Regions of the World

Second Edition

David Burroughs and
Norman Bezzant

Published on behalf of the
Wine and Spirit Education Trust

Winner of the 1980 International Promotion Prize (Books Category) given by the Centre Internationale de Liaison des Organismes de Propagande en faveur des Produits de la Vigne, Paris

Heinemann Professional Publishing

Heinemann Professional Publishing Ltd
Halley Court, Jordan Hill, Oxford OX2 8EJ

OXFORD LONDON MELBOURNE AUCKLAND SINGAPORE
IBADAN NAIROBI GABORONE KINGSTON

First published 1979
Reprinted 1981, 1984, 1985, 1986, 1987
Second edition 1988
Reprinted 1989, 1990

British Library Cataloguing in Publication Data
Burroughs, David
Wine regions of the world. – 2nd ed.
1. Wines
I. Title II. Bezzant, Norman III. Wine
and Spirit Education Trust
641.2'2

ISBN 0 434 90174 1

Maps by Reginald Piggott
Drawings by Diane Tippel and Martin Cottam

Typeset by TecSet Ltd., Wallington, Surrey

Printed in Great Britain by
Billing & Sons Ltd, Worcester

Contents

List of maps	*viii*
Introduction to the First Edition by David J. B. Rutherford	*x*
Introduction to the Second Edition by Grahame D. McKenzie	*xii*
Authors' Preface to the First Edition	*xiii*
Authors' Preface to the Second Edition	*xv*
Part One – Vines and Wines	1
1 **Viticulture**	3
2 **Vinification**	18
Part Two – Wines of the European Community	39
Introduction to Part Two	41
3 **France – General**	43
South-West France	48
Northern France	71
Central and Eastern France	93
Mediterranean France	112
4 **Germany – General**	135
German Quality Wine Regions	145
5 **Italy – General**	161
Northern Italy	168
Central and Southern Italy	178
6 **Spain – General**	190
Spanish Quality Wine Regions	194
7 **Portugal**	211
8 **Other Community Wines**	228
Greece	229
Luxembourg	231
England and Wales	232
9 **Aromatized Wines**	238

Part Three – Wines of Third Countries 243
10 Central and South Eastern Europe 245
Switzerland 246
Austria 248
Czechoslovakia 251
Yugoslavia 252
Hungary 254
Romania 257
Bulgaria 258
11 The Levant and North-West Africa 260
The Levant 260
North-West Africa 265
12 North America 269
The United States 270
California 271
North-Western States 277
North-Eastern States 277
Canada 278
Mexico 279
13 The Southern Hemisphere 281
Australia 282
New Zealand 286
The Cape 288
South America 295

Part Four – Spirits 299
Introduction to Part Four 301
14 Fruit Spirits 302
15 Grain Spirits 315
16 Vegetable Spirits 328
17 Flavoured Spirits 333

Part Five – Legal Aspects 339
18 Effects on the British Trade of National and EEC Legislation 341

Appendixes

1 Notes on EEC Directives and Legislation 355
2 The Annual Cycle of the Vine and Vineyard Work 362
3 Maladies of the Vine 366
4 Maladies of Wine, their Symptoms, Prevention and Cure 370
5 Bibliography 374

Glossary 379
Index 386

Maps

EEC Wine-zones	27
General Map of France	44
Bordeaux and Neighbouring Regions	49
Bordeaux (1)	55
Bordeaux (2)	59
Bordeaux (3)	67
The Heart of Champagne	75
Loire (1)	82
Loire (2)	84
Loire (3)	87
Alsace	90
Burgundy (1)	97
Burgundy (2)	98
Burgundy (3)	101
Burgundy (4)	106
Côtes du Rhône (north)	115
Côtes du Rhône (south)	118
Provence	122
Central Southern France	126
Rhineland	139
Rheingau	146
Rheinhessen	148
Rheinpfalz	151
Nahe	152
Mosel-Saar-Ruwer	154
General Map of Italy	162
North-West Italy	170
North-East Italy	174
Central Italy	180
The Iberian Peninsula	191

Rioja 195
Cataluña 198
Levante 202
Sherry 204
Madeira 216
The Douro 221
The Principal 'Districts' of English Wines 233
Central and South-Eastern Europe 246
Central Europe 249
Cyprus 261
California 272
South-East Australia 284
New Zealand 287
Cognac 303
Scotch Whisky 316

Tables

1 *Sparkling Wine Styles* 79
2 *Planting of Grape Varieties in West Germany by QbA regions, in percentages* 138
3 *Elements of German labelling regulations* 142
4 *Spanish Denominaciónes de Origen* 192
5 *Elements governing authorization of Corgo vineyards* 224

Introduction to the First Edition

by
David J. B. Rutherford
(Chairman of Trustees: Wine and Spirit Education Trust, 1980 to 1984)

Never in its 3000 years of recorded history has the wine industry faced such dramatic change and such world-wide interest as it does today. Until the Second World War, each of the wine-producing countries was master of its own destiny; not only were there no groups of trading nations, but there were few countries whose governments monitored the quality and quantity of production of vineyards, wineries, and distilleries. However, the Common Market and the cult of the consumer have changed all this.

Realizing this, the Wine and Spirit Association of Great Britain, with financial help from the Worshipful Company of Vintners, formed the Wine and Spirit Education Trust from the chrysalis of their Education Subcommittee. The objects of the Trust, governed by the two bodies mentioned with the Worshipful Company of Distillers and the Institute of Masters of Wine, are to educate entrants to the industry and to upgrade the knowledge of those already in it – particularly in the retail sector.

Today, over 5000 students sit each year for the Trust's examinations, at Certificate, Higher Certificate and Diploma levels. Three years ago, the Trust commissioned *The Wine Trade Student's Companion*★ as a guide to learning, setting out for beginners the fundamentals of growing grapes, making wines, and distilling spirits. Now, following the success of the *Companion*, the Trust

★ Burroughs and Bezzant. Republished as *The New Wine Companion* (London: Heinemann, 1980).

has commissioned the same authors to write this book for more advanced students. *Wine Regions of the World* explains the modern scene in greater depth and explores the wine- and spirit-producing regions of the world in a way that will not only encourage the student to learn, but must also be of help and interest to every wine and spirit lover.

One theme running through the book is the importance placed by the Trust on a responsible attitude to the use of alcohol. Sensible drinking is healthful and praiseworthy, but abuse by excessive consumption can only lead to personal misery and a bad name for the trade.

This book, which contains a wealth of interesting facts and information on practically all wines or spirits the wine lover is likely to meet, should be a 'must' for every wine library.

Introduction to the Second Edition

by
Grahame D. R. McKenzie
(Chairman of Trustees: Wine and Spirit Education Trust,
1984 to 1988)

During the nine years since *Wine Regions of the World* was first published, it has been the manual for over twenty thousand Higher Certificate students in the trade. The number of students taking the Trust's examinations annually has increased since 1979 by over 60 per cent, and the book has been reprinted over six times with only minor corrections; that so few have been required is a credit to its authors.

However, recent years have seen great changes in the European Community and technical advances in the production and marketing of wines and spirits; these variations have been reflected in changes to the Trust's syllabuses and incorporated in this new edition of *Wine Regions of the World*. Several new maps and illustrations and two new chapters on viticulture and vinification have been added and the whole book has been brought up to date.

As an enjoyable work of reference, and particularly for the footnote questions and answers which run through the book, I warmly commend this second edition not only to all students, but also to the wine and spirit trade and the general public.

Authors' Preface to the First Edition

As with our other publication, *The New Wine Companion*, the student will get the best value from *Wine Regions of the World* if he or she understands how it is intended to be read, and in this there is a basic difference between the two works. In *The New Wine Companion*, each chapter progressively builds up the understanding necessary to the next chapter; thus the reader first understands the essential part played by nature in determining the world's wine regions; there follows the culture of grapes, and their conversion into wine involving the chemical conversion of sugar to alcohol; and so on through the distillation of spirits to the flavoured and sweetened drinks called liqueurs.

Wine Regions of the World has two major objectives. It outlines (in its Introduction to Part Two and Chapter 18) the more important intricacies of EEC and British law affecting the wine and spirit industry. And secondly it treats in turn each of the main wine- and spirit-producing regions of the world, its history, its products and the manner of their making. But each chapter is separate: to learn about the wines of Puglia or Rioja, for instance, it is not necessary first to have read the chapters on the French wine regions. Nevertheless, the book does introduce the world's regions roughly in their order of importance to the British trade, and so a continuous narrative again unfolds.

The student should therefore read straight through the book quickly to see how the whole hangs together; *Wine Regions of the World* should then be studied in detail, and with each chapter the student should read as many as can be readily obtained of the works mentioned in the bibliography relating to that chapter.

In passing, we would caution readers not to confuse 'Third Countries' – an official term to identify countries which are not members of the EEC – with 'The Third World', a common term to describe countries not included in the NATO or COMECON blocs.

As in the *Companion*, there is a running series of questions at the foot of each right-hand page to which the answers are given at the foot of the following left-hand page. These questions assist recall at the time of reading, and are a valuable aid to revision before examinations.

The authors again acknowledge the help given by experts from many countries, and by the personnel of the Wine and Spirit Education Trust, in bringing together so much information in palatable form. Above all the authors hope that those who read this book will enjoy it, for 'if learnin' ain't fun, people won't'.

Authors' Preface to Second Edition

A book about wine, once written, reflects a point in time. Once published, it is history, for many months must elapse in its production, and the world does not stand still to await publication. Not only does vintage succeed vintage, each different in quality and quantity, but changes in the market, consumer tastes and Regulations also occur. So that authors of books such as this one might well despair.

Nevertheless we do *not* despair, and urge our student readers to accept this edition as our view of the summer of 1987. Between 'sealing' our work and writing this preface we have seen the appearance of four new Appellations Controlées of the Corbières to mention only one of the edicts which flow continually from the pens of INAO. France, always in our minds as the leader in wine matters, is only indicative – changes are taking place all over the world.

It must therefore be the responsibility of the student to update the information in this book with the aid of the wine trade press – particularly that of wine-producing countries, new publications by other authors, and their own observations. Until the next edition arrives!

PART ONE

Vines and Wines

1
Viticulture

The vine which produces fruit for making into wine is a deciduous climbing shrub of the family *Ampelidaceae*, genus *Euvitis*, and species *Vitis Vinifera*. Other species of the sub-genus Vitis also produce fruit: *V. Labrusca, V. Rupestris, V. Riparia,* and *V. Berlandieri* are generally referred to as 'American Vines' from their origin, but the wine made from their fruit has a pungent flavour often described as 'foxy'. They are, however, essential in providing rootstocks on to which scions of *V. Vinifera* may be grafted, as their roots are resistant to attacks of *Phylloxera* and nematodes, while the roots of *V. Vinifera* are not.

If you have not read Authors' Preface on pages xiii to xv, you should do so now: it will help you to use the book to your best advantage.

1 Why will I find the answer to this question at the foot of the next page?

2 If the answer to a question cannot be found in the preceding text, where else could it be found?

The aim of viticulture for wine-making is to produce a crop of fruit with sufficient sugar, acids, proteins, and other compounds to yield stable wine with pleasant flavour characteristic of the region; within the natural constraints of climate and soil of their region and the legal constraints of local, national, and international regulations, growers can vary their rootstocks, planting density, grape varieties, and treatment of the soil and of the vine.

New varieties are continually being produced by crossing. This process involves first removing all the male parts from the flower of one of the two varieties to be crossed (because the vine is self-fertilizing), and then fertilizing the female parts remaining with pollen from the other variety. The grapes resulting from this union yield pips which may be grown into vines containing the genes of both the original varieties. It is a long process – the variety Müller-Thurgau was first produced by crossing a century ago, but has only in recent years attained full recognition – though just as unreliable as the mating of humans to produce the ideal.

Not only are *V. Vinifera* crosses, such as *Riesling* × *Silvaner*, possible, but also crosses between different species. For instance, the cross *Berlandieri* × *Rupestris* yields a number of varieties of rootstock which satisfy different climatic and soil conditions. Crosses between species are called hybrids and hybrids of *V. Vinifera* with other species are known as 'direct producers'; many experiments have taken place to evolve the ideal direct producer vine which can resist phylloxera and yet produce good wine, but even though crossed and recrossed until the non-*vinifera* content is less than one-eighth, the wine is no more than ordinary. For this reason, the wine of direct producers is not generally admissible in the EEC. Control is made easier by the fact that wine from direct producers (and other hybrids) is detectable by the presence of the diglucoside compounds which do not occur in wine from pure *V. Vinifera*.

New clonal varieties can also be created by grafting a bud of one variety of *V. Vinifera* on to the stock of another; pips from the fruit can then be grown to fix the clone.

Varieties may also change their character naturally after transplantation from one region to another: the Traminer vine, originating from the village of Tramin in the region of Alto-Adige (Süd-Tirol) in Italy, was transplanted to the region of Alsace in France. It was

1 Because all questions appearing as footnotes are answered at the foot of the following page.
2 In the Appendices or Glossary, which commence on page 355, or in *The New Wine Companion*.

then noticed that the wine from some vines had a spicier quality than most Traminer; the canes from these vines were chosen for grafting when planting new vines, and where it was noted that the characteristic of the scion's parent persisted these clones became established as a new variety, which was called Gewürztraminer (Spicy Traminer). Gradually, this variety excluded the original, which is now no longer permitted in Alsace. Crosses between different clones of the same variety, for example *Riesling* × *Riesling*, provide new varieties also.

It is not yet fully clear whether climatic or geological factors are responsible for the initial clonal separation, but it is clear that transplantation is a cause, as evidenced by the different wines produced from clones of Syrah, Chenin Blanc, and Cinsaut after transplantation to the Southern Hemisphere. While Trebbiano (Ugni Blanc) and Pinot are subject to clonal variation, Cabernet Sauvignon and Chardonnay seem less so, retaining their original characteristics wherever they are planted.

CLIMATE

The ideal climate for the wine-producing vine is found between the latitudes of 30° and 50° North, and 30° and 40° South – on the equatorial side of the temperate zones in each hemisphere, where the winter temperature does not generally fall below −16°c. Temperatures below this level will freeze the sap in the trunk and roots, causing them to split and the vine to die. Some success in keeping the vine alive in such conditions is achieved, for instance in Canada, by burying the whole vine. Some frost is desirable, however, to harden the wood and to kill pests and fungi in the soil.

The temperature is not entirely ruled by latitude, however; proximity to the sea has a moderating effect on temperature both in winter and summer: distance from the sea has an opposite effect, creating in effect 'maritime' and 'continental' sub-climates whose rainfall patterns also differ. Moreover, as temperature decreases by about 5°c for every 300m altitude, sub-climates in which cultivation of the vine is possible occur in the high uplands of Chile, Peru, and Mexico which lie in the equatorial zone. The cold polar sea-currents

1 Why are 'American' vines essential to the winemaker?

which flow along the western coast of the American land-mass have a similar effect of moving the vine-growing climate nearer to the equator. On the contrary, the extreme winter colds and summer heat of the low-lying Euro-Asian landmass between latitudes 35° and 50°N, coupled with the extremely low rainfall in this part of the world, render vine growing impossible.

In order that the vine can create enough sugar in its fruit, 1500 hours of sunshine are required between the beginning of April and the end of October; the summer temperatures must not be too hot, or the leaves will become scorched and unable to perform their task of assimilation. Furthermore, there must be enough rain – 700 mm – mostly falling before April.

The leaves, growing each year from the shoots emerging from the canes, are the powerhouse of the vine. They contain chlorophyll, which is light-sensitive; when light falls on the leaves, a process known as photosynthesis occurs, by which carbon dioxide gas absorbed by the leaves is combined with water and minerals (phosphorus is particularly important) drawn into the leaves via the canes, trunk, and roots from the soil. This combination yields acids, proteins, and sugars, which the vine uses for its own growth and also stores in the fruit. The by-product is oxygen, which escapes to the atmosphere. At night, the process is reversed to some extent, releasing carbon dioxide to the atmosphere in exchange for oxygen. These two processes are also called assimilation and respiration.

Nevertheless, hours of sunlight are not the limiting factor; the degree of assimilation will vary with the intensity of light, but it can only take place if there is sufficient leaf growth. The shoots which bear the leaves will not grow until the temperature has reached 10°C, and growth is directly related to the amount of heat received by the vine during the growing season from April to October. The heat-sum is found by multiplying the average number of degrees over 10°C by the number of days in each month to give the day-degrees.

Successful commercial wine-production needs between 1000 and 3000 day-degrees, although fine wines have been made in England with as little as 800 day-degrees. Maps with isothermal lines defining these limits will indicate possible vineyard areas.

1 For rootstocks, because their roots are resistant to *Phylloxera*.

When temperatures are cooler, the photosynthesis process in the leaves produces more acids; when they are hotter, more sugar. So that the lower the heat-sum figure is, the lesser will be the quantity of sugar in the grape, and the greater the quantity of fruit acids, producing at the extreme a thin, acid wine with too little alcohol: at the other end of the scale, with a high heat-sum, the wine will tend to be fat and flabby, with an excess of alcohol. Wine-makers are permitted to compensate for lack of sugar in the grape by adding sugar during fermentation. More details of this process are given in the chapter on vinification. It is strictly controlled, and the maximum amounts have been assessed by reference to the heat-sum of the region concerned. The map on page 27 shows zoning of some western European countries.

Within sub-climatic regions, local topographical features, natural or man-made, can produce minor changes to the regional pattern, known as microclimates; for instance, the presence of large lakes or rivers in the vineyard area increases the amount of sunlight by reflection, and flowing water decreases the risk of frosts. Damming rivers for hydro-electric schemes, as has been done in the Mosel and Douro valleys, has influenced the microclimate; the grower can produce the same effect on a small scale by planting belts of trees.

An anomalous condition of the vine exists in India, where vineyards have been established in the Poona Hills 128 km east of Bombay in Maharashtra Province; these vineyards lie at latitude 19°N at a height of only 750 m. Here, the vine is no longer deciduous: carrying leaves the whole year round, it would, if permitted, bear two crops each year and rapidly exhaust itself. However, by cutting off the summer crop as soon as the flowers appear, and pruning appropriately, the Chardonnay variety grown on trellises is induced to give an excellent crop in February.

ASPECT

The situation of the vineyard, and its aspect to the sun and to the prevailing weather, is important. The ideal for the vineyard in

1 What is 'crossing'?
2 What is 'cloning'?
3 What is the minimum temperature at which the vine can survive?

temperate climes is to face south-east, so that it gets the benefit of early morning sun and warms up quickly before the obscuring dust and mists rise; at this time also there is more carbon dioxide lying close to the ground, produced by respiration of the vines during the night, which the vines can now assimilate with the light.

A south aspect is the next best; a south-west aspect is not as good, lacking the early morning sun. An eastern aspect, although getting the morning sun, loses it shortly after midday; this applies even more to a north aspect, although this is a benefit in the sub-equatorial climates where too much exposure to the sun's rays might scorch the leaves.

Ideally, the sun's rays should strike the ground at right angles, which will explain why some of the best vineyards in low latitudes are planted on steep slopes.

Prevailing winds must also be taken into consideration, particularly if they are strong or salt-laden; vineyards at the top of hills are particularly at risk, and windbreaks may have to be planted to give protection near the sea – the Landes Forest of Bordeaux and the reed screens of the Portuguese coast are excellent examples.

SOIL

Above all, the soil needs to be well-drained, but with enough humus in it to retain moisture. The vine grows best in regions that have little summer rain, so the soil must retain sufficient water from the previous winter's rainfall, or irrigation will be necessary. Gravel, stony, or even rocky soils are often suitable, as they have the added quality of retaining the sun's heat, which is then radiated to the vines during the night. The soil should also have the right mineral balance; the growth elements phosphorus, potassium, and nitrogen must be present, but the amount of nitrogen – provided by humus and organic manures as well as by artificial fertilizers – must be carefully controlled, for too much will result in over-production of foliage to the detriment of the crop.

The minerals in the subsoil are also of great importance in determining the character of wines; the greatness of the Grands Crus

1 Propagation by pollination of the female flowers of one variety or species with the pollen of a male.
2 Asexual development of a variety by selective grafting.
3 –16°c.

of the Médoc is in large measure due to the presence of a layer of *alios* ironstone in the subsoil, and the Grands Crus of the Côte de Nuits in Burgundy owe much to the underlying marl.

The vine's roots can extend 15 m or more in deep fertile soils, but the vine yields crops of better quality where the depth of the soil is restricted by a sub-stratum of hardpan, rock, or clay which will also force the roots to spread out under the surface. While it is important that the topsoil should be kept loose and friable, encouraging aeration and the suppression of weeds, machinery should be used sparingly between the rows, lest the soil become too compressed or the shallow roots of the vine damaged.

THE VINE

'The life-span of a vine is as that of a man' ran the old saying, but in the old days a man did not live very long; the life of a vine is now reckoned to be 30 years, possibly 45 for some varieties. During the first year the roots form: in the second, wood and some berries that do not reach maturity: and during the third and fourth years the roots and trunk continue to develop, and the crop of berries in the fourth year may be used for quality wines in France, though not generally in the EEC. From the fifth year onward, the vine produces a crop of mature grapes, but does not reach its full productive potential until about the tenth year, which it will maintain until it is about 30 years old. Although the amount of crop will then decline, the quality may continue to improve.

The constituent parts of the vine are the roots, the trunk and branches, the shoots or canes, the leaves, and the flowers which, when pollinated, become fruit. The trunk and branches (old wood) support the canes (previous year's growth), their shoots (new year's growth) and crop. Fruit is only borne on shoots from the previous year's growth: shoots from old wood will not bear fruit, although *their* shoots in the next year (when they have become canes) will bear fruit.

The roots continue to grow throughout the life of the vine, and bring water and minerals in solution to the rest of the plant.

1 What temperature must be reached before the vine will grow?
2 What are the degree day limits for successful commercial wine-production?

Rootstock varieties differ in their characteristics – some plunge, some spread: they also differ in their ability to tolerate, for instance, chalk – so it is important for the grower to select rootstocks to suit the soil, the climate, and the *V. Vinifera* variety grafted onto them.

As may be allowed by the regulations for their region, growers can decide the length of the trunk and how many branches there should be to suit the climate. Danger of frost dictates that the trunk shall be long; where it is desired to get the maximum reflected heat from the soil, a short trunk is required. In hot climates it may be desirable to grow the vine on a high trellis with a very long trunk and, where wind is a problem, there will often be several branches on a short trunk.

Grafting

The vine-louse, *Phylloxera Vastatrix*, which so nearly destroyed the true vine for ever, can never be eradicated. It changes form during its life, existing as *radicole*, in which it attacks the roots and causes most damage; *gallicole*, in which form it eats the leaves and lays its eggs; and a winged form in which it can move to attack vineyards up to 48 km downwind. It has been excluded from certain parts of the world by natural features and rigorous quarantining of imported vines. The most important of these are Chile, Cyprus, and Australia (with the exception of the State of Victoria). Moreover, if the vine is planted in pure sand, the pest cannot reach the roots to assume its radicole form, so the vineyards of Colares in Portugal and of the Rhône Delta, as well as those of the Great Plain in Hungary, are also undamaged by the pest.

Generally, however, the only remedy against it is to graft scions of the true vine *V. Vinifera* on to rootstocks of the American vines on which the pest originated. This works only because the roots of these vines have the ability to heal over the wounds caused by Phylloxera with scar-tissue; *V. Vinifera* lacks this ability and bleeds to death. So, except in the areas mentioned and a few isolated apparently immune vineyards, the vast majority of wine-producing vines are *V. Vinifera* scions grafted onto American rootstocks.

In warmer climates such as that of southern Italy or of Jerez in Spain, the American rootstock vines are planted out in their final

1 10°C.
2 1000 and 3000.

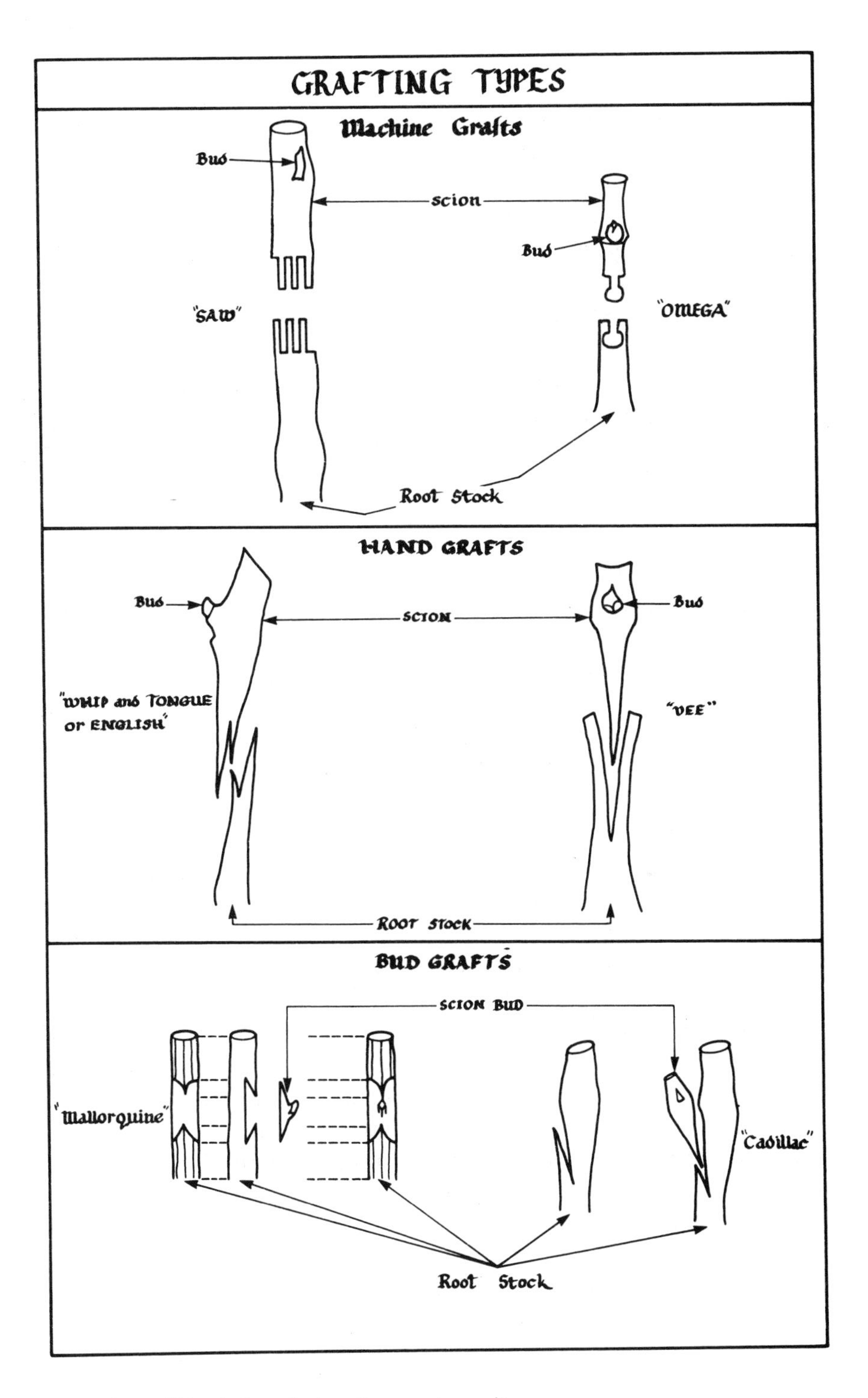

1 Which would be the best site in a German vineyard?
2 Which growth elements should be present in vineyard soils?
3 Will the vine yield better crops on loose subsoils?
4 What are 'canes' and their purpose?

positions. When they have established themselves, all the foliage is cut off and a slit cut in the top of the stump, into which a scion of the desired *V. Vinifera* variety is inserted as a V-graft. Another method is to insert a bud beneath the bark of the rootstock. These methods are known as field grafting.

More generally, grafts are prepared in nurseries. During the winter, matched short lengths of canes of rootstock and *V. Vinifera*, each containing one or more buds, have matching machine-made mortice cuts made in their ends and are joined together, the join being sealed with wax. The grafted canes are then packed in a nutrient compost and kept in a hothouse until the weather becomes warm enough to plant them out. Those which have not formed a callus over the graft are discarded, and the remainder are planted out closely in rows in the open. After a year, the rootstocks will have formed roots, and the scions will have formed buds from which shoots will grow; they are then ready for lifting, packing and despatch to their eventual vineyards.

When ordering from nurseries, growers will be careful not only to order the *V. Vinifera* variety to give them the wine they desire, but also a rootstock that will suit their soil and climate, as varieties differ in their tolerance to soil temperature, acidity, and moisture.

Pruning and Training

As the process of assimilation takes place entirely in the leaves, it follows that the capacity of a vine to produce varies with the total leaf surface area; it would therefore seem that the more leaves there are, the more fruit there would be; this is not so, because the more shoots there are, the less vigorously will they grow. The canes from last year may have grown as long as 5 m, and if left at that length would produce a multiplicity of spindly growth and poor fruit.

Although cutting back last year's growth will reduce the capacity of the vine, fewer shoots will grow more vigorously. Seeing that any vine can only bring to perfection a certain number of bunches of grapes, depending on its environment and history, it pays to cut back in order to produce a decent crop. Moreover, if too big a crop is taken one year, the future production capacity of the vine will be

1 Half-way down a steep slope facing SE.
2 Phosphorus, Potassium, and Nitrogen.
3 No. It will do better where it is restricted by an impermeable substratum.
4 The wood of the previous year's growth, from whose buds the new year's growth will shoot.

reduced. The canes are therefore cut back each year so that a limited number of buds remain, depending on local tradition and practice, evolving from the above principles. What must be achieved is a balance of these principles, so that there is sufficient leaf area to support moderate plant growth and to produce a moderate crop of ripe grapes of good quality without exhausting the vine.

The time of pruning will vary from region to region, but does not generally take place during the coldest part of the winter. Very good results are obtained by pruning immediately after the leaves have fallen, before the onset of winter; otherwise, the best results come from pruning in March (Northern Hemisphere). In spring, pruning will delay the bud-burst by three weeks, which will be an advantage in regions susceptible to spring frosts. Canes should be cut about 25 mm above the last bud to avoid frost damage or drying-out, but no more extra wood should be left, lest it should become infested with insects. Sharp instruments must be used to avoid bruising the cane.

Training the vine, either by tying the canes to wires or posts, or letting the vine assume a bush form, is a response on the part of vinegrowers to their local climate and soil, and to the characteristics of their chosen vine varieties; and as such, is often governed by regulation. The principles involved are that the more vertical a shoot grows, the more vigorously will it do so, and that a large cane can produce more than a small one and can therefore carry more buds.

Systems of training can be classified as long (cane), or short (spur). In cane training, one to four canes are selected and pruned to the permitted length of up to 18 buds – any remaining canes being cut out entirely. In spur training a number of canes are retained, but pruned to no more than two to four buds. The systems can be mixed: the vines may be grown low or high, according to the length of the trunk: and the method of support may vary also.

The actual method used will depend on local tradition and practice, which is to say that it has evolved to suit the climate and soil conditions, and the local grape varieties. Some basic training styles are described below; as illustrated on pages 14 and 16, the vines are shown at the end of the season's growth with an indication of where pruning cuts might be made. Later chapters on individual regions will mention the styles of training most commonly used.

1 Why are vines planted in sand immune to *Phylloxera*?
2 In the Index, page numbers appear in **bold type**, *italics*, or ordinary type. What does each imply?

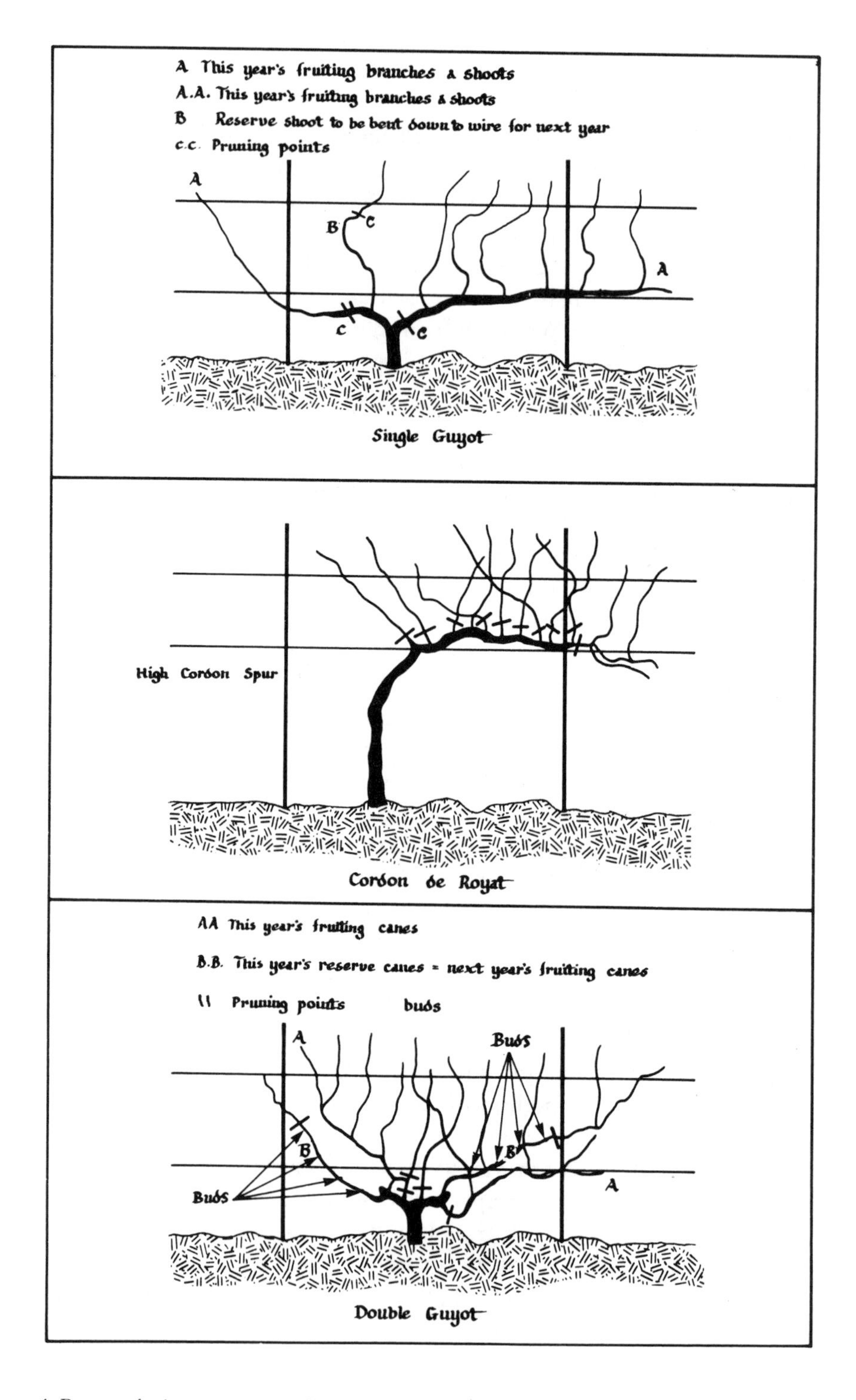

1 Because the insect cannot reach the roots to damage them.
2 For your answer, please *now* read the legend at the beginning of the Index on page 386.

Guyot/(cane): This may be single, double, or quadruple. For single Guyot, two canes only are retained, one on either side of the trunk; one, to bear the current year, is pruned to three, four, or more buds; the other, which will provide the cane for the following year, and may issue from old wood, is pruned to two buds only. Bearing canes are selected from different sides of the trunk in alternate years. For double Guyot, two canes are left on either side of the trunk, one trimmed short, the other long; the long canes may be trained in a bow to the opposite side. Quadruple Guyot is used for trellis training on a long trunk, the same principles applying.

Cordon/(spur): The trunk is grown to a certain height, bent over, and tied to a horizontal wire (Cordon de Royat); or two branches may be allowed to form, trained in opposite directions. Canes have grown from these branches, and are trimmed short, so that each branch carries three to five spurs, each with a spur pruned to two or three buds.

Espalier/(spur): This system is used where vines are grown against a wall, and consists of a central trunk, with horizontal branches at intervals, each bearing spurs as in the Cordon system.

Post/(spur or cane): This system is characteristic of steep-slope vineyards as in the Mosel valley; either one cane is pruned long, or several spurs on a vertical trunk are pruned short. Posts are also used initially in Bush and Gobelet styles.

Gobelet/(spur): A short trunk is allowed to have five or six spurs pruned to one or two buds each; the shoots are tied together after they have grown to about 1 m length, making the vine adopt a shape resembling a goblet.

Bush/(spur): The short trunk is allowed to form several branches; on each branch, one or more spurs are left, the canes of which are pruned to one to three buds. This style is particularly suited to regions where winds are strong enough to break off shoots tied to wires or posts.

1 Will cutting back the previous year's growth reduce the vine's vigour?
2 Describe the two classifications of vine training.

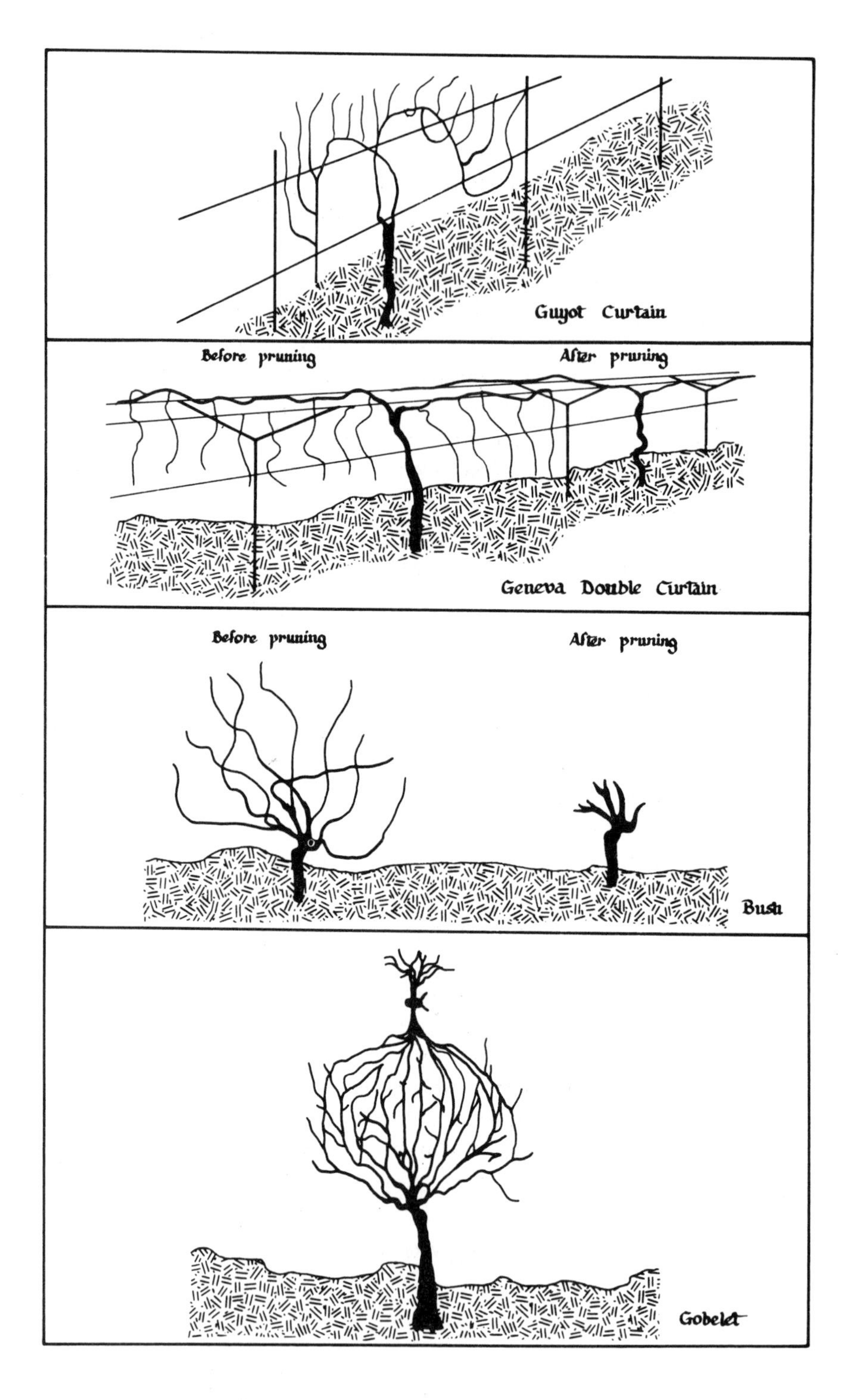

1 No. Fewer shoots will grow more vigorously.
2 Long (cane), and short (spur).

HARVESTING

Grapes are normally ready for harvesting 100 days after flowering, which should give the vineyard owner enough time to organize the considerable workforce necessary; traditionally in Europe, they are itinerant workers from the south who work their way north as the grapes ripen; gauging the right date is important, for once the workers have arrived they have to be housed, fed, and paid. The ideal is to get the harvest in quickly – but the weather does not always (if ever) co-operate. Vineyard owners will have been measuring the sugar content of the grapes with a refractometer, to give them an indication of when the harvest should start. Local regulations may impose some restrictions.

Rain may interrupt the proceedings, for rainwater adhering to the grapes will dilute the sugar content; on the other hand, if the vigneron waits, he may find rot setting in. All in all, the vintage is an anxious time, and one when he is most likely to be pestered by unthinking visitors!

Labour is getting more and more expensive, and more and more owners are turning to machine harvesting; the design of the machines is nowadays becoming better adapted to the delicate work of harvesting grapes. The early machines straddled the rows and beat the vines with rods, breaking off the bunches to fall onto a conveyor belt and be conveyed into a trailer; this tended to shorten the life of the vine, but later improvements have eliminated this. A new type of machine has fingers which stroke the ripe grapes from the stalks, leaving unripe grapes on the vine.

The most important factor in harvesting is to avoid oxidation of the juice. This means getting the bunches of grapes to the winery as quickly as possible without breaking the skins, for which tractors with rubber tyres, or closed gondolas filled with carbon dioxide, may be used. With headlights, machinery can be used at night, when the vines are exhaling carbon dioxide and the air is cool, and the harvest in hot regions may therefore be done more safely then. In cooler climates where rain and hail are the risks, the machines will work both day and night whenever weather permits. Small baskets may be used to feed the tractors.

1 In regions of high winds, would you expect to find cane or spur training, and why?
2 Where would you normally train vines to (a) the Espalier system, and (b) the Post system?

2
Vinification

Vinification is the process by which wine is made from grapes, so the first consideration must be of the ripe grape.

STRUCTURE & COMPOSITION OF THE GRAPE

Wine grapes vary from spherical to ovoid in shape, and from 10 to 20 mm in diameter, according to the variety. All are attached to the stalk of the bunch by a pedicel which continues into the grape centre and to which the seeds (two to four in number) are attached; few wine varieties are seedless. The seeds are enclosed by the juicy pulp,

1 Spur training in bush form, because the winds would break shoots tied to wires.
2 (a) Against a wall. (b) On steep slopes.

and the pulp by a skin, which in ripe grapes may be green to golden or roseate to blue-black in colour, again depending on the variety.

The pulp (which is practically all juice) constitutes about 86 per cent of the grape by weight, the stem 4 per cent, the seeds 3 per cent, and the skin 7 per cent. The pulp divides into three zones: Zone I is a thin outer layer inside the skin which yields juice with difficulty, but contains much flavouring material; Zone II, the largest, is the middle zone from which the juice is released easily; and Zone III is a small inner layer surrounding the pips, retentive of its juice. Zone III has the most acidity and the lowest sugar, and Zone II the most sugar and the least acidity. The detailed constituents of the various parts of the grape are:

Stem: Tannin, minerals, acids, and cellulose.

Skin: Tannin, pigments, flavouring matter, and cellulose.

Pulp: Zone I Flavouring, sugar, and acids,
Zone II Sugar and acids,
Zone III Acids and sugar.

Seeds: Tannin, oils, resinous material, and cellulose.

The main constituent of the grape is water, amounting to 70-80 per cent by weight; sugars account for 12-25 per cent, acids about 1 per cent, tannin 0.2 per cent, and there are lesser quantities of pectins, minerals, and flavouring elements. To the wine-maker the most important of these are the sugars and acids, for ripeness of grapes is gauged not only by the sugar content but also by the sugar/acid ratio (which will vary for different types of wine, but for light wines may be of the order of 40 to 1).

Sugar content of grapes is measured in the field by a refractometer as mentioned earlier, but this only gives an indication from a very small sample; a more accurate method is to take a larger sample of juice for measurement with a hydrometer (saccharometer) calibrated in one of the various scales employed. France uses Beaumé, the Germans Oechsle, the Americans Balling or Brix: they are all related to Specific Gravity (SG).

1 If eight canes on a vine are each pruned to three buds, is this spur or cane pruning, and what form of training would you expect?
2 How do vineyard owners know when to start their harvest (after the permitted date)?
3 And what should be avoided, if possible, during harvest?
4 What *must* be avoided during harvest?

Oechsle is probably the easiest scale to understand, as it consists of the last three significant figures of the Specific Gravity. Divide degrees Oechsle by 4 to get degrees Balling/Brix, and divide degrees Oechsle by 7.143 to get degrees Beaumé.

Thus, for example, an SG of 1.080 ('ten-eighty')
= 80° Oechsle
= 11.2° Beaumé
= 20° Balling or Brix

Acids are measured by titrating samples with an alkali of known strength, and are usually expressed as grams/litre of tartaric acid, the principal acid of the grape at maturity; malic acid is also present, and is the principal acid in unripe grapes, but is converted to sugar during the final stages of ripening; about 10 per cent of the acid content is of citric acid, and traces of twenty other organic acids have been identified by scientists. Acidity is sometimes quoted as grams/litre sulphuric acid: to convert to grams/litre tartaric multiply by 1.53.

Tannin is found in the pips, skin, and stalk of the grape, and is a preservative of wine. Too much, however, will make the wine astringent and 'hard'. The skins of black grapes contain three times as much tannin as those of white grapes, but their pips and stalks about the same amount.

Yeasts and Bacteria

The skin of the grape is covered by a waxy layer known as the 'bloom', to which adhere a very large number of micro-organisms. These are mostly yeasts, of which over 150 different species have been identified including over forty *Saccharomyces*, of which only two or three are useful to the wine-maker. Generally, the micro-organisms can be categorized as 'wild yeasts' (of which there are about 10 million on every grape), 'wine yeasts' useful to the wine-maker (100,000 per grape), and bacteria (100,000 per grape).

The chief wine yeast is *Saccharomyces Cerevisiae* (often called *S. Ellipsoidus*), while *S. Uvarum* and *S. Oviformis* also play some part; these are anaerobic species which can ferment sugars in the absence of oxygen (although oxygen is needed for them to reproduce). The most common 'wild yeast', *Kloeckera Apiculata* (lemon-shaped), can

1 Spur pruning in Bush or (possibly) Gobelet form.
2 By measuring the sugar content of the grapes with a prism saccharometer, and by assessing the weather prospects.
3 Picking during, or immediately after, rain; this dilutes the sugar content.
4 Oxidation of the grape-juice, which will give off-flavours to the wine.

also ferment sugars up to about 4 per cent alcohol, but only in the presence of oxygen. Most of the other 'wild yeasts' and all bacteria give off-flavours to the wine, and it is usual to dispose of them by sulphiting before fermentation.

There are probably more strains of *S. Ellipsoidus* than there are varieties of *V. Vinifera*, and each one differs in its make up of enzymes and its capability of creating flavouring esters. As strains are peculiar to their locality, this is another reason why wines from the same grape made in different regions can differ so much in aroma and flavour.

FROM VINE TO VAT

Pickers may have been told to cut out any rotted grapes from bunches, and to leave unripe bunches unless the must sample has shown a lack of acids; alternatively, bunches used to be sorted at the end of the rows – *épluchage* in Champagne.

The grapes must be got to the winery as quickly as possible, and handled carefully on their way, for it is important that the juice should not become oxidized before fermentation. For this reason, machine harvesting is often done at night, when it is much cooler – for heat aids oxidation. During the day, grapes may even be loaded into closed 'gondolas' filled with CO_2 gas.

On arrival at the winery, the grapes will usually be crushed and destalked before pumping to the vat in the case of red or rosé wines, or to the press in the case of white wines. For certain red wines (e.g. Beaujolais) the grapes may be put straight into the vat and for some white wines (e.g. Champagne) into the press, without either crushing or destalking. The springy stalks provide runnels for the juice to escape from the press, and also provide extra tannin which can help in clarification. Some stalks may be left with crushed grapes for white wine.

There are various types of presses: the oldest is the basket press, where the mass of grapes is placed in a 'basket' from which the juice can escape freely, and is pressed vertically by a hydraulic ram. This is still used in Champagne and in small old-fashioned wineries, particu-

1 By what is the grape connected to the stalk?
2 What percentage by weight of a grape does the pulp consist?
3 Apart from water, what are the other constituents of the grape?

larly for extracting the press wine from black grapes used for red wine. Several gentle pressings are necessary to avoid crushing the pips and releasing bitter oils, and the pomace (*marc*) must be broken up with spades between each because it cakes.

This task is avoided in the rotating presses, of which there are three main types. In the Vaslin, plates joined by chains advance from either end of a slatted cylinder; when the rotation is reversed, the plates separate, and the chains break up the pomace. The Bucher-Guyon is similar, except that it is of far greater diameter, and there is only one plate; on its retirement, the tumbling of the pomace is sufficient to break it up. The third rotating press, the Willmes, is much favoured for delicate white wines; it contains an elastic rubber bag which can be inflated with air to press the grapes against the perforated walls of the cylinder. On deflation, the pomace being more evenly distributed, breaks up easily.

These are sometimes referred to as cylindrical presses because of their shape. Another type of cylindrical press, used only for ordinary wines and called a Continuous press, has a screw rotating in a fixed barrel which pushed the grapes forward on the Archimedean principle against a plate balanced with a heavy weight to give great pressure. As the distance between flanges of the screw diminishes from beginning to end, so the pressure increases, and it is possible to collect the juice separately from various sections of the press. The pomace emerges almost bone-dry, many pips being crushed in the process, so juice from the final section is sent for distillation.

Juice extractors or dejuicers (*égouttoirs*) separate the free-run Zone II juice from the mass of crushed grapes, and in their basic form consist of a tray or drum with perforations over or through which the crushed grapes pass; in one variety, the grapes are brushed along the tray. Another is a cylinder placed at an angle with a tapering screw – like the continuous screw press, but without the weight at the end; gentle pressure is exerted by the taper, and by the weight of the grape mass on a gentle slope. As with the screw press, juice of various qualities is collected at different points.

The latest development is the Serpentine press, which consists of two endless belts running in a channel, the upper impervious, the lower perforated. Crushed grapes are fed between them and carried along; exterior rollers are placed at intervals, each exerting more

1 The pedicel.
2 About 86%.
3 Sugars, acids, tannin, pectins, minerals, and flavouring elements.

pressure than the last; as in other continuous presses, the juice is collected at various points.

The must is normally treated with sulphur dioxide (sulphited) before it reaches the vat in order to kill bacteria and the unwanted wild yeasts; rarely, these are held to impart characteristic flavour to red wines, and the must is not sulphited. Must for white wines is always sulphited, and may stand for 24 hours in settling tanks (*débourbage*) to allow solid particles – skin fragments, pips, pieces of stalk, earth, etc. – to fall to the bottom. Otherwise, the must may be centrifuged or filtered to exclude these gross impurities, but filtration may take too much out of the juice.

Sulphiting is the commonest of the few treatments which may be used to preserve musts and wines against contamination and oxidation. The sulphur is applied either by burning a 'candle' of elemental sulphur in the fermentation barrel (a process seldom used nowadays), or by adding SO_2 dissolved in water or compressed in liquid form, or by adding potassium metabisulphite with which the fruit acids will react to release SO_2. The amount of sulphur to be added depends on the type of wine to be made and the condition of the must. Healthy musts with a high proportion of acid need less sulphur.

WINERY EQUIPMENT AND HYGIENE

Materials: All winery equipment, even the rollers in the mill, must be of materials, or coated with materials, that will not react with the must or wine.

Traditionally, all equipment, including the mill rollers, were made of wood; but modern machinery demands the use of metals and plastics. If metal is used, all surfaces to come in contact with must or wine must be corrosion-proofed. Glass and stainless steel are suitable and easily cleaned, so are the preferred modern materials for presses, pumps, filters, vats, and bottling machines. Vats may be lined with ceramic tiles or concreted, but the rough surfaces left make them difficult to clean; steel tanks lined with epoxy resin are suitable if the resin is periodically renewed.

1 How do you convert degrees Baumé to degrees Oechsle?
2 What is the most important measure of grape ripeness?
3 How do you convert acidity expressed in g/l sulphuric acid to g/l tartaric acid?
4 When might unripe grapes be included in the harvest?
5 Why must grapes be taken quickly and carefully to the winery after picking?

Cleaning: The need for cleanliness in the winery cannot be too highly stressed. Walls, floor, roof, and all equipment should be thoroughly sterilized before the harvest, and equipment cleaned and sterilized after each use. Stalks, and pomace from the presses, must not be left to lie about the winery – for they contain sugar which will attract the fruit fly *Drosophila* which is a carrier of acetic bacteria *Acetobacter*. Equipment is generally cleaned with steam, which is also effective in removing accretions of tartrates from the walls of vats, but for complete sterility a proprietary brand of detergent or a hypochlorite solution will be used.

FERMENTATION

This is the process whereby yeasts act on the sugars in the must and convert them into ethyl alcohol and carbon dioxide gas; fermentation gets its name from the bubbling and seething of the gas escaping from the liquid. This simplistic description conceals a series of twelve different reactions, each using a different set of enzymes and requiring different catalysts.

The process was summed up by Gay-Lussac in the theoretical formula

$C_6H_{12}O_6$	$\rightarrow$	$2C_2H_5OH$	+	$2CO_2$
glucose		ethyl		carbon
or	$\rightarrow$	alcohol	+	dioxide gas
fructose				
180 g		92 g		88 g

In practice, the process yields only 84 g of alcohol from 180 g of sugar; the remainder of the sugar is converted into other products or consumed by the yeast in growth. Additionally, the process releases much energy in the form of heat, a factor to be considered.

1 Multiply by 7.143.
2 The sugar/acid ratio.
3 Multiply by 1.53.
4 If analysis of the must shows too little acid.
5 To avoid oxidation.

Factors Affecting Fermentation

Temperature: Fermentation can take place at temperatures between 18–20°C for white wines and 25–30°C for red wines depending on the yeast strain. If the temperature rises or falls outside the limits of the particular yeast strain, fermentation will cease ('stick') and drastic measures have to be taken to restart it for, if overheated, the sugars may change into unfermentable mannitose, which gives an unpleasant mousey aroma to the wine.

Alcohol and Sugar Concentrations: With musts over 120° Oechsle fermentation slows down, and will stop before all the sugar is consumed. Yeast strains vary considerably in their tolerance to concentrations of sugar and of alcohol, but high sugar levels lower their alcohol tolerance. Once the alcohol level has reached the amount that the yeast tolerates (9–18 per cent by volume) or the fermentable sugar has all been consumed, fermentation will cease.

Aeration: Although fermentation by *S. Cerevisiae* is anaerobic, and is most efficient in producing alcohol under these conditions, yeasts need a limited amount of oxygen to be able to reproduce efficiently and continue the fermentation. With too little oxygen, fermentation slows down and the dead yeasts may contribute off-flavours: with too much, the wine may become flabby and oxidized.

Pressure: High pressure slows fermentation, and if high enough will stop it; this has been used by German winemakers to preserve *Süssreserve* wines under a pressure of 8 atmospheres of SO_2, now more commonly used than CO_2.

Permitted Treatments in Fermentation

Sulphiting: This has been discussed already as a means of eliminating wild yeasts and bacteria. It may also be used to stop fermentation, but is seldom used for this purpose nowadays. Its main use is as an anti-oxidant.

1 Which type of rotating press is most favoured for delicate white wines and why?
2 What treatment is usually given to musts before fermentation, and why?

Must Improvement (Chaptalisation): Named after its proponent, Dr Chaptal, it consists of adding sugar to fermenting musts in order to increase their potential alcohol. The practice is strictly controlled, and in the EEC the basic amounts permitted vary in prescribed Zones (see Map on facing page). There must be a minimum of potential alcohol before any chaptalisation is permitted; the permitted additions in each Zone are listed in the Glossary under 'Chaptalisation'. For Spain and Portugal Zoning had not been prescribed at July 1987.

Refined crystallized cane sugar is generally used. In regions of Italy and Spain where no addition of sugar is permitted, concentrated grape must may be used instead; there are moves to extend this requirement. Addition of sugar or concentrated must softens the wine and gives it more weight by reducing the acidity and increasing the glycerol content.

Acidification: This is sometimes necessary in hotter climates, and is achieved by adding tartaric acid or, as is frequently done in Jerez, by the addition of calcium sulphate (gypsum).

De-acidification: This may be needed in cold climates and in bad years elsewhere, and is achieved usually by the addition of calcium carbonate (chalk) which combines with the acids, mainly tartaric; it can lead to excess tartrates in the wine. As a natural de-acidification may take place in the wine by malolactic fermentation later, the treatment must be applied carefully.

Refrigeration: This process may be used to arrest fermentation, as in production of Asti Spumante. Fermentation is likely to restart when the wine warms up, if there are any yeasts left.

Aeration: As already mentioned, anaerobic yeasts need oxygen to be able to reproduce, and controlled aeration of fermenting musts may therefore be necessary to stimulate the yeasts if fermentation is failing.

Fortification: If spirit is added during fermentation to raise the level above that which the yeast can tolerate, fermentation will stop. This process is used in production of Port and Vins Doux Naturels, for example.

1 The Willmes, because its pressure is gentle and even.
2 Addition of sulphur dioxide to kill bacteria and unwanted 'wild' yeasts.

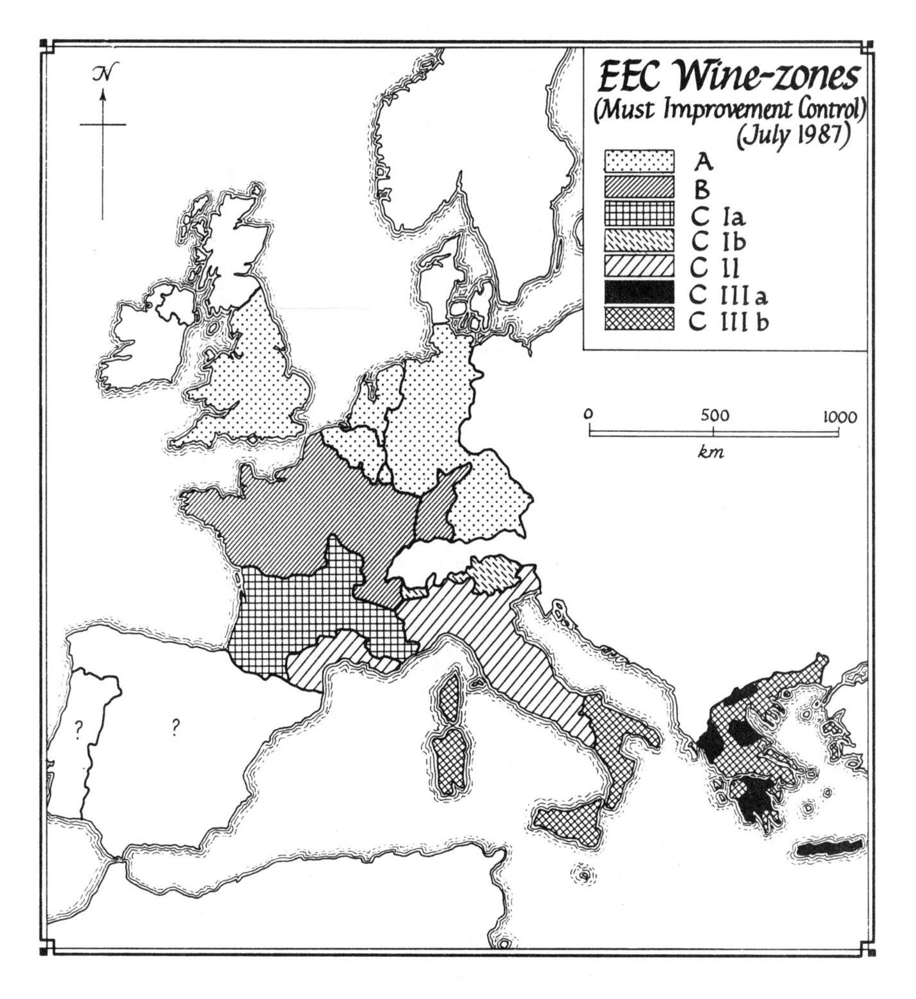

Pasteurization: Raising the temperature to 54°C will kill the yeasts and stop fermentation, but will have an adverse effect on the quality of the wine.

Centrifuging and Filtration: This procedure may also be used to produce wines of low alcohol content, but effective filtration will remove much of the body from the wine.

Inoculation: Used to replace yeasts killed by pasteurization or removed by centrifuging, or to supplement natural local yeasts. This may also be referred to as leavening.

1 What percentage of the grape sugar is converted into alcohol during fermentation?
2 Between what temperature limits can alcoholic fermentation take place?
3 Over what degree Baumé will must not ferment naturally to dryness?

VINIFICATION

Red Wines

The first stage in the vinification of red wines is the extraction of colour from the skins. The anthocyanine compounds in the skin which provide colour are soluble in alcohol and can also be extracted by pressure or by heat. The traditional method of extracting colour for red wines is to macerate the skins in the fermenting must for a period whose length will depend on the degree of colour required.

The must is pumped from the crusher-destalker directly to the vat, up to 100 g/litre sulphur being added according to the healthiness of the grapes and the pH of the must (more sulphur is needed for open vats than for closed). The first tumultuous fermentation generates much heat which assists colour extraction, but the wine-maker will watch the temperature readings closely, and prevent the temperature exceeding limits by pumping the must through cooling radiators; with closed vats of stainless steel, cooling can be achieved by running cold water down the sides of the vat, or by the use of internal or external cooling coils. A temperature between 27°C and 30°C is desirable.

Carbonic Maceration: Pressure is also brought into play with this method, of which there are variations. The tank may be closed with a device that limits the escape of CO_2 gas so that a controlled overpressure exists in the tank. However, in Beaujolais the term is applied to the traditional method of wine-making in which uncrushed bunches complete with stalks are tipped into open or closed vats. The weight of grapes crushes those at the bottom, thus starting fermentation; with the generation of CO_2 gas the pressure builds up even more, forcing yeasts through the skin of unbroken grapes, so that they start to ferment inside then burst. After 48 hours the vat is emptied to presses, sufficient colour having been extracted.

The normal unpressurized maceration period is 10 to 12 days, depending on the amount of colour and tannin required. During this period, the skins are forced to the top by CO_2 pressure and form a cap. With open vats, it is important that this cap should be kept immersed lest it become infected with acetic bacteria; in the old days

1 Usually about 47%.
2 Between 5°C and 35°C, depending on the strain of yeast used.
3 Approximately 17° Baumé.

men were employed continually to push it down with poles but, more efficiently, a grating is fixed just below the level of the liquid in the vat.

On the second day of fermentation the juice is run out of the vat into a basin, whence it is pumped to the top and sprayed over the cap; this not only cools the must and runs alcohol over the skins to extract colour, but also aerates the must and energizes the yeasts. The same may be done with closed vats purely for aeration purposes, because modern closed vats have internal Archimedean screw apparatus to circulate the skins, and interior and exterior cooling coils to maintain the temperature.

Siphon Vat ('Autovinificator'): This vat, illustrated on page 30, is used in hot countries to extract the colour quickly. With the central tube removed, the vat is filled with crushed grapes from the crusher-destalker; the central tube is then replaced (the central ball-valve covering the hole in the bottom of the tube is unscrewed to make this possible and then resealed) and fermentation starts in the sealed vat. CO_2 pressure forces the must up the side tube into the open part of the vat at the top, and at the same time acts to push water in the separate hydraulic valve from one part to another. When the pressure reaches a preset level, it trips the hydraulic valve open to release all the CO_2 in the vat, so allowing the wine in the open part to rush down the inner tube, up again to hit the roof of the vat, and spray down with considerable force upon the cap of skins, beating the colour out of them. This vat is efficient in controlling temperature and extracting colour, but needs a rapid fermentation to keep the aeration to a minimum. The process is noisy, resembling the roar of waterfalls in an underground cave.

Thermotic Extraction: A method of extracting colour *without* fermentation using the principles of pressure and heat can be employed for ordinary bulk wines. The crushed grapes are sulphited to remove oxygen and pumped into a series of closed vats under nitrogen pressure to prevent fermentation; they are then heated to 87°C for 40 minutes, which extracts all the colour from the skins and kills all the yeasts and bacteria. After centrifuging to remove all the solid matter including dead yeasts, the sweet black juice is pumped to nitrogen-

1 How can the vinegrower compensate for lack of sugar in the grape?
2 Is this process controlled, and what is it called?
3 How may alcohol levels in light wines be improved in regions where addition of sugar is not permitted?

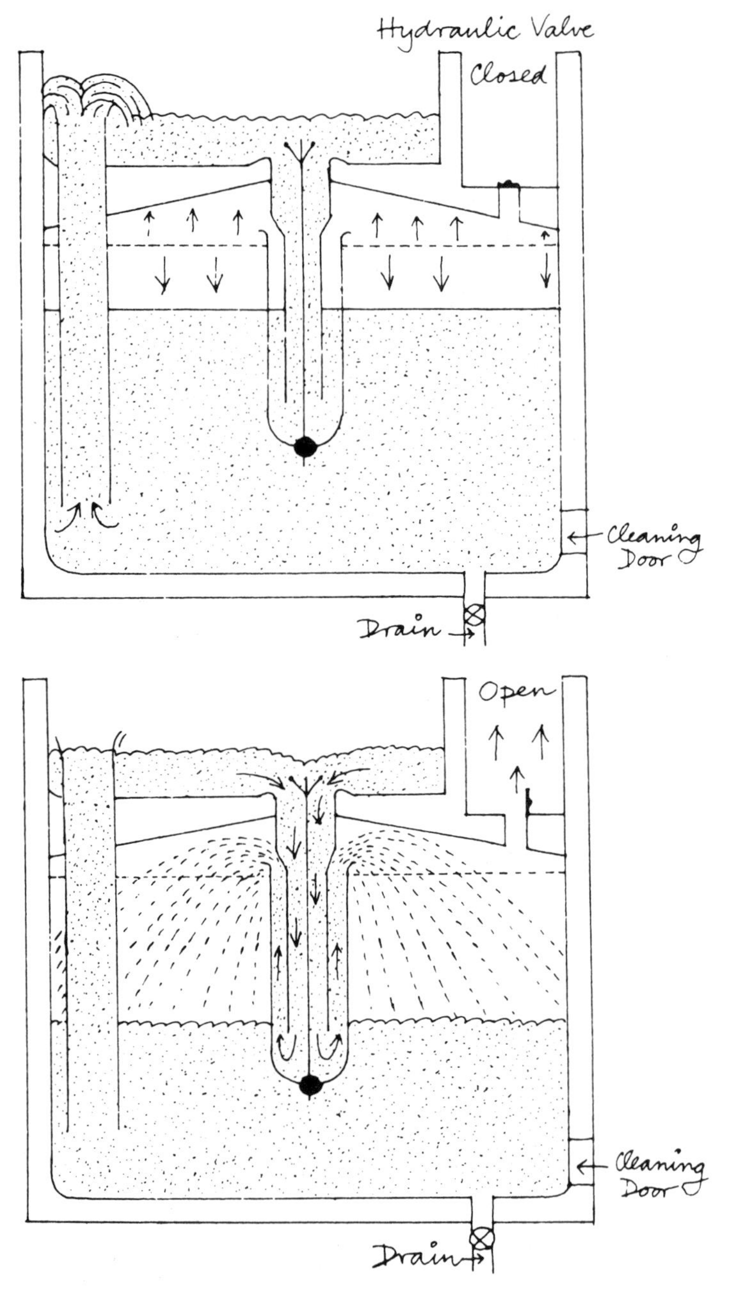

AUTOVINIFICATION - A Siphon Vat

1 By adding sugar during fermentation.
2 Strictly, according to Regional regulations: *Chaptalisation*.
3 By addition of concentrated grape juice.

filled storage tanks where it will remain until required either for sale as natural grape juice or for inoculation with cultured yeasts to produce wine.

After the required amount of colour has been extracted, the 'free-run' wine (consisting of all the juice from Zone II, and most of the juice from Zones I and III of the grape) is run from the bottom of the vat and pumped into closed vats to complete fermentation at a slightly lower temperature – usually about 25°C. This will take four to five weeks, at the end of which the yeasts die and the wine must be racked off them with aeration, for in decomposing (autolysis) they generate hydrogen sulphide. This gas (poisonous in anything but minute concentrations) smells of bad eggs, and itself reacts with other wine constituents to form mercaptans, the compounds used to give an unpleasant smell to natural gas.

Rosé Wines

The traditional method is exactly the same as that for red wines, but the free-run wine is removed from the skins after a very limited time – between 24 and 48 hours only. Also traditionally, rosé wine can be made by vinifying black and white grapes together as for red wine; wines made by this method attain a bronze colour likened to that of an onion skin or the eye of a pheasant.

Some wines called '*vins gris*' by the French are made by pressing black grapes with enough force to press colour out of the skins, and fermenting the juice as for white wine: the coloration is very pale. Such wines are called 'Blanc de Noir' or 'Blush wines' in the United States.

White Wines

The Zone II juice free-running from crushing and/or the first light pressing will, after sulphiting and clearing by settling or centrifuging, be pumped to the fermentation vat. It may also be cooled to about 5°C, for the ideal fermentation temperature is 18–20°C, much lower than for red wines. The juice from later pressings may be vinified separately, as it will produce wines of lesser quality. The control of temperature is more important with white wine, and the

1 What methods may be used to extract colour from the skins of black grapes?
2 What is the desirable temperature range for the fermentation of red wines?

ideal is to maintain a steady rate of fermentation until all the sugars have been converted – this will take about six weeks.

When fermentation is complete, the wine will normally be racked off the lees *without* aeration and sulphited again (for white wines oxidize more easily than red) before being pumped to the maturing vats.

SULPHURING

When SO_2 is added to wine, some of it combines with the wine constituents (principally acetaldehyde); this 'binds' it. The sum of the 'bound' and the 'free' SO_2 is the 'total SO_2': it is the maximum amount of this which is regulated by law.

The normal limit of free SO_2 before bottling is 20–30 mg/litre for red wines and 40–50 mg/litre for white, but care must be taken that bound SO_2 does not take these levels over the maxima permitted. These depend on whether the wine is 'dry' (less than 5 g/litre sugar) or not:

Red Wines	Dry	160 mg/litre
	Other	210 mg/litre
White Wines	Dry	210 mg/litre
	Other	260 mg/litre
	Sauternes and Barsac	400 mg/litre

In the EEC, there is constant legislative pressure to reduce these levels, particularly the last.

COMPOSITION OF WINE

The chief constituents of wine are water: 80–90 per cent, and the ethyl alcohol produced by fermentation: up to 15 per cent. There are also constituents of the original grape-juice which remain unaltered by either the alcoholic or malolactic fermentations: fruit acids, anthocyanins, tannins, and unfermented sugars. Moreover, during

1 Pressure; heat; maceration in the alcohol formed by fermentation.
2 25°C to 30°C.

the alcoholic fermentation about 3 per cent of the original sugar was converted into 'by-products giving flavour': higher alcohols, aldehydes and ketones, esters, and acids; of these, the principal aldehyde is acetaldehyde, and the principal acid is succinic. If there has been a malolactic fermentation there will be a quantity of lactic acid.

This mixture is not stable and if oxygen is freely available will degenerate to acetic acid. If exposure to air is strictly controlled, in a full and closely-sealed container, the reactions will be slower and different, softening the harshness of tannin and developing ester combinations giving bouquet.

Fixed and Volatile Acidity

Acids in wine are classed as 'volatile' or 'fixed'. The chief volatile acid is acetic, and the measure of volatile acidity is expressed as grams/litre acetic acid; it is an important measure of the health of a wine, as body temperature is of human health.

Fixed acids may combine with metallic ions to form salts, for example, calcium tartrate. Fixed and volatile acidity together make total acidity which is measured in grams/litre tartaric acid.

MATURATION

Maturation has been called 'controlled oxidation', but is in reality a much more complicated series of inter-reactions between the various different constituents of the wine. Acids react with alcohols to produce esters (organic salts) which contribute to the aroma and flavour of wine. Water, the principal constituent of wine, has no smell or taste, and ethyl alcohol, the next most important, has little.

One reaction involved in maturation is malolactic fermentation – fermentation of malic acid by lactic bacilli, principally *leuconostoc*, the enzymes of which convert the malic acid into softer lactic acid, releasing CO_2. It does not take place in hot climates where there is little malic acid, and can be prevented if it is desired to retain the acidity. On the other hand, it is often promoted, and the ideal conditions are warmth, a reasonably low level of total acidity, a low

1 What are 'blush wines', and how are they made?
2 What is the ideal fermentation temperature range for white wines?

SO_2 level, and the presence of lactic bacteria. As these conditions occur during normal alcoholic fermentation, the two fermentations often naturally take place at the same time.

Wines are not always matured at the vineyard where the grapes were grown; often they are sold to a shipper (through the intermediary of a broker) who will mature them in his cellars, probably in the local town: hence his title in French, *Négociant-éleveur*. Besides maturing wines that he buys, he may blend wines from different vineyards or districts to balance their faults and emphasize their qualities; the label of the blend must be of a district, region, or country that encompasses all the constituent wines of the blend.

Red Wines: These should be matured in wood; the wine evaporates through the pores of the wood, and the casks must therefore be kept topped up to avoid excessive oxidation and conversion of alcohol to acetic acid. Oak is generally used, but chestnut, redwood, and satinwood are also used, particularly for the larger vats. In turn the wood, particularly new oak, imparts flavour from its own constituents such as vanillin, lignin, and tannins.

The length of maturation will depend on the type of wine and the size of cask used; wine matures more rapidly in the Bordeaux or Burgundy cask of 225 litres (2.25 hl) than in the Piedmontese vat of 25–50 hl. So Bordeaux reds are traditionally matured for two years in wood, while Barolo is matured for at least three. During maturation, the tannins and anthocyanins (colouring) combine to form insoluble compounds which fall to the bottom of the cask or vat; every three months or so, therefore, the wines are racked off the sediment into fresh casks.

The 'press-wine' obtained from the pressing of the pomace after maceration will have finished its fermentation and been matured separately. During maturation, wine-makers have the choice whether to add it or not; if the wine lacks tannin, they will add some or all of it; otherwise they may blend it with wine from vines too young to count for quality wine or wines from lesser properties for sale as table wine, or for home consumption by their workers.

White and Rosé Wines: Lacking the tannin that preserves red wines, these wines must have the air excluded; a higher level of sulphiting is

1 Very pale rosé wines, made by pressing black grapes forcibly, and then fermenting the juice without the skins.
2 18°C to 20°C.

used; racking is kept to a minimum, and may be carried out under a blanket of CO_2. Frequently, white wines will not be racked after initial separation from the yeasts until they are prepared for bottling, and some Muscadet, for example, is bottled directly off its lees. It follows that the maturation period for white wines is much less than for red, and most will be bottled within six months of the end of fermentation. Sweet white wines such as Sauternes are an exception, needing up to three years in oak, but many white wines are matured in stainless steel vessels.

Maturation in Bottle

For wine to mature in bottle it must have an airtight seal of cork, always kept moist and flexible by contact with the wine. If the cork is sealed with wax, as with Vintage Port, the amount of air which can get in is negligible and maturation will be very slow. In bottles which are not so sealed, the wine will mature more quickly in small bottles than in large, as the area of cork exposed to the air differs little with the size of the bottle.

PREPARATION FOR SALE

Until recently much wine used to be exported from the region of origin in bulk, and much was bottled in England; nowadays, more and more is matured and bottled by the grower or shipper (*négociant*) in the region; less and less is imported to the UK in cask or bulk container.

FINING

When the wine has matured, it should have 'fallen bright', but will yet not have the perfect clarity required because of protein matter deriving from yeast autolysis or pectins, which hangs as a haze in the wine. This is removed by fining: that is to say, the addition of substances which will stick to the protein and be heavy enough to settle or be filtered out. These substances are usually albumen from fish or egg-white, or earths.

1 What level – free, bound, or total – of sulphur dioxide is subject to legal control?
2 What are the present legal limits of sulphuring?
3 Which are the two principal acids found in wine, and are they fixed or volatile?
4 What is leuconostoc, and what is its connection with wine?

If albumen is used, it is mixed to a froth with some of the wine, and stirred thoroughly into the cask; the protein is fixed by the albumen and falls to the bottom as a gel. After a week the wine is racked direct to the bottling plant, after checking its SO_2 level. Earths are of two types. Kieselguhr, which is a diatomaceous earth, is only used in cases where the wine is very cloudy, and is a filter medium, not a fining. The most commonly used fining agent is montmorillonite clay, which is finely powdered and sold under the trade name Bentonite.

Bentonite is generally preferred because of its ease of use and because it does not take too much out of the wine. Casein, in the form of milk, once used as a fining agent, was abandoned because it took too much character from the wine. It is still used in particularly stubborn cases which resist other treatments.

Refrigeration

Tartaric acid combines with metallic salts to form tartrates; these are more soluble in water than they are in alcohol, and so deposit during fermentation on the walls of the fermentation vessel to a depth of up to 20 mm. However, in warmer regions particularly, they do not all precipitate, and may do so in the bottle. To avoid this, the wines may be refrigerated before bottling, which will cause the tartrates to crystallize out.

Packaging

Bottling operations must be conducted with great care, first of all to see that no contamination enters the wine, and second to reduce the amount of contact with air. Wine may be filtered with an EK or micropore filter and bottled cold under completely sterile conditions; or it may be pasteurized by heating to 54°C and filled into heated bottles.

Although the term 'bottling' is still applied to all methods of packaging wine for sale, the methods have proliferated in recent years to include tins, screw-top and crown cork closures, large-

1 Total.
2 160–210 mg/l for red wines; 210–260 mg/l for white other than Sauternes.
3 Acetic acid (volatile); tartaric acid (fixed).
4 A lactic bacillus causing 'malolactic' fermentation of malic acid to softer lactic acid.

mouth flasks, Tetrapacks of lined waxed paper, and plastic liners of 'bags-in-boxes'. While these are at present used for wines to be consumed without long keeping, some wines so treated are of good quality. The same rules of hygiene govern the new forms of packaging; as they are not cheaper, their success and merit must lie in attracting new customers to the beneficial practice of drinking wine.

1 How is the controlled name of a quality blended wine determined?
2 Does the size of the cask have any effect on maturation?
3 What is the purpose of racking?
4 Why do white wines require a higher degree of sulphuring than red?
5 How will a fine Sauternes be matured?

1 By the name of the district, region, or country that encompasses all the constituent wines.
2 Yes; wines mature more rapidly in smaller casks.
3 To separate the wine from its lees, which may otherwise affect its flavour.
4 Because they lack the preservative tannin present in red wines.
5 In oak, for up to three years.
(Questions continued on page 41.)

PART TWO

Wines of the European Community

Introduction to Part Two

The European Economic Community (also called the European Community) was set up in 1957 between France, Germany, Italy, Belgium, Holland, and Luxembourg, and was joined by Great Britain, Denmark, and Ireland in 1973, by Greece in 1981, and by Spain and Portugal in 1986. Its objects are to develop resources jointly, to protect trade, and to promote fair competition. In agriculture, which includes the cultivation of the vine and the production of wines and spirits, the objectives are to increase productivity, to ensure a fair standard of living for agricultural workers, to stabilize markets, to assure availability of supplies, and to ensure that these reach consumers at reasonable prices. Unfortunately these laudable objectives often operate against each other.

The Community set up a permanent Commission to govern its affairs, and this Commission, with a Council of Ministers of the Member States, issues Directives and Regulations designed to achieve the Community's objectives, and these are binding on Member States.

The wine trade makes representations to the Commission through various channels, the most important of which are the Wine Management Committee (on which the UK is represented by the Ministry of Agriculture, Fisheries and Food), the EEC *Comité Consultative Viti-Vinicole*, the International Federation of Wine and Spirit Merchants (FIVS) and its *Comité Vin*, and the EEC Wine and Spirit Importers Group, on the last four of which the Wine and Spirit Association of Great Britain is represented.

The main regulations applying to wine issued by the Commission deal with:

- Control of vineyard plantings, in area and vinestock type.

1 What substances are generally used in 'fining'?
2 What is the purpose of refrigerating wines before bottling?
3 To what temperature would wine be raised during 'hot-bottling'?

- Regulation of viticultural practices.
- Control of vinification practices including pasteurization.
- Control of minimum prices for both EEC and non-EEC wines.
- Support of wine exports.
- Control of wine imports.
- Control of bottle sizes, and the amount of their contents.
- Control of quality for both EEC and foreign wines.
- Control of labelling and advertising.

These controls at present apply only to wines, but similar controls for spirits and liqueurs are under discussion. The references and purposes of the most important Regulations to date affecting the wine trade in the UK are listed in Appendix 1.

The definition in Regulation 337/79 of 'Liqueur Wine' as 'wine fortified with spirit during fermentation' may seem an over-simplification. It is forbidden to add alcohol to fully fermented wines, but Sherry is deemed to have traces of residual sugar and therefore be capable of further fermentation. The British can translate this as '*Fortified Wine*'.

The quality wines referred to in Regulation 338/79 comprise the Appellation Contrôlée and VDQS wines of France, and the DOC and DOCG wines of Italy, the Qualitätsweine and Qualitätsweine mit Prädikat of Germany, and wines bearing the Marque of the Grand Duchy of Luxembourg. Also included now are the Denominacion de Origen (DO) wines of Spain and the Denominação do Origem wines of Portugal. Each of these countries is responsible for making national rules for its own wines.

1 Albumen from eggs or fish, or absorbent earths.
2 To precipitate tartaric acid salts which might otherwise form crystals in the bottle.
3 54°C.

3

France – General

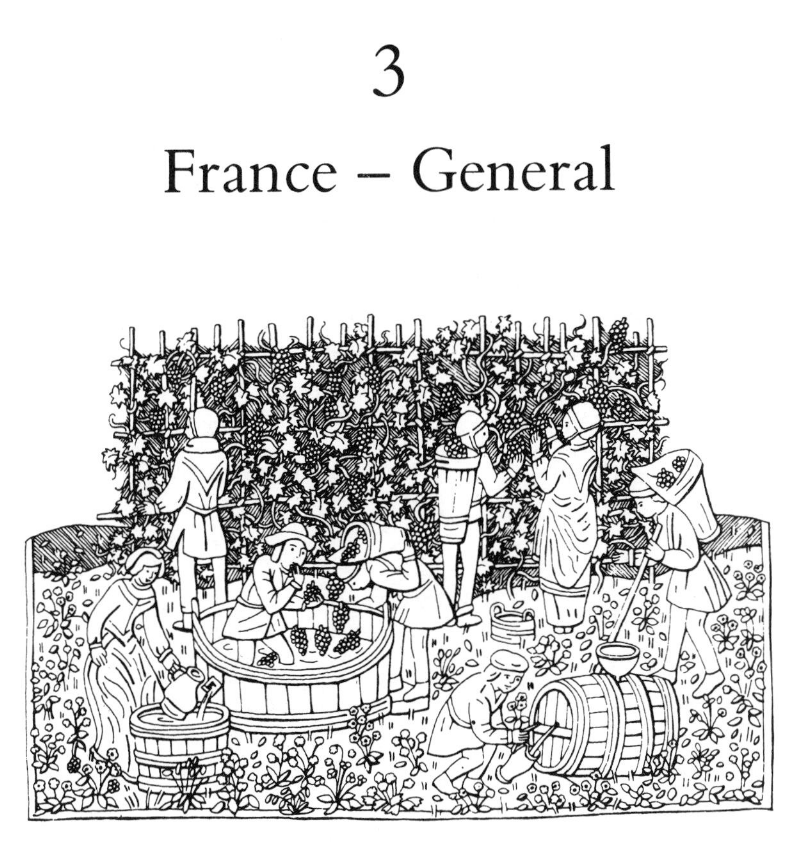

France is the foremost country in the world today for wine production. Sometimes it produces more wine than any other country, although in most years this doubtful honour goes to Italy. But France is remarkable for the diversity of its wines, deriving from the variety of its climate, its soils and grape types, and its long unbroken experience of cultivating the vine and making wine.

The naming of wines in France is strictly regulated and controlled by Government bodies in order to prevent fraud. The principal authority, responsible for recommending decrees for the grant of quality denominations, is the National Institute for Names of Origin of Wines and Spirits (INAO).

1 Which are the twelve Member States of the EEC?
2 Which Ministry represents the British Wine Trade to the EEC?

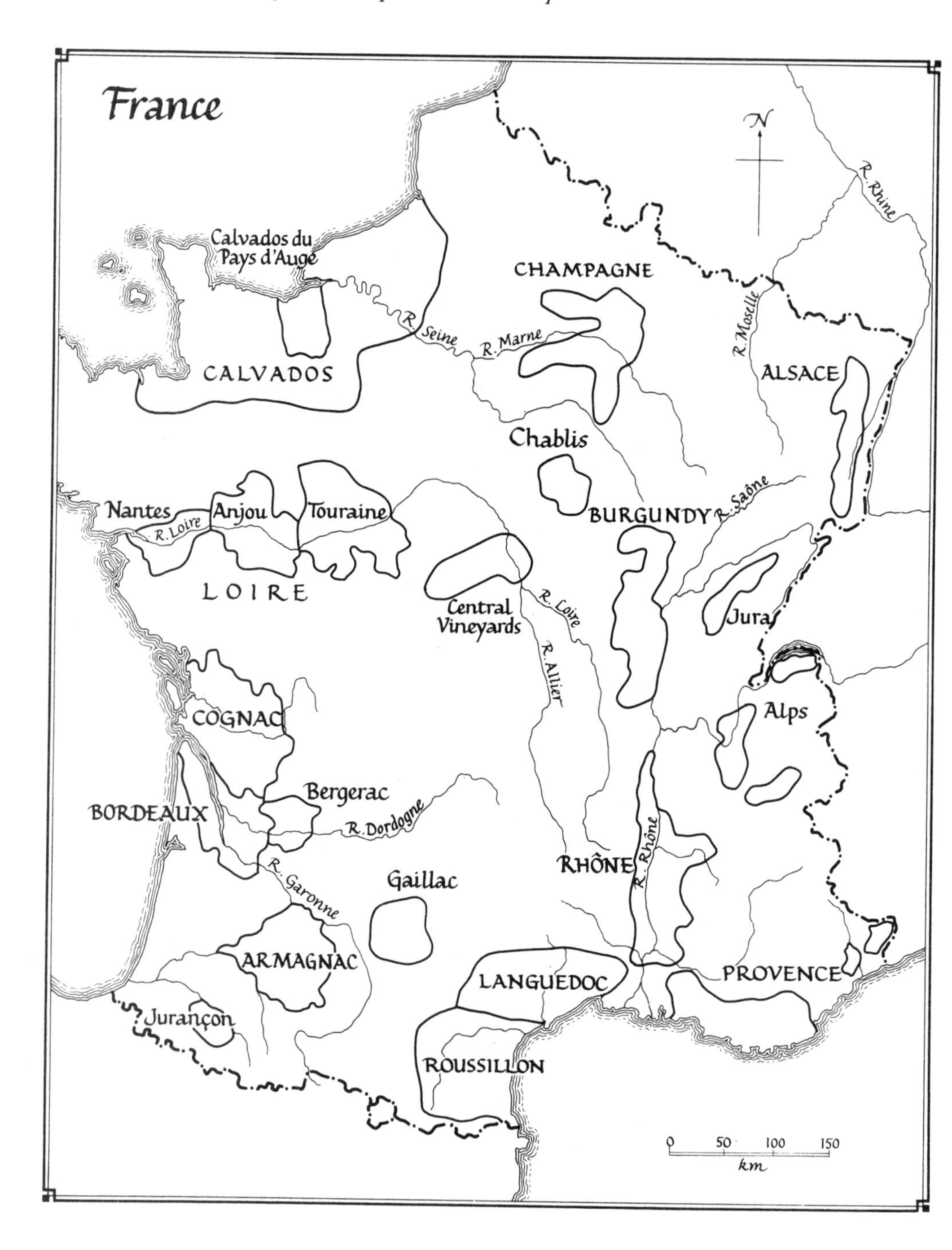

1 France, Germany, Italy, Belgium, Holland, Luxembourg, Great Britain, Denmark, Ireland, Greece, Spain and Portugal.
2 Ministry of Agriculture, Fisheries and Food.

About 21 million hl, or about 28 per cent of France's annual wine production, is qwpsr with *Appellation (d'Origine) Contrôlée* – AC or AOC for short. Another one million hl (1.3 per cent) is qwpsr (quality wine produced in a specified region) with the description *Vin Délimité de Qualité Supérieure* (VDQS). A further 60 per cent of French wines are Table Wines (*Vins de Table*) formerly known as AS (*Appellation Simple*) or VCC (*Vins de Consommation Courante*). The remaining 10 per cent goes for distillation into brandy.

An increasing proportion of Vins de Table are submitting to stricter controls on yield, minimum alcohol level, and other analysable qualities for closer geographical description as *Vins de Pays* (13% vol. alcohol). There are at present three levels of delimitation – regional, by Département, and local – with increasing strictness of control as the delimitation narrows. The three regions are Jardin de la France, comprising the Départements either side of the Loire as far up as Orléans: Comté Tolosan, comprising SW France south of Gironde: and Pays d'Oc, comprising Provence, Languedoc and Roussillon, and part of the Rhône valley.

The Département Vins de Pays are those most often encountered, for instance Vin de Pays de l'Aude. The local Vins de Pays, which in 1985 numbered over ninety, vary in size from two Départements, as Vin de Pays Charentais (*Charentes* and *Charentes Maritimes*), to a single commune, as Vin de Pays des Hauts de Badens just east of Carcassonne. Large or small, all have to be approved by a tasting panel as characteristic of their area.

AC wines are subject to many controls which vary considerably in their strictness and method of application. All AC wines now have to undergo a tasting test, for instance. Vineyard areas are strictly delimited, but a wine from a given vineyard may qualify for three or even four Appellations of increasing merit, depending on its qualities. The alcoholic degree is always controlled, as is the selection of grape varieties from which the wine may be made. In some areas the method of pruning and of training, the date of harvesting, and even the price paid to growers for their grapes may be controlled. The yield in hectolitres per hectare (hl/ha) is always controlled, and the finer the AC, the less the yield permitted.

Up to 1975 growers could operate the '*cascade*' system, that is to say, if in Pauillac they produced 50 hl, they could sell the maximum of 40 hl as AC Pauillac, a further 3 hl as AC Haut-Médoc, a further

1 Which two main Regulations control wine and quality wine?
2 What government body controls the naming of French wines?

2 hl as AC Médoc, and the balance of 5 hl as AC Bordeaux (or Bordeaux Supérieur), in decreasing order of excellence. They could put the name of their château (or vineyard) on all of them, for this does not, in Bordeaux, form part of the Appellation. All would have had to pass a tasting test, for that is the rule for red Bordeaux wines although it is not yet the rule in Burgundy. The 'cascade' system was open to criticism, if not to abuse, and new laws have been introduced.

The yield, as laid down in the law for each AC, is termed the Basic Yield (*rendement de base*). This may now be varied annually by a local Commission to an Annual Yield. Within this figure, which varies with the quality of the AC, growers can market their wine, with or without a tasting test, as the particular Appellation requires. Production over this Annual Yield may be admitted by an allowance (but fixed for each region, and invariable from year to year) up to a 'Ceiling Yield' (*Plafond Limite de Classement* or *plc*). Production above Annual Yield but not exceeding Ceiling Yield may be admitted to AC by a tasting panel, but if production exceeds Ceiling Yield, all quantities up to the Ceiling Yield must pass a tasting trial before admission to AC, and the excess above Ceiling Yield must be converted to spirit or to vinegar. If any wine fails to pass at the trial to which it has been submitted, it must all be converted to spirit or vinegar and none of it may be declassified, as in the 'cascade' system.

Consider a Burgundian example, from a one-ha plot in the Grand Cru vineyard of le Corton, in a particular (and extraordinary) year when all the Annual Yields were 5 hl/ha above the Basic Yields. The possible AC are:

	Yields in hl/ha		
AC	*Basic*	*Annual*	*Ceiling*
Corton	30	35	42
Aloxe Corton	35	40	48
Bourgogne, or Bourgogne Grande Ordinaire	50	55	66

The grower producing 47 hl/ha has three choices:

1 337/79 and 338/79 respectively.
2 National Institute for Names of Origin (INAO).

1 Sell all 47 hl as AC Bourgogne at a low price, OR
2 Sell 40 hl as AC Aloxe-Corton, and submit 7 hl for trial as Aloxe-Corton, OR
3 Send 5 hl for conversion to spirits or vinegar, and submit 42 hl for trial as AC Corton – a risky business, as the standard required of AC Corton is much higher.

In this book, the wine regions of France have been grouped clockwise broadly in relation to climate; the south-west, comprising those regions and districts watered by the Gironde and its far-reaching tributaries and those to the south of it whose rivers flow to the Atlantic: those north, comprising the Loire Valley, Champagne, and Alsace: followed by the east, comprising Burgundy, Jura, and Savoie: and finally the south, comprising the Rhône Valley, Provence, Languedoc & Roussillon. Each of these groups has variations in climate, in soil, and in grape variety. These differences merely serve to underline the remarkable diversity of French wines.

1 What are the four categories of French wine in ascending order of quality?
2 Are Vins de Pays 'Quality Wines'?
3 What are the levels of delimitation of Vins de Pays?
4 What are the names of the three Regional Vins de Pays?

South-West France

BORDEAUX

Bordeaux, at once the most important, most complicated, and most variable fine wine region in the world, lies in the Département of the Gironde, 45° north of the Equator. The wine-lover's comment: 'know Bordeaux and understand all' may sound encouraging, but Henri Bertrand has said, 'In Bordeaux everything happens in half tones: men's behaviour, the hierarchy of the soils and the vineyards, the diversity of the intrinsic qualities of a multi-faceted production, and even the meaning given to words, depending on the speaker.' So each must allow his opinion to evolve as his own knowledge of Bordeaux becomes more refined. He will find the wines of Bordeaux to be among the best in the world; Hugh Johnson describes them as

1 Vin de Table, Vins de Pays, VDQS, and AC.
2 No: they are 'table wines'.
3 Regional, Départemental, and local.
4 Jardin de la France, Comté Tolosan, and Pays d'Oc.

being 'indescribably delicate in nuance and complexity.' The red wines are unquestionably superlative; the dry white wines are very good, but only rarely exceptional, while the sweet golden wines of Sauternes and Barsac are outstanding.

Topography and Soil

The general map of the Bordeaux wine region shows the main districts into which the region is divided, and the geographical reasons for those sub-divisions. The region is first trisected by the

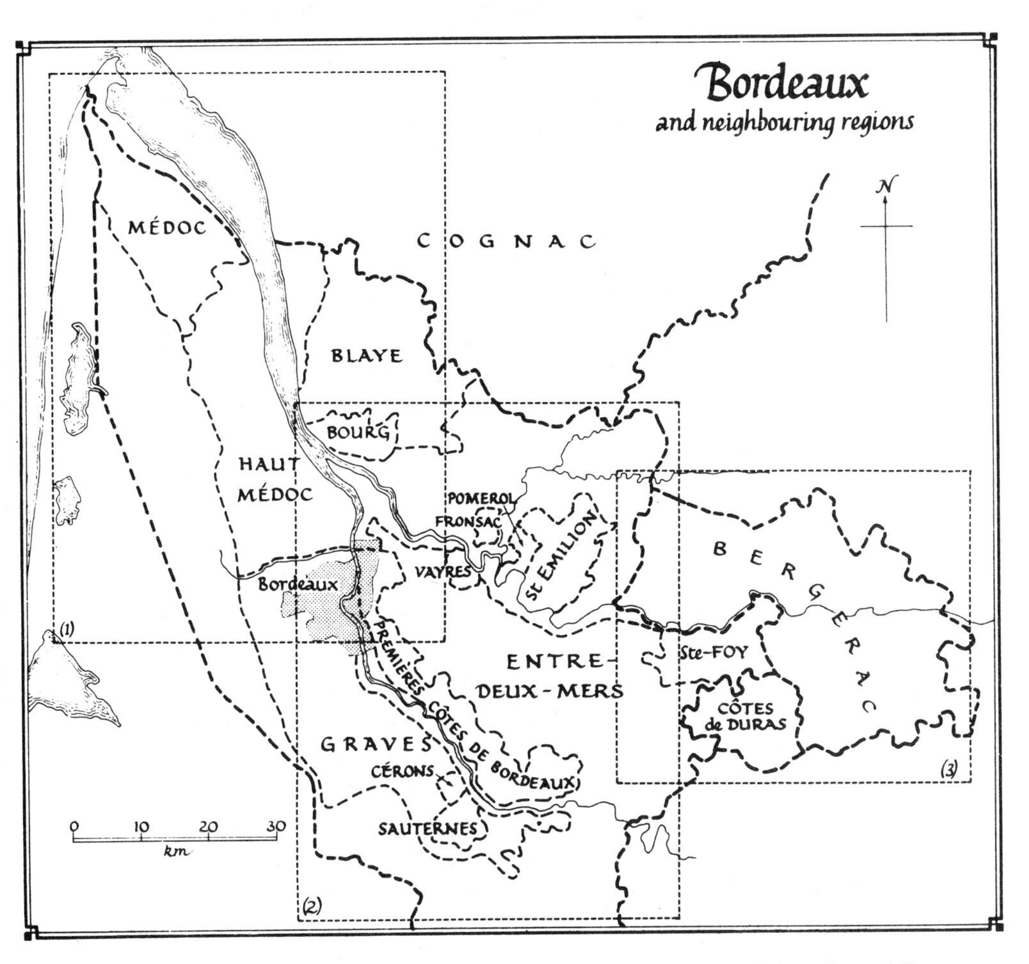

1 May quantities of wine produced above the AC ceiling yield be downgraded to a lower AC, or sold as vin de table?
2 What natural variations between French regions lend diversity to the wines?

rivers Garonne and Dordogne, which meet 24 km north of the city of Bordeaux and flow on as the Gironde to the sea. These rivers, with a network of tributaries, give constant irrigation to the region. The nearby Atlantic Ocean stabilizes the climate, so that summer and winter temperatures vary less than inland; drought is rare; thunderstorms in the summer heat are sudden and frequent and provide much-needed moisture; hail, however, is an enemy to be reckoned with, particularly along the Garonne, and one for which the grower, or *vigneron*, has little answer.

To the west and south of the Garonne forested sand dunes by the sea protect the district from prevalent south-westerly salty winds. Inland, gravel predominates in the 112 km sweep from Langon to the Pointe de Grave at the mouth of the Gironde estuary. In the Haut-Médoc the gravel bank runs in a long mound along the Gironde. Between the Garonne and the Dordogne is an area of sandstone hills with outcrops of clay; and over the Dordogne, to the north and north-east of the region, is a large area of sand and limestone hills, bounded on the north side by an undulating clay plateau. It can be seen that the soil varies considerably, and the subsoils, which also vary, include lime, marl, clay, and sandstone rich in iron. Gravel subsoils abound to the west of the Garonne. The alluvial lands (*palus*) close to the rivers, being rich are unsuitable for fine wines.

Historical Background

From Phoenician times, Bordeaux has been the premier Atlantic port of south-west France. Long before the Gironde countryside was put down to vines, the merchants of Bordeaux had been exporting wines from the 'Haut Pays', Provence, and the Rhône Valley. After the Roman edict reducing vineyards was relaxed in AD 280, however, a local wine industry developed. But for eight hundred years Saxons, Franks and Normans pillaged and sacked the region in their turn. It was only in the tenth, eleventh, and twelfth centuries that a measure of law and order was achieved. Encouraged by Henry of Anjou (Henry II of England), the wine industry prospered; red wines mainly were produced, and a light red wine, *clairet*, the forerunner of

1 No: they must be converted into vinegar or industrial alcohol.
2 Climate, soil, and grape variety.

claret, was made by blending red and white wines – an illegal practice nowadays. Graves was principally a red wine district in the medieval era; only in the last three hundred years has white wine production increased.

The influence of the Angevin suzerainty lasted for three hundred years, but when Gascony reverted to France in 1453, English trade fell away, and with the growth of Calvinism trade between the Protestant Huguenots and the Hanseatic states developed in its place. These new markets favoured sweeter white wines, and the vignerons of Graves consequently switched much of their production from red to white wine. But the historical links with England lasted so that she still now takes 15 per cent of Bordeaux wine exports, mostly in fine claret.

The American taste for Bordeaux wines was doubtless influenced by Thomas Jefferson who in 1792 was extolling the virtues of the Rozan family, makers at that time of the wines of Château Margaux and Château Latour. 'Rozan Margau,' said Jefferson, 'is made by Madame de Rozan – this is what I import for myself.' The market in America for Bordeaux château and regional wines has prospered and continues to increase in volume through to the present day. During the nineteenth century the Bordeaux region suffered – as did most other continental wine regions – from the ravages of *oïdium tuckerii* and *phylloxera*; and during the twentieth century the region has suffered from war. The vintages from 1905 to 1918 were very poor, and during the Great War much wine was requisitioned for the French Army. In 1940, Bordeaux suffered occupation though strangely, perhaps, the Bordeaux merchants reached agreement with the Germans on quotas for ordinary wine sales, and most of the fine wines escaped confiscation and pilfering. Today, in addition to Britain, Belgium, and America, two new markets – Japan and the Soviet Union – feature in the list of countries importing Bordeaux wines.

Pattern of Trade

In the Gironde, the pattern of trading varies. On the one hand is the vigneron who produces and ships his own wine, e.g. Rothschild; on

1 In which Département do the vineyards of Bordeaux lie?

the other hand is the small grower who sells his wine through the offices of a broker (*courtier*) to a Bordeaux shipper (*négociant*, often called a *négociant-éleveur* because he buys young wines and matures them or 'brings them up'). Today, some twenty firms are responsible for three-quarters of the business.

Characteristic of Bordeaux is the château, but in fact the word 'château' here means vineyard rather than castle, and has no relation to size. Bottling at the property is becoming generalized, stimulating greater interest in individual châteaux among consumers. However, while a 'château-bottled' description may be an additional guarantee of quality from the great châteaux, it must be recognized that the facilities at many of the smaller properties fall far short of the high standards obtaining in the cellars of shippers in the cities of Bordeaux and London.

Types of Wines and Spirits

The annual Bordeaux crop of AC wines in the last decade averaged 2.75 million hl, of which 75 per cent was red and rosé wine, and 25 per cent was white. In addition, nearly the same quantity of table wine (but in reverse proportion of colour) was produced. No VDQS is produced in the region.

Red wines are predominant in the northern districts of the region, and white wines in the southern. This, according to experts, is connected with the rainfall, which is greater in the red-wine-producing north, and less in the south, where the drier conditions favour development of benevolent *botrytis*, undesirable in black grapes.

The ACs require minimum strengths for different qualities of wine, varying from 10% vol. alcohol in the case of Bordeaux Rouge up to 13% for Sauternes. For most Bordeaux AC, checking of the wine's characteristics by analysis must be supplemented by official INAO tasting. AC Bordeaux Rouge at present fetches three or more times the price of AC Bordeaux Blanc. AC Bordeaux Mousseux must be made by the *méthode Champenoise*; non-AC sparkling wines can be made by the cheaper tank method (Eugène Charmat, inventor of the tank method, came from Bordeaux). Every wine region has its

1 Département of Gironde.

spirit production: here Eau-de-vie-de-vin-originaire-d'Aquitaine and Eau-de-vie-de-marc-originaire-d'Aquitaine are produced, and in the city of Bordeaux itself, the Cusenier firm produces liqueurs.

Grape Varieties

The regulations limit the variety of grapes, both black and white, that may be used for table wines, and a lesser number for AC wines, in the Bordeaux region. The important black varieties are Cabernet Sauvignon, Cabernet Franc and Merlot. The Cabernets are the backbone of Bordeaux red wines and Merlot is noted for its softness and early ripening. The important white varieties are Sauvignon (a heavy cropper), Sémillon, and Muscadelle. Care must be taken not to confuse the black Cabernet Sauvignon with the white Sauvignon. Bordeaux quality wines are made from blends of grape varieties.

Viticulture

In France, viticultural methods are strictly controlled by law. Not only the selection of vine varieties but also the density of planting and the methods of pruning and training are decreed. Generally in Bordeaux the vines are planted one metre apart in rows which have one metre between them. The vines are trained on two wires, close to the ground to help in their ripening. They are most often pruned according to the method known as Guyot Simple, in which three or four buds are trained from one main cane, with a reserve bud for future use. There are no special frost precautions in Bordeaux, but the system of *buttage* (in which soil is ploughed up to the roots in winter) is usually adopted. As the young shoots develop, the vines are sprayed with Bordeaux mixture (copper sulphate, lime and water); this spraying is continued, after flowering, at intervals and particularly after rain, until three or four weeks from harvest; from July on, they are also dusted with sulphur powder against mildew.

1 What are the two main rivers of the Bordeaux region?
2 What do the forested sand dunes of the Landes do for the vineyards of Bordeaux?
3 What is palus? Does it connect with fine wines?

VINIFICATION

Red Wines: The grapes are destalked and crushed by hand or machine, and the must is pumped into fermenting vessels varying from open-topped wooden vats to sealed stainless steel vats, depending upon the property. Fermentation continues at a temperature of 25–30°C for about fourteen days. The must is not generally sulphured to kill the wild yeasts as is done in some regions. After this time, the young 'running wine' completes its fermentation and is matured in oak barriques of 225 litres (48 gallons) capacity for 18–24 months. The casks are topped up weekly and the contents racked at three-monthly intervals, starting in December or January. After maturing for from one and a half to two years, the wine is fined, usually with egg white, and is bottled one to three months later. The 'press wine' is kept for blending at one year if required, or if not, for consumption on the property.

Dry and Medium Dry White Wines: The grapes are pressed without destalking, and usually sulphured; the third and fourth pressings are fermented and matured separately. Fermentation takes place in tiled, concrete, or stainless steel vats, and the wine is matured in oak vats for six to nine months. Racking takes place every three months, and the wine is fined with isinglass or bentonite, an absorbent clay (montmorillonite).

SOUTH-WEST OF THE GARONNE

This area includes the districts of Médoc, Graves, Sauternes & Barsac, and Cérons; their climates are similar, all being protected by the pine forests of the Landes to the West. Gravel predominates in the soil, but peters out to the south. The subsoils contain *alios* (ironstone) and calcareous deposits, again contributing to the firm characteristics of the wines.

1 Garonne and Dordogne.
2 Protect them from salty south-west winds.
3 Alluvial land close to rivers. No.

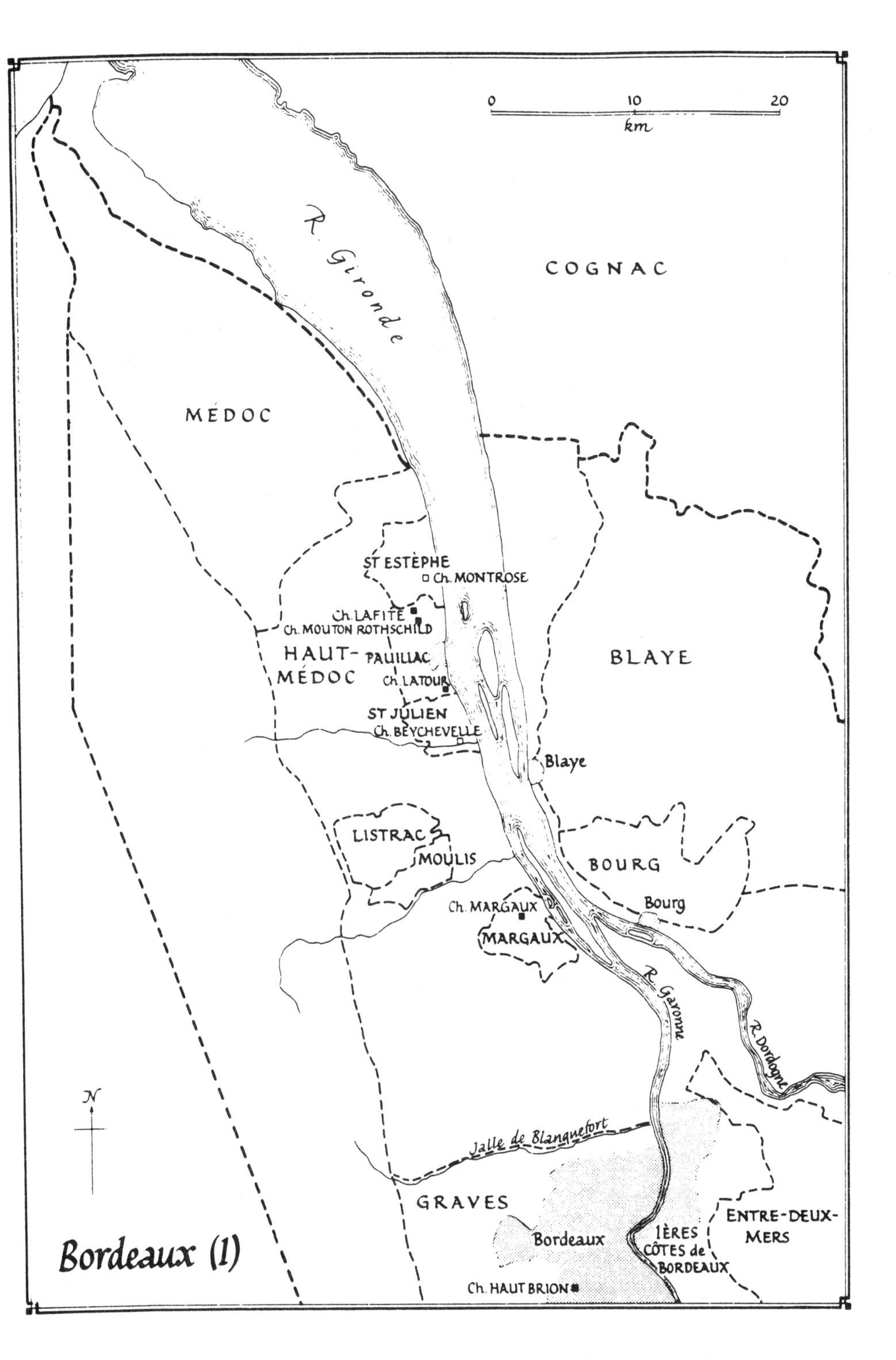

1 What do the following Bordeaux terms convey? (a) Courtier, (b) Négociant-éleveur.
2 Where do red and white wines predominate in the Bordeaux region?
3 What range of minimum strength is required for Bordeaux ACs?
4 What are the three most important white grapes of Bordeaux?
5 What is Bordeaux Mixture?

MÉDOC

The Médoc lies between the Jalle (stream) de Blanquefort, north of the city of Bordeaux, and the Pointe de Grave. The southern part of this district, from St Seurin de Cadourne, is the Haut-Médoc, and it is here that the gravel bank rises to its maximum height of 48 m above the Gironde. Only red wines produced in the Médoc have AC Médoc or Haut-Médoc, though some white wines are produced there under AC Bordeaux. The classification of 1855, since modified very slightly, remains a simple and faithful guide to the wines of the Médoc. The classification is not to be confused with the AC system, however. The latest EEC labelling regulations allow the terms 1er cru and 2nd cru only; quoting '3rd', '4th', or '5th' growth has fallen into disuse – they are labelled simply as 'crus classés'. In the Haut-Médoc there are twenty-seven communes, six of which – Margaux, St Julien, Pauillac, St Estèphe, Listrac, and Moulis – have their own Appellations. The Médoc has many internationally famous châteaux; also there are several wine-making co-operatives, and a great many lesser-known châteaux.

1855 Classification: Under the 1855 Classification there were sixty-two classed-growth châteaux. Later they were followed by seven 'Exceptional Growths' and the remainder were classed as 'Bourgeois' or 'Artisan'; under current EEC rules, these are all classed as 'Bourgeois'. The supremacy of Margaux, St Julien, Pauillac, and St Estèphe is a matter of simple geography. They occupy supreme positions on the gravelly mound of the Haut-Médoc, above the alluvial deposits in the lower areas. Here the small gravel pebbles (*cailloux*) retain heat from the sun to continue ripening the low-trained grapes by night, and also provide good drainage.

Vines: Predominantly, the vines of the Médoc are Cabernet Sauvignon, with Merlot, Cabernet Franc, and Malbec in lesser quantities. The Petit Verdot and Carmenère may still be found in some properties. The Cabernet Sauvignon gives wine which can be hard when young but is long-lasting, while Merlot gives wine of more supple balance. The vignerons constantly ponder the dilemma of replanting their vineyards; the grape on the older vine improves in

1 (a) Wine broker, (b) A Bordeaux shipper who stores and matures young wines.
2 Red in the north and white in the south.
3 From 10% for Bordeaux rouge to 13% for Sauternes.
4 Sauvignon, Sémillon, and Muscadelle.
5 Copper sulphate, lime, and water.

quality but reduces in quantity, while the young vine entails an initial labour cost (it bears little or no harvest in its first two years, and cannot be used for AC wine until the fourth year after grafting). Usually the grapes are ready to harvest by the third week of September. Because there is some danger at the time, the vintage can vary considerably according to the time of picking; 2.5 cm or one inch of rain adhering to the grapes can lower the sugar by one degree Baumé, and wet grapes are in danger of rotting.

Chaptalisation, the addition of sugar to fermenting must in order to increase alcohol production, is permitted in the Médoc, and often used. Here the *chais* (wineries) are above ground, because the hillocks are so low, and the water table so high, that cellars would be flooded. The *vin nouveau* is fermented in a separate part of the chai from the *vin vieux* or wine of the previous year, not yet fully cask-matured. The average period of maturation is one and a half to two years in cask, followed by a longer period in bottle, depending upon the hardness of the wine: the higher the tannin content, the longer the wine will need to mature.

Graves

The Graves district lies along the left bank of the Garonne and runs south-east from the Haut-Médoc almost to the Département boundary east of Langon; it includes the city of Bordeaux. In all it is some 65 km in length, and varies in width from 6 to 19 km.

The sandy gravel plain which gives Graves its name has a subsoil of ironstone, clay, or chalk. The gravel gives a taste, '*goût de terroir*', to the wines of the Graves, which tend to be drier than those of the Médoc. The climate is influenced by the Atlantic, and the warm clear autumn weather in the south induces mists from the woodlands surrounding the vineyards. There is more rain in the north of the district, but the risk of hail increases in the south. The wines vary from the fine red Graves in the north, through dry white wines in the centre, to mostly ordinary wines at the southern end of the district.

Classification: Graves has fifteen châteaux classified in 1959, of which six are classified for both red and white wines, a further seven for red

1 To what AC are the white wines of the Médoc entitled?
2 What is the approximate capacity of a barrique?
3 What is the usual fining agent for red wines in Bordeaux?
4 What is bentonite and what is it used for?

wines only, and two for white wines only. Château Haut-Brion, one of those classified for red wines only, was also classified in 1855 with the finest growths of the Médoc. The six communes containing classified châteaux all lie in a 15 km strip south-east from Bordeaux. Near the City is the new AC Pessac-Léognan, west of the Garrone. There are no grades within the 1959 Appellation. In addition to the classified Graves vineyards, Féret lists some three hundred 'Principal Vineyards', of which rather more than half produce white wine only, having the characteristic delicate and fruity flavours of the district.

For red wines, the Cabernet Sauvignon, Cabernet Franc, and Merlot are planted; for white wine the vines are Sauvignon, with Sémillon and some Muscadelle.

Sauternes and Barsac

This small district forms an enclave of Graves next to the Garonne, about 30 km south-east of Bordeaux on low gravel hillsides facing north. The subsoil is of clay and sandstone, with more limestone occurring in Barsac. The microclimate is an interesting one. The atmosphere is moister than in the Médoc and Graves, yet receives more heat in the autumn after morning mists have cleared, encouraging the development of *botrytis* as 'noble rot' and also as the hazard 'grey rot'. There is a high risk of hailstorms from August onwards which can damage the grapes. Rockets or aeroplanes are used to seed hail-clouds to disgorge their load before reaching the vineyards.

The 1855 Sauternes classification singles out Château d'Yquem as the sole 'First Great Growth', and goes on to list eleven vineyards as Premiers Crus or First Growths, and twelve as Second Growths. The AC Sauternes is reserved only for white wines with high alcohol content at 13% vol. Barsac, which has its own AC, enjoys AC Sauternes also, but not vice-versa. If Sauternes fails to reach 13% alcohol, the wine is declassified to AC Bordeaux.

Great care is taken in selecting the over-ripe and shrivelled grapes, involving successive pickings of individual grapes by local womenfolk over a long period, stretching even into December. The vines are grown only on the gravel slopes, and not on the clay plain or in the alluvial valleys. For classic Sauternes, grapes are used in the

1 AC Bordeaux Blanc.
2 225 litres (48 gallons).
3 Egg-whites (four to seven per barrique).
4 An absorbent clay used for fining.

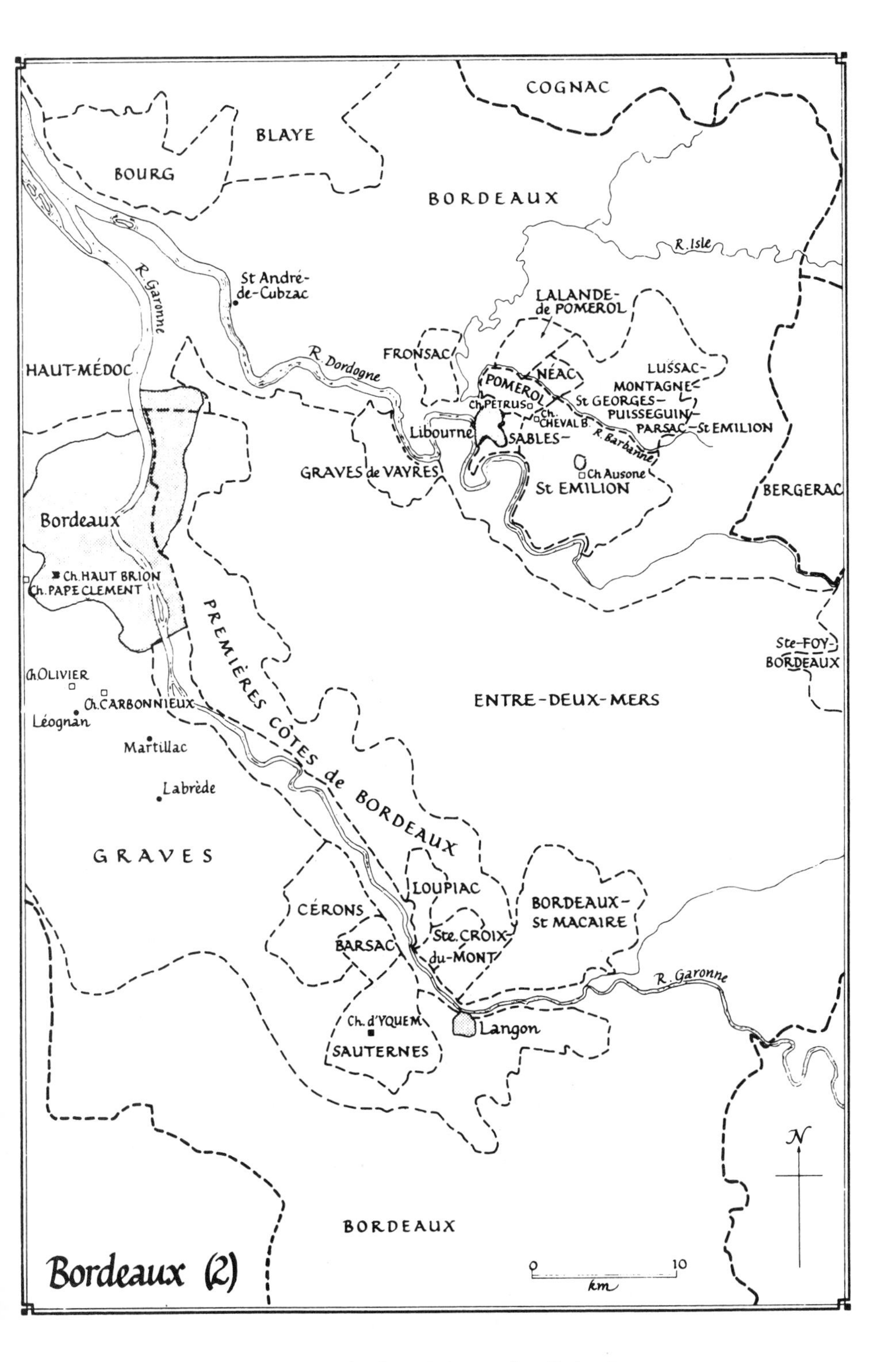

1 Which six communes in the Médoc have their own Appellations?
2 Why are the chais of the Médoc sited above ground?
3 Are fine red Graves found in the north of the district or in the south?

proportion of two-thirds Sémillon to one-third Sauvignon, pressed five times (there are no stalks as the grapes have been picked individually). The first pressing yields the best must. Wild yeasts are killed by the addition of SO_2, and the must is fermented in oak at 18° to 20°C by natural yeasts.

The wine is racked every three months, and the finest wines will be left to mature up to three years in cask before bottling, after fining with ox blood. In years when the Sauvignon vine over-produces, the excess is vinified with some Sémillon to produce AC Bordeaux dry white wines. Sauternes will improve in bottle for a long time provided the storage conditions of constant temperature and protection from light and oxygen are observed. Exposed to air or light, Sauternes maderizes rapidly. The dry white wines are matured only for a short time, to bring out a fresh fruitiness in the lighter wine.

Cérons

This small district adjoins Barsac on the left bank of the Garonne. The soil is sandy and gravelly on a chalk subsoil. Cérons produces semi-sweet wines from Sémillon and Sauvignon grapes, occasionally affected by noble rot. AC Cérons requires a strength of 12.5% vol. alcohol and applies only to white wines. If the strength is lower, the wines may be declassified to AC Graves Supérieure or AC Bordeaux.

BETWEEN THE GARONNE AND THE DORDOGNE

The district lying 'between the rivers' is hilly, with great local variations of soil and subsoil on a theme of chalk clay. On its south-west border, steep slopes down to the Garonne produce excellent dry and sweet white wines, and lower down the river sound red wines are made. Altogether about six times as much white wine as red is produced, and the white is of better quality. The

1 Margaux, St Julien, Pauillac, St Estèphe, Listrac, and Moulis.
2 Because the hillocks are low and the water table high, which would flood cellars.
3 In the north.

principal vine for red wine is the Merlot and for white wine, the Sauvignon; both are heavy-cropping varieties.

Here high training of vines may be seen in some vineyards. The vines are grown to two metres in height on wires, in rows two to three metres apart. This makes spraying, cultivation, and harvesting easier, and there is no need for straddle tractors. Another feature of this southern district of the Bordeaux region is the occasional use of *macération carbonique*, a fermentation process in which CO_2 is retained under slight pressure. This removes colour from the skins more quickly in red wine production, and helps to produce a fresh light wine needing a short maturation period.

Premières Côtes de Bordeaux

This district, on the right bank of the Garonne, stretches up-river for some 60 km from a point opposite Bordeaux to Langon. In width it is rarely more than 2 km. The produce conforms to the pattern in Graves – more red wine in the north, more sweet white wine in the south. Both share the AC Premières Côtes de Bordeaux, and provided the alcohol level meets the legal requirement, thirty-seven villages (e.g. Langoiran) may add their name to the Appellation.

Loupiac, Ste Croix-du-Mont, and St Macaire

To the south, opposite Cérons and Barsac, are several enclaves on the limestone hillside, each entitled to its own AC, but only for sweet white wines. These wines are not as good as Sauternes, but generally they are better than other Premières Côtes, and improve with bottle age.

Entre-deux-Mers

This undulating, hilly territory is the largest in the Bordeaux region, but it is also the least heavily planted. Clay soils predominate over gravel subsoils. The Guyot Double method of training is used quite

1 When were the vineyards of Graves classified?
2 What element is essential to Sauternes and Barsac production?

extensively. In the Entre-deux-Mers district, a considerable proportion of the harvest goes to large co-operatives and wine-making firms.

Only the dry white wines of 11.5% vol. alcohol have AC Entre-deux-Mers, the red wines only meriting AC Bordeaux or Bordeaux Supérieur according to strength (10% and 10.5% respectively).

GRAVES-DE-VAYRES AND STE-FOY-BORDEAUX

The small district of Graves-de-Vayres borders the south bank of the Dordogne opposite Libourne, in the north-west of Entre-deux-Mers. Its soil is gravel, in contrast with the surrounding clay, and its wine is accordingly superior. Its tiny production of red wine, averaging only 7000 hl per annum, is near in quality to the wines of Pomerol and St Emilion, being hard in youth but maturing quickly. The white wines are of bourgeois quality.

Also on the south bank of the Dordogne, but at the north-eastern tip of Entre-deux-Mers adjoining Bergerac, is Ste-Foy-Bordeaux. On its soil of chalky clay rather better than ordinary white wines are produced. A small production of red wine shares the AC Ste-Foy-Bordeaux but is of lesser quality.

NORTH OF THE DORDOGNE AND GIRONDE

As with the Médoc, the wine-producing districts to the north of the Dordogne are dedicated to the production of fine red wines. Generally the territory is hilly, with sand or gravel soils over limestone or clay, which sometimes appear as outcrops. Here stands Libourne, the second port of the Bordeaux region, at the confluence of the Dordogne and its tributary, the Isle. It was named after Roger de Leyburn, the English Seneschal of Guyenne, who is said to have built the town and its fortifications. Libourne conducted a lively trade in wines from the Haut-Pays (Languedoc) and Rhône long before the Bordeaux vineyards were developed. It remains to this day the trading centre for St Emilion, Pomerol, and Fronsac wines.

1 In 1959, except for Château Haut-Brion which was included in the 1855 classification.
2 *Botrytis* – as 'noble rot'.

In the hilly country to the east is Castillon-la-Bataille which, like Hastings in Sussex, commemorates a French victory over the English. It was here in 1453 that the French finally regained Guyenne and Bordeaux after some three hundred years of English rule. Its rather better than ordinary red wines have AC Bordeaux-Côtes-de-Castillon.

The climate north of the Dordogne starts to have more continental and less maritime characteristics; being further from the sea, there is less rainfall, and the variation between hot summer and cool winter is more marked. The main districts in this third of the Bordeaux region are St Emilion and Pomerol with their subsidiary districts, and Bourg and Blaye. In the red wine districts the Cabernet Franc (known locally as Bouchet) and the early-ripening Merlot are the principal vines.

St Emilion

St Emilion lies east of Libourne on the slopes of limestone hills bordering the Dordogne, and on the sand and gravel plateau beyond. The finest properties are divided between these two parts, and there is a confusing use of the term 'Graves' for the wines of the plateau as opposed to 'Côtes' for the wines of the slopes bordering the old town of St Emilion itself. Ch. Ausone, the most famous Château on the Côtes, has a cellar under its growing vines on the edge of the limestone escarpment round the town. The equally famous Ch. Cheval Blanc lies out on the gravelly plateau which is more prone to the effects of spring frost.

The vines are trained low (Guyot Simple), and the earliest date of harvesting is fixed by the *Ban de Vendange* and instituted with great ceremony. Many chais are underground, for here there is no risk of flooding as in the Médoc. The 1954 Classification divided the properties into Premiers Grands Crus, of which there are twelve, and Grands Crus, of which there are seventy-three. Chx. Cheval Blanc and Ausone head the list, but thereafter the list is alphabetical and of little help in judging the relative merits of each. Eight communes are entitled to AC St Emilion and five more are entitled to add St Emilion after their AC name. These AC are for red wine only. They

1 What happens if a bottle of Sauternes is exposed to air or light?
2 What is macération carbonique?
3 What sort of wines are produced in the Premières Côtes de Bordeaux?
4 What is the soil in Loupiac, Ste-Croix-du-Mont, and St Macaire?

are fuller-bodied than the clarets of the Médoc, mature more quickly, are less hard, and are sometimes likened to Burgundies.

Pomerol

Pomerol lies to the west, between St Emilion Graves and the river Barbanne. Its best property is Ch. Pétrus, only some 600 m from Ch. Cheval Blanc. The soil is a continuation of the sand and gravel plateau stretching out from St Emilion, with some clay outcropping and, beneath the best properties, an ironstone subsoil. There has been no classification for Pomerol, for which the AC is for red wines only. The wines of Pomerol come from the same grape varieties as those of St Emilion; they are, however, harder in character than the St Emilions, more nearly resembling the clarets of the Médoc, though experts will distinguish them from both. To the north and west of Pomerol are the small sub-districts of Lalande-de-Pomerol and Néac with the Barbanne stream separating them. Lalande is hillier than Pomerol with a more fertile topsoil of clay over the gravel. Generally the wines of Lalande are of a lesser quality than those of Pomerol, resembling more nearly the Médoc bourgeois growths. There are some good wines, however, and Ch. Bel Air is the best of its vineyards.

Fronsac (Fronsadais)

Beyond the river Isle, to the north-west of Pomerol, is Fronsac, with the high bluff of Canon Fronsac facing Vayres across the Dordogne. It is necessary to distinguish between the two AC – Côtes Canon Fronsac and Fronsac: wines with the first AC are made on the limestone bluff overlooking the Dordogne and are better than the AC Fronsac from grapes grown on richer soil. The Fronsadais are heavy, meaty wines, which can be very appealing, especially if allowed to age in bottle for a few years. As they are less expensive than St Emilion and Pomerol wines, this can be well worthwhile. Only red wines are entitled to the Appellation.

1 It will maderize rapidly, turn brown in colour and bitter in flavour.
2 Retention of CO_2 under pressure to accelerate coloration of red wine and reduce maturation period.
3 Red wine in the north and sweet white wine in the south.
4 Limestone hillside with clay topsoil.

Bourg and Blaye

These districts are situated on rolling hills for forty km along the right bank of the Gironde, from the confluence of the Garonne and Dordogne. Each has its own AC. The soils are marl over gravel with occasional ironstone, and are too rich to give great quality. Bourg and Côtes de Bourg, Blaye and Premières Côtes de Blaye are all ACs for both red and white wines, but the reds are superior to the whites. The reds of Bourg are in turn superior to those of Blaye, being fuller-bodied, generous wines improving with age, like the Fronsadais. White wines may be made from up to 30 per cent of heavy-yielding grapes, such as Colombard and Ugni Blanc. As an exception to the normal for this part of the region, Blaye grows more white grapes than black, and the white wines of AC Côtes de Blaye are of higher quality than those of AC Blaye.

REGIONS AND DISTRICTS ADJOINING BORDEAUX

Bergerac

This AC region lies immediately to the east of the Bordeaux region and the river Dordogne runs through the centre of it. The climate is a little more continental than that of St Emilion which its northern part adjoins, though the grapes are similar – Cabernet and Merlot for red wines, and Sauvignon and Sémillon for white wines. The soil is generally sandy clay with some lime; the best areas have gravelly subsoils with some iron. Thus it seems that the region possesses Bordeaux blessings in diminished scale, which is quite true of the wines also.

The region is smaller than Bordeaux and can be divided generally into two parts, north and south of the river Dordogne. North of the Dordogne, white wines are made in the AC districts of Montravel, Côtes de Montravel and Haut-Montravel in the west. The best red wines come from the AC district of Pécharmant. These are light in bouquet but nevertheless fine, made mostly from the Malbec grape,

1 Is AC Entre-deux-Mers for red wine, white, or both?
2 What are the four main districts of Bordeaux north of the Dordogne?

for which the official name is Cot. Other AC wines are Bergerac, Bergerac sec, Côtes de Bergerac, and Côtes de Bergerac Moelleux.

South of the Dordogne river, on clay soil, the AC district of Monbazillac produces light, sweet white wines of high alcohol, similar to Sauternes but with less bouquet, from the same grapes. The AC district of Saussignac adjoins it to the west. A further AC, adjoining Ste-Foy-Bordeaux, is Rosette, which is a lighter, sweetish white wine with a pronounced Muscat flavour. Because they are little known, these wines tend to be good value for money.

Côtes de Duras

This small AC region lies immediately south of Ste-Foy-de-Bordeaux between the southern half of Bergerac and the Entre-deux-Mers district of Bordeaux. The soil is generally sandy, with a little lime; the region produces red wines primarily, although the Appellation extends to white also. The Côtes de Duras lie on the banks of the little river Dropt and its tributaries. Both red and white wines are made from the Bordeaux grape varieties, and perhaps the only reason they are not included in Entre-deux-Mers or Ste-Foy is that they lie in a different Département, Lot-et-Garonne.

OTHER DISTRICTS

Although there are no VDQS in the Gironde, one VDQS district lies near the river Garonne a few kilometres outside the Département on alluvial lime soil. Côtes de Marmandais VDQS produces sound red wines and sweet wines of some quality from Bordeaux grape varieties, also some brandy. AC Côtes de Buzet lies further upstream, adjoining the brandy region of Armagnac. This district produces red wines from the Malbec grape which are often blended with the wines of the Gironde. Sweet and dry white wines are made. A new VDQS, Côtes de Brulhois, lies upstream between Buzet and Agen. The study of light wines of south-west France would be incomplete without reference to those grown in the foothills of the

1 Only the white.
2 St Emilion, Pomerol, Bourg, and Blaye.

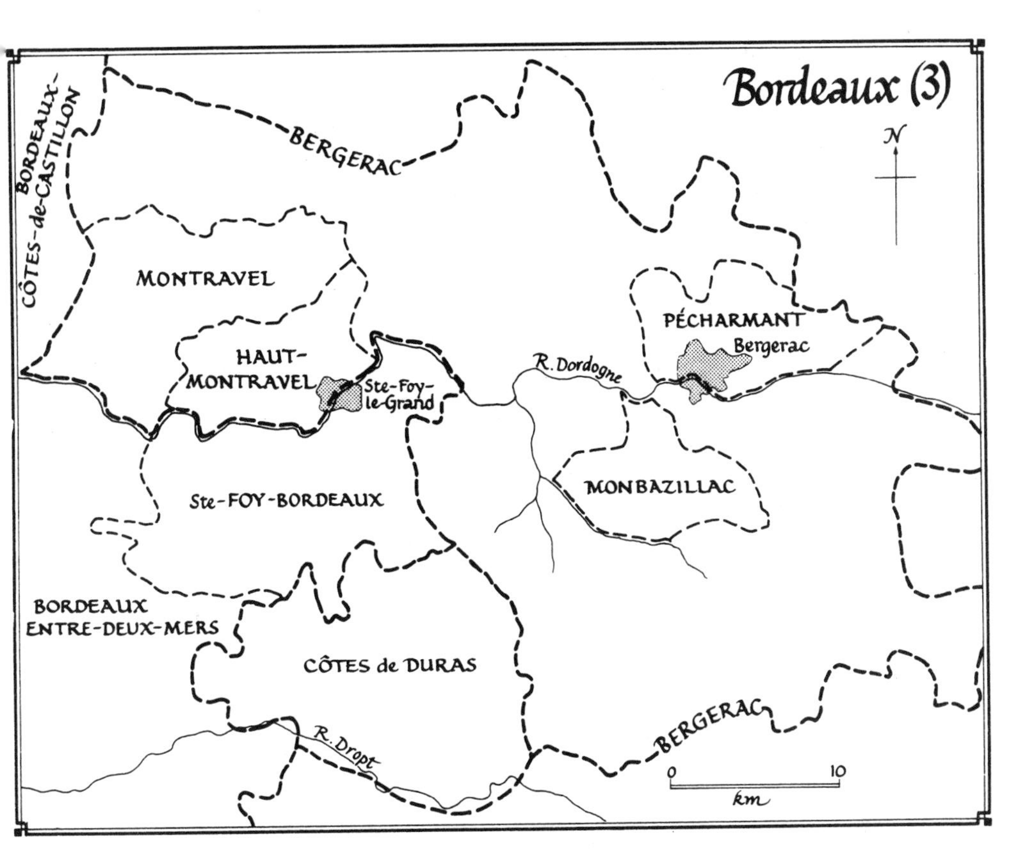

Pyrenees, in the Départements of Hautes-Pyrénées, and Pyrénées-Atlantiques. This misty land beyond the plains was well known to the British between the wars. Pau had its tennis and cricket clubs; nor were these only for the gilded rich escaping the English winter, but also for the poor wine merchants seeking the good AC wines of Madiran and Jurançon for their customers.

The soils are of sandstone in the foothills, becoming sandy in the plain; the grape varieties are often strange to the remainder of the south-west. The vines are frequently trained on 2 m high trellises.

Madiran

The vineyards were formerly planted with Cabernet, but now with Tannat, a local grape which gives a full-bodied red wine of good

1 What is the best vineyard of Pomerol?
2 What are the wines of the Fronsadais like?
3 How do the red wines of Bourg and Blaye compare with the white wines?
4 Which is the best red wine of Bergerac?

flavour. In this district are also found the sweet or dry white AC wines of Pachérenc-du-Vic-Bilh, Pachérenc being the local name for the Manseng grape, and Vic-Bilh the name of the village.

Jurançon

Further south again are the white AC wines of Jurançon, whose principal town is Pau. These are sweet or medium-dry wines according to the year, and have a distinctive flavour which has been likened to peaches, camomile, and even carnations. The grape is Manseng, as for AC Pachérenc-du-Vic-Bilh.

Vins de Béarn

These AC wines from the valley of the Gave de Pau take their name from the province. Like sauce Béarnaise, the red, white, or rosé wines are piquant yet velvety.

Irouléguy AC

Irouléguy, centred round the border village of St Etienne-de-Baïgorry, produces red wines from local grape varieties – Tannat chiefly – in the most romantic setting one could hope to find, nestled in the deep valleys of the Pyrenees. It also produces a white wine of quality.

VDQS

Tursan covers about thirty villages in the Landes Département on sandy soil, and produces velvety red wines (and some whites too) from Picpoul, Malbec and Tannat grapes.

Côtes de St Mont VDQS, lies in Armagnac to the north of Madiran; red, white, and rosé wines are made from local grape varieties.

1 Ch Pétrus
2 Heavy, red, meaty, and appealing – especially when a few years old.
3 The red wines are generally superior.
4 AC Pécharmant.

EAST AND SOUTH OF BORDEAUX

North of Toulouse there are several AC and VDQS districts which lie on tributaries of the Garonne, yet are quite different from the other wines of the south-west. Nor, although some of them use the same grapes as Languedoc districts, do they fit in there either. They are a link joining the circle of the French vineyards.

AC Gaillac

This AC district lies in the Tarn valley some 40 km upstream to the east as far as the little town of Albi. Here, on clay-lime soil, fine white wines are made from the Mauzac grape. An inner delimited district, AC Gaillac Premières Côtes, gives a wine with more body. As with the wines of Limoux, the wines are best in youth and become bitter after only a few years. Also like Limoux and by the 'méthode rurale', sparkling and semi-sparkling wines are made; these are entitled to AC Gaillac Mousseux. There is also an AC Gaillac Doux. The AC extends to red wines also, and some sound examples are now appearing on the British market.

AC Cahors

70 km to the north-west of Gaillac lies the little town of Cahors on the Lot, another tributary of the Garonne. Here the valley is much steeper and the soil is of harder limestone. The wine, made from the Malbec grape (Cot), used to be known as 'the black wine of Cahors' and was astringent and tannic. Nowadays, fermentation on the skin is not allowed to go on so long, and the resultant wine is more noted for its fruitiness.

AC Côtes du Frontonnais and la Villedieu VDQS

These districts lie between the Garonne and the Tarn, and are noted for pleasant red wines, full-bodied, rich in colour and alcohol, from

1 Where is Bergerac? Is it part of the Bordeaux region?
2 What is AC Monbazillac wine like?
3 (a) Around what river do Côtes de Duras lie? (b) Are their wines AC?
4 Name two VDQS on the Garonne upstream from the region of Bordeaux.

the Negrette grape grown on sandy-clay soil; the best vineyard is that of Pech-Langlade in the Côtes du Frontonnais, whose wines improve with age. Some white wines made from the Mauzac grape are not as good as the red.

Moving up the Lot into the mountains of the Auvergne, the vineyards get fewer and fewer, but 100 km from Cahors, three VDQS districts, Vins d'Entraygues et du Fel, Marcillac, and Estaing, are found around the junction of the Lot and the Truyère. Light and fruity red and rosé wines are produced from a local grape called Fer. Entraygues and Estaing also have VDQS for white wines, made from the Chenin grape of the Loire with the Mauzac of the Midi. Comté Tolosan is the regional Vin de Pays.

LOCAL VINS DE PAYS

Hautes-Pyrénées: Bigorre
Gers: Côtes de Gascogne (includes Côtes de Montestruc and Côtes de Condomois)
Lot-et-Garonne: Agenais
Lot: Coteaux de Glanes
Tarn-et-Garonne: Coteaux de Quercy, Coteaux et Terrasses de Montauban, and Saint Sardos
Tarn: Côtes du Tarn
Aveyron: Gorges et Côtes de Millau

This completes our study of the wines of South-West France; although the wines of Limoux are included under this heading by the EEC authorities, these lie on the river Aude which flows to the Mediterranean, and we have therefore included them with Roussillon under 'Mediterranean France'.

1 Immediately east of the Bordeaux region but not part of it.
2 Sweet white wine, similar to Premières Côtes de Bordeaux.
3 (a) The river Dropt. (b) Yes.
4 Côtes de Marmandais and Côtes de Brulhois.

Northern France

CHAMPAGNE

Champagne is one of the crossroads of Europe; invaders through the ages have traversed it, and yet it remains the same. The Romans were there, made wine, and built towns which were destroyed by the barbarians. The towns were built up again, and the potential for wine attracted the Church, which established great abbeys and cathedrals; St Rémi, patron saint of the region, gave his name to Reims, and both baptized and crowned Clovis as King of the Franks in its cathedral – with the wine of Champagne – in AD 496. Many other Kings followed in their turn, and the abbeys grew in power and knowledge: that at Hautvillers took special care in wine-making,

1 Name two AC wines made from the Manseng grape.
2 Where is Madiran? What sort of wine does it produce from what grape?
3 (a) Where is the district of Gaillac? (b) For what wine is it famous; what is the soil; what grape is used?
4 Why was 'the black wine of Cahors' so-called?

and in blending the different *cuvées* of its vineyards into a harmonious whole, a process perfected by its cellarmaster in 1670–1715, Dom Pérignon.

He is also credited with bringing to the north the idea of sealing bottles with cork, from his studies in Catalonia whence the cork at that time came – and as the cold winters of Champagne came early, inhibiting fermentation, and the acid wine had a strong malolactic fermentation, this was a desirable object.

The power of the abbeys ended with the Revolution and with the wave of anti-clerical feeling that followed; however, the fame of the now sparkling Champagne had become established, and there was no lack of *citoyens* ready to take on production. There were already many firms making the glorious wine – Ruinart, Möet & Chandon, Clicquot, Lanson, Roederer, Heidsieck and others, all founded before the Revolution; and by this time Champagne was a very popular drink in England.

Napoléon Bonaparte was a great friend of Champagne, and Champagne of him; yet, in spite of the almost complete blockade between England and France during the wars, Champagne still got through to England. It must be a great peacemaker, for Reims, occupied by the Russians after Waterloo, suffered not at all, and trade with Russia increased tremendously.

CONTROL

The wars of 1870, 1914, and 1939 raged over the fields of Champagne, and each damaged the vineyards, countryside, and cellars more; yet out of each the Champenois gathered strength. In 1941, Comte Robert-Jean de Vogüe of Möet & Chandon was Mayor of Épernay; not only did he resist the occupying Nazis in their attempts to loot the cellars – to the point of being sent to a concentration camp – but also, in 1942, founded the *Comité Interprofessionel du Vin de Champagne* (CIVC) to regulate the growing of vines and the production of wine in the region.

The interests of growers and of makers of Champagne (the 'Houses') are opposed: the one is a seller of grapes, the other a buyer; but many small growers combine the functions of growing grapes and making Champagne. The CIVC is composed of a number of growers and an equal number of makers who have to agree; if they

1 Pachérenc-du-Vic-Bilh and Jurançon.
2 Foothills of the Pyrenees. Full-bodied red AC wine from the Tannat grape.
3 (a) In the Tarn Valley near Albi. (b) White wine (AC Gaillac) and sparkling wine (AC Gaillac Mousseux); clay-lime soil; the Mauzac grape.
4 Lengthy fermentation on the skins of the Malbec grape produced an astringent and tannic wine.

cannot, which is rare indeed, the Government-appointed Head of the CIVC arbitrates. The service has a small permanent staff, financed by a tax on each kg of grapes and each bottle of wine; this income also provides for publicity and education.

Climate

Climate is one thing that the CIVC cannot control, although it takes note of it in deciding the dates each year when the harvest should begin. The region has cold winters, lying as it does between the 49th and 50th parallels of north latitude; but, fortunately, it usually enjoys long warm summers giving an average temperature of 10°C (50°F) – the extreme lower temperature for the full ripening of grapes. Frost is a problem in spring, and grafts must be done in greenhouses, not in the field.

However, the topography of the region helps, in that the wooded tops of the hills seem to absorb the worst of the winter frosts, and moderate the humidity in summer. The nature of the soil also helps.

Soil

The principal base soil of the region is belemnitic and oolitic chalk, like the South Downs of England; this layer was raised up from the Ile de France in a past age by earthquakes, which also spewed out a thin layer of tertiary deposits on top. The chalk, under such a thin topsoil, acts as a night storage heater in summer, conserving the heat to help ripen the grapes: it provides easily worked accommodation for cellars underground and gives the exact composition of minerals required for the best wine. The occurrence of chalk controls the choice of vineyard sites, for it is on this type of soil that they were originally planted. This soil does not occur at all places throughout the region, and this is taken into account when permitting new plantations. Furthermore, the law of Champagne grades the vineyards in relative quality from 100 per cent downwards to determine the basic price of their crop to the grower. Formerly, the gradings reached down to 50 per cent, but in 1978 the lowest was 77 per cent, and in 1986 it was 80 per cent. Seventeen Crus were graded

1 Name the vines used in the VDQS districts of the Auvergne mountains.
2 Why have the wine areas of Limoux been excluded from South-West France in this book?
3 Name the patron saint of the Champagne region.

100 per cent in 1986; in some Crus, the white grapes were graded differently from the black.

Districts

Champagne is divided into three main districts from north to south: Montagne de Reims; Vallée de la Marne; and Côte des Blancs. The Montagne has an outcrop to the north-west of Reims which is of lesser quality – la Basse Montagne; the Haute-Montagne surrounds the northern and eastern parts of the Montagne de Reims for 25 km from Verzenay to Tauxières. Wines from the Montagne give vinosity and backbone to the Champagne blend; in 1986 the Montagne contained ten 100 per cent communes. The Vallée, whose wines give fruit and body to the blend, has but one 100 per cent commune, Ay. The Côte de Blancs, whence the wines give finesse and elegance, had in the same year six 100 per cent communes, while a further forty-four were graded 90 per cent or higher. In the Département of the Aube, the still wine AC Rosé-des-Riceys has been made for many years from the Pinot Noir grape; the produce of this district is also admissible for Champagne. The districts, and some of the more important villages in them, are shown in the map opposite.

Grapes

The varieties are strictly controlled, and the only varieties now permitted are the black Pinot Noir, an early-ripening grape susceptible to frost damage, the black Pinot Meunier, less fine but later-ripening and therefore a good insurance against frost, the white Petit Meslier, rarely grown successfully in this northern climate and understood to be related to the Sémillon, but producing wine with great bouquet, and the white Chardonnay, which alone fills the 2500 ha of the Côte des Blancs and gives the district its name. The Chardonnay is a better bearer than the Pinot Noir, but is even more susceptible to frost, and takes longer to ripen. The last permitted grape is the white Arbanne, grown in the southern vineyards of the Aube, whose wines are usually excluded from the Champagne cuvées, and where the soil is Kimmeridgian clay.

1 The Fer, Chenin, and Mauzac.
2 Because they lie on the river Aude which flows into the Mediterranean, hence 'Mediterranean France'.
3 St Rémi.

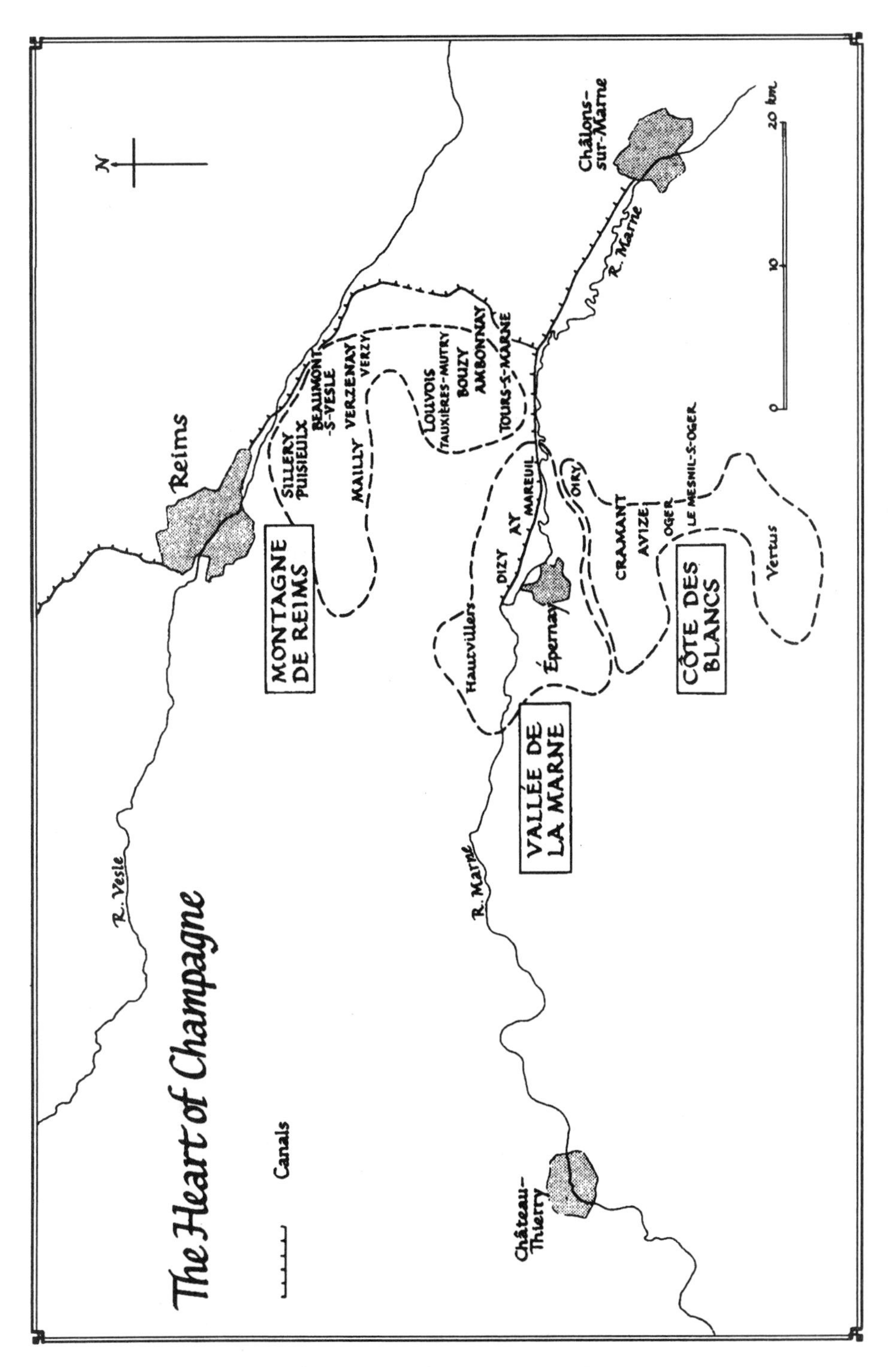

1 Who is claimed to have first perfected Champagne, and where did he work?
2 Name six firms already making Champagne at the time of the French Revolution.
3 What groups have conflicting interests in the Champagne region?
4 (a) In what latitudes does the region of Champagne lie? (b) What is the principal feature of its soil?

Viticulture

This is a particularly hard labour in Champagne, for the soil is very hard, making ploughing difficult; and the topsoil so thin that it must be replenished every three to five years from the black tertiary deposits on the hilltops. In years when this is not necessary manure must be put down, for these vineyards are extraordinary in that they need nitrogen. Moreover, the cool damp climate encourages pests and diseases, necessitating frequent spraying. The Champagne law controls the density of planting, depending on the variety of grape; the Chardonnay needs more room than the Pinot. The law also controls the training method and the extent of pruning. Only three methods of training are allowed in Champagne: the *Taille Chablis*, in which, three, four, or five main branches trained parallel to the ground are allowed, each pruned to four (Pinot) or five (Chardonnay) buds; the *Cordon de Royat*, in which a main branch is trained 0.6 m high with two to three buds on each shoot; and the *Guyot*, either *Simple* (one branch with ten buds), or *Double* (two branches with ten buds each). In all cases, a replacement shoot is retained for use in future years. The pruning is best done in March, although the work may take more than a month; done earlier, frost may damage the cut shoots; done later, the shoots may 'bleed' too much and weaken the vine. It is the same with roses.

Harvest

The date for the start of the harvest is controlled as in other areas (and may vary by as much as a month from year to year), but in Champagne separate dates are fixed for white and black grapes, and for different Crus; furthermore, the date by which the harvest must finish is also stipulated. The permitted yield per hectare is fixed each year, depending on the generosity of the crop. The price to be paid to the growers for their grapes is also fixed each year by negotiation within the CIVC, as are all other variables: the price each year is related to the sales of Champagne in the previous year.

Vinification

The amount of the harvest that may be used for making Champagne is limited to 100 litres from 150 kg of grapes – quoted as 2666 litres

1 Dom Pérignon, as cellarmaster at the Abbey of Hautvillers.
2 Clicquot, Heidsieck, Lanson, Moët & Chandon, Ruinart, and Roederer.
3 The growers of grapes and the producers of wine.
4 (a) Between latitudes 49° and 50°. (b) Chalk subsoil.

from a standard press-load of 4000 kg (4 tonnes). However, now that the Vaslin and Wilmes-type presses are more widely used in Champagne, with varying loads, it is easier to relate the yield to the law, which also gives the percentage of each pressing of the permitted yield, as follows:

First pressing of 150 kg yields 38 litres of must Second pressing of 150 kg yields 22 litres of must Third pressing of 150 kg yields 15 litres of must	Vin de Cuvée 75 litres
Fourth pressing of 150 kg yields 8 litres of must Fifth pressing of 150 kg yields 7 litres of must	Premières Tailles 15 litres
Sixth pressing of 150 kg yields 10 litres of must	Deuxièmes Tailles 10 litres

Another 10–12 litres of juice (*Rebêche*) is extracted to be fermented and distilled for industry. The pomace (*marc*) is soaked, fermented, and distilled to produce *Eau-de-Vie – Eau-de-Vie de Marc de Champagne* or *Marc de Champagne*. Some wine made from the *tailles* is distilled to make *Eau-de-Vie-de-Vin de la Marne* or Fine Marne. The better Houses use only the cuvée, and sell the tailles to lesser manufacturers. The juice does not always run clear from the press, usually having a little colour pressed from the black grape skins; while the juice is settling (*débourbage*), a small amount of SO_2 is added to kill the wild yeasts, and this also effectively oxidizes the colouring matters, thus bleaching them. After the first fermentation, the wine must have at least 9.5% vol. alcohol. Then follows the great mystery of Champagne as explored by Dom Pérignon – the blending. The wines from different villages, often from different vineyards, are vinified separately, then blended together; the qualities that each district of the region brings to a blend have already been mentioned, but the biggest question is – shall a Vintage be declared? A Vintage is a fine advertisement, but the House must have a wine of sufficient quality to warrant such an advertisement and must have it in sufficient quantity. For the bread and butter of the House is its blend, and it relies on the fine wines of former years to make good the qualities (and quantity) lacking in later poor years. If cultivation of the vineyard is hard for the grower, construction of his blend (*marque*) is just as hard for the manufacturer; that both realize this makes the CIVC both possible and efficient.

1 How does the law of Champagne affect grape prices?
2 (a) Name the three main districts of Champagne. (b) What qualities does each contribute to the wine of Champagne?

The wines so far described are still wines, capable of being made into Champagne. Some of these wines are sold as still wines with AC – Rosé-des-Riceys has already been mentioned – and the wines of the Montagne de Reims, notably the white wine of Sillery and the red of Bouzy, are also popular. Formerly these could be sold as 'Vin Nature de Champagne' but, as EEC regulations confine the use of the word Champagne to sparkling wines, they are now sold under AC Coteaux Champenois. The red wine of Bouzy is also used, by blending with white wine, to produce the basic still wine for making into pink Champagne. It should be noted that this is the *only* qwpsr rosé wine in the EEC that may be made by blending red and white wines.

The still wine for Champagne, whether white or rosé, is mixed in large tanks equipped with paddles, with the *liqueur de tirage* of cultivated yeasts and sugar. The normal method of corking the bottles into which the wine is removed for its second fermentation is with a crown cork; however, this crown cork is of stainless steel and very strong, since it has to withstand pressures of up to six atmospheres. Inside the cork is a special plastic seal, which extends down into the bottle like the inverted crown of a top hat and will serve to receive the sediment shaken down into it by *remuage*.

Mechanization of the work of the *remueur* is fast being developed, but special machines needed in Champagne are very expensive. In addition to hand remuage, many producers now use *giropalettes* – machines holding metal baskets (in which 300–500 bottles are stacked) which rotate and tilt the bottles at regular intervals – with the same effect. With the crown cork, automatic *dégorgement* (after freezing the sediment and the small amount of wine lying in the cup, now above the cork) is possible, and is generally practised. However, in the case of pink Champagne, the *dégorgeur* will wish to check that the colour of the wine has not been adversely affected by the second fermentation, so will remove the closure by hand – *Dégorgement à la volée*. Different amounts of wine may escape from different bottles during dégorgement, so extra wine is removed from every bottle until they all contain the same amount; a measured amount of sweetening *dosage* is then added to refill them completely before corking.

The resulting Champagne will conform to one of the French sparkling wine styles in Table 1.

1 By the grading of vineyards from 100% (dearest) downwards, and by annual fixing of the 100% price.

2 (a) Montagne de Reims, Vallée de la Marne, and Côte des Blancs. (b) Vinosity and backbone; fruit and body; finesse and elegance, respectively.

Table 1

Sparkling wine styles

Style	*Sugar content*
Extra brut	0 g/l to 6 g/l
Brut	0 g/l to 15 g/l
Extra dry	12 g/l to 20 g/l
Dry	17 g/l to 35 g/l
Semi-dry	33 g/l to 55 g/l
Sweet	Over 50 g/l

The final corks used for Champagne must be of high quality. With the oversize cork needed for Champagne it would be prohibitively expensive to use pure corks, so those used for Champagne are composite: two or three thin discs of the finest cork about 5 mm thick are glued to the main body of the cork, which is made from compressed cork fragments glued together. The fine discs must be on the inside of the bottle next to the wine, and this part of the cork must, by law, be branded with the word 'Champagne'. Until recently this meant that the corks all had to be loaded by hand, but this can now be done mechanically. The middle portion of the corks is waxed to facilitate extraction.

THE LOIRE

The vineyards of the Loire Valley were probably developed by the Romans in the first and second centuries; but the wine districts grouped around the river Loire and its tributaries have little political history to touch the imagination.

At the end of the fourteenth century, Rabelais (born at Chinon) extolled the food and drink of the Loire and the virility that these gave, in stories of the legendary Gargantua and Pantagruel.

It is clear that the early wine trade of the Loire was with England, until the English switched their taste to the red wines of Bordeaux. However, the Dutch had a developing wine trade with Nantes, Angers, and Tours, which remained steady until the Franco-Dutch

1 What are Taille Chablis and Cordon de Royat?
2 State the proportion of must that may be taken from grapes for making Champagne.
3 What is débourbage?

war of 1672. It was then that the vignerons of the Loire turned inward to serve France with their considerable variety of red, white, and rosé wines. Under the conditions of those days, marketing was an arduous task over bad roads beset by highwaymen. Nevertheless the home market was established, and has remained ever since.

Improvements in vinification have enabled the wines of the Loire to travel well so that markets in England, Belgium, and the United States have become important outlets for them. The river Loire is more than 960 km long – the longest in France. It rises at the northern tip of the Cévennes Mountains and runs north for 480 km; then at Orléans it turns west and runs a further 480 km to the sea at St Nazaire. The river has many tributaries, notably the Cher, Indre, Vienne, Layon, and Sèvre, along which the vineyards were established, for in the early days, rivers were the only practical method of transport for heavy and bulky cargoes such as wine. The main vineyards lie in the Départements either side of the river Loire downstream from its confluence with the Allier at Nevers; upstream from Nevers, and in the upper reaches of other tributaries, lie several VDQS districts.

Four important wine districts lie along these river banks, from the Nantais in the west to Pouilly-sur-Loire in the east. Taken in order upstream from the mouth of the river Loire they are: the Nantais, Anjou, Touraine, and Central Vineyards, each of which has sub-districts. There are also several outlying VDQS districts.

The Loire products cover a wide range of red, white, and rosé wines, both still and sparkling, those coming from Anjou and Touraine tending to be sweeter than those from the estuary or the Upper Loire. There are fifty-nine Appellations Contrôlées of the region; all may add the words 'Val de Loire' to their name except AC Crémant de Loire; surprisingly, AC Rosé de Loire, the only other AC covering the whole region, may do so. The reds are strong and highly perfumed. The still white wines are of good quality and great variety; the dry ones are ready for drinking when between one and five years old, while the sweet improve for fifteen years or more; sparkling and semi-sparkling white wines are another speciality; Anjou, Saumur, Touraine, and Montlouis have separate ACs for both Mousseux and Pétillant wines, Vouvray for Mousseux only, and Rosé d'Anjou for Pétillant only. The Loire is famed for its rosé wines above all.

1 Permitted vine training methods in Champagne.
2 100 litres per 150 kg of grapes (2666 litres from a standard pressing of 4 tonnes).
3 Allowing solids to settle out of the must.

Nantais

Nantes has a mild climate which is shared by neighbouring Anjou, and the best vineyards are on the gentle slopes bordering the river. The soil in the Nantais district is of clay with a granite subsoil rich in minerals; this contributes to the AC wine of the district, Muscadet, which has become justifiably popular as a dry white wine superb for accompanying fish dishes. Muscadet, named after the grape from which it is exclusively made, ages quickly, however, and is best drunk young. Within the area of the general AC lie two AC sub-districts of better quality – Muscadet-de-Sèvre-et-Maine (better), and Muscadet-des-Coteaux-de-la-Loire.

Some of the wines are matured in cask on their lees (Muscadet-sur-lie), which gives a vinous taste to the wine. The wines are not heavy in alcohol, and the AC is unusual in decreeing a maximum alcoholic degree each year. Also within the Nantais are two VDQS districts: Gros-Plant-du-Pays-Nantais, producing a white wine made from the Folle Blanche grape (Gros Plant); and Coteaux d'Ancenis with red and rosé wines from the Gamay, Cabernet, or Pinot Bearot, and white wines from the Chenin Blanc (or Pineau de la Loire), or Malvoisie grape varieties – the name of the variety used must be stated on the label. This last district lies to the north of the Loire river, and connects with Anjou. In the Département of Vendée to the south VDQS Fiefs Vendéens are found; these also are light dry white wines, but some red and rosé are produced.

The Vins de Pays of Retz and Marches-de-Bretagne lie in Loire-Atlantique south of the river.

Anjou and Saumur

The soil of Anjou varies on each side of a line running north and south through the district. In the west the soil is schistous with a shallow warm topsoil ideal for storing heat and ripening the grapes in summer. To the east, there is limestone, out of which have been carved famous underground caves. The soil is ideal here for the fresh light wines, both still and sparkling, of Saumur. The Anjou district, covering an area of about 2700 ha, produces in all one million hl

1 Name two AC still wines made in Champagne.
2 What is a 'giropalette'?

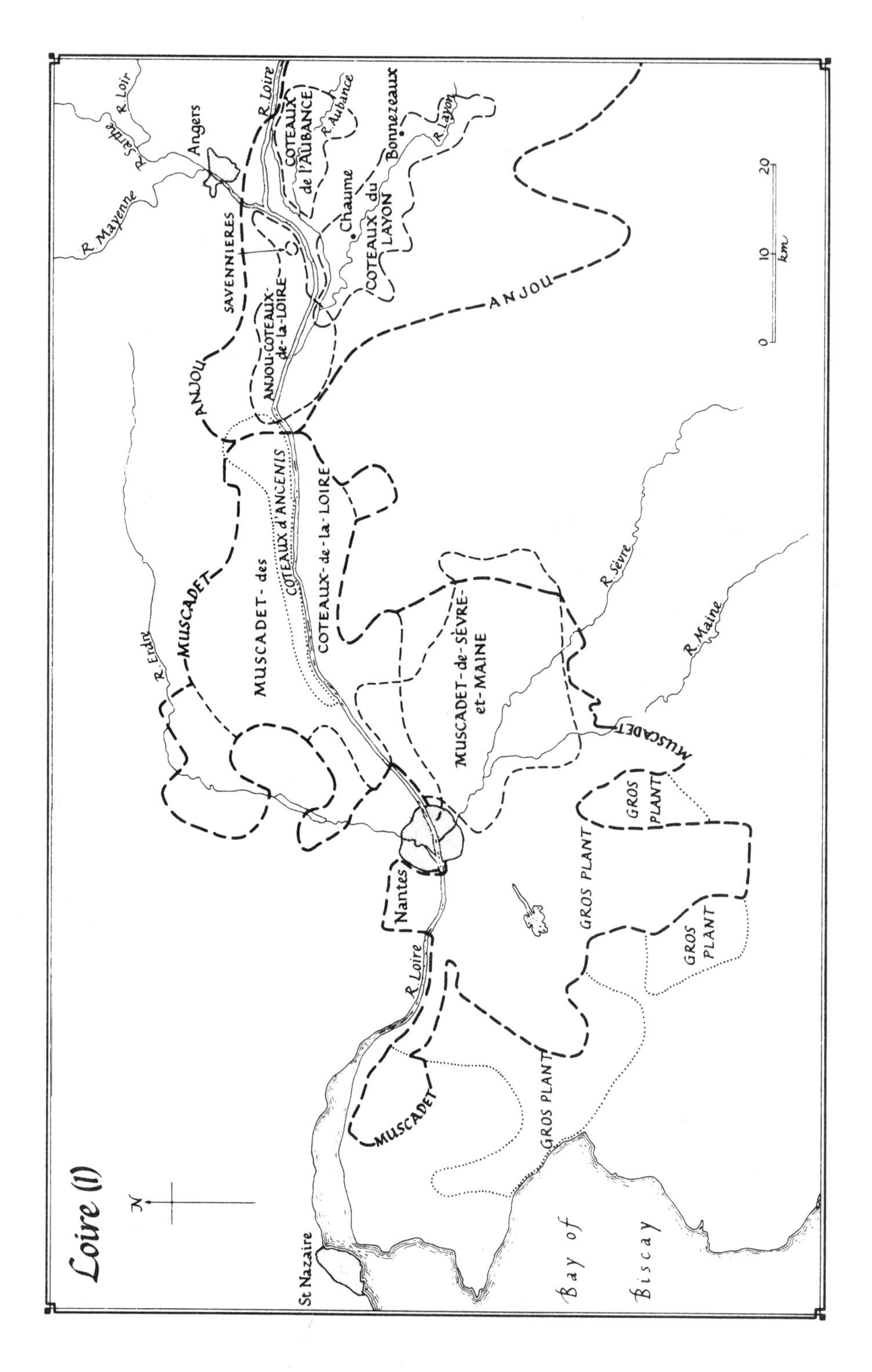

1 AC Rosé-des-Riceys and AC Coteaux Champenois.
2 A machine for automatic remuage in batches of 300–500 of wines made sparkling in bottle.

annually, of which 30 per cent is white, 15 per cent is red, and 55 per cent is rosé.

AC Anjou which covers the whole district includes white wine from the Chenin Blanc (Pineau de la Loire) grape, with a small proportion of Chardonnay and Sauvignon; some is made *pétillant* (prickly). The AC also covers red wine made from Cabernet Franc, Cabernet Sauvignon, and Pineau d'Aunis grapes. There is a separate red wine AC, Anjou Gamay – which is self-explanatory.

AC Rosé d'Anjou, a medium-dry rosé with an attractive pink colour, is made from the same grapes as the red with the addition of Gamay, Côt, and Groslot. A finer, sweeter, rosé wine, made from Cabernet grapes alone, has AC Cabernet-d'Anjou-Rosé, or AC Cabernet-de-Saumur-Rosé. Similarly, red and white wines from the Saumur sub-district are entitled to AC Saumur.

North of the Loire lies the AC sub-district of Anjou Coteaux de la Loire (not to be confused with the adjoining Muscadet-des-Coteaux-de-la-Loire), producing delicate dry or medium-dry white wines from the Chenin Blanc grape. Within this lies the little commune of Savennières where, on a bank of volcanic debris brought down the river from the Massif Central, the wines are of much better quality and age slowly; the four vineyard slopes are Roche-aux-Moines, Coulée-de-Serrant (each with its own AC), Château d'Epire, and Clos-du-Papillon.

On the southern side of the Loire Valley, facing Angers and the Coteaux-de-la-Loire, lie two further sub-districts – AC Coteaux du Layon, to which seven commune names may be appended, and AC Coteaux de l'Aubance, centred round the rivers of those names. They produce fine white wines from the Chenin Blanc grape which are sweeter than those of Savennières. In the Coteaux du Layon, and in particular in its AC vineyards of Quarts-de-Chaume and Bonnezeaux, *pourriture noble* occurs, producing fine Sauternes-like wines, but with more delicacy and acidity.

Saumur lies 40 km east of Angers on the southern side of the river and is the home of the famous cavalry school. The sub-district extends 30 km to the south. Here fine AC white wines, both still and sparkling, are produced, which may vary from dry to medium-sweet. Just to the east of the town of Saumur, and immediately adjoining that part of Touraine producing similar wines, lies the

1 Name the four main districts of the Loire, in order upstream from the sea.
2 What labelling rules govern the use of the words 'Val de Loire'?
3 For what type of wine is the Loire best known in the UK?
4 Describe Muscadet wine.
5 Name three VDQS districts near Nantes.

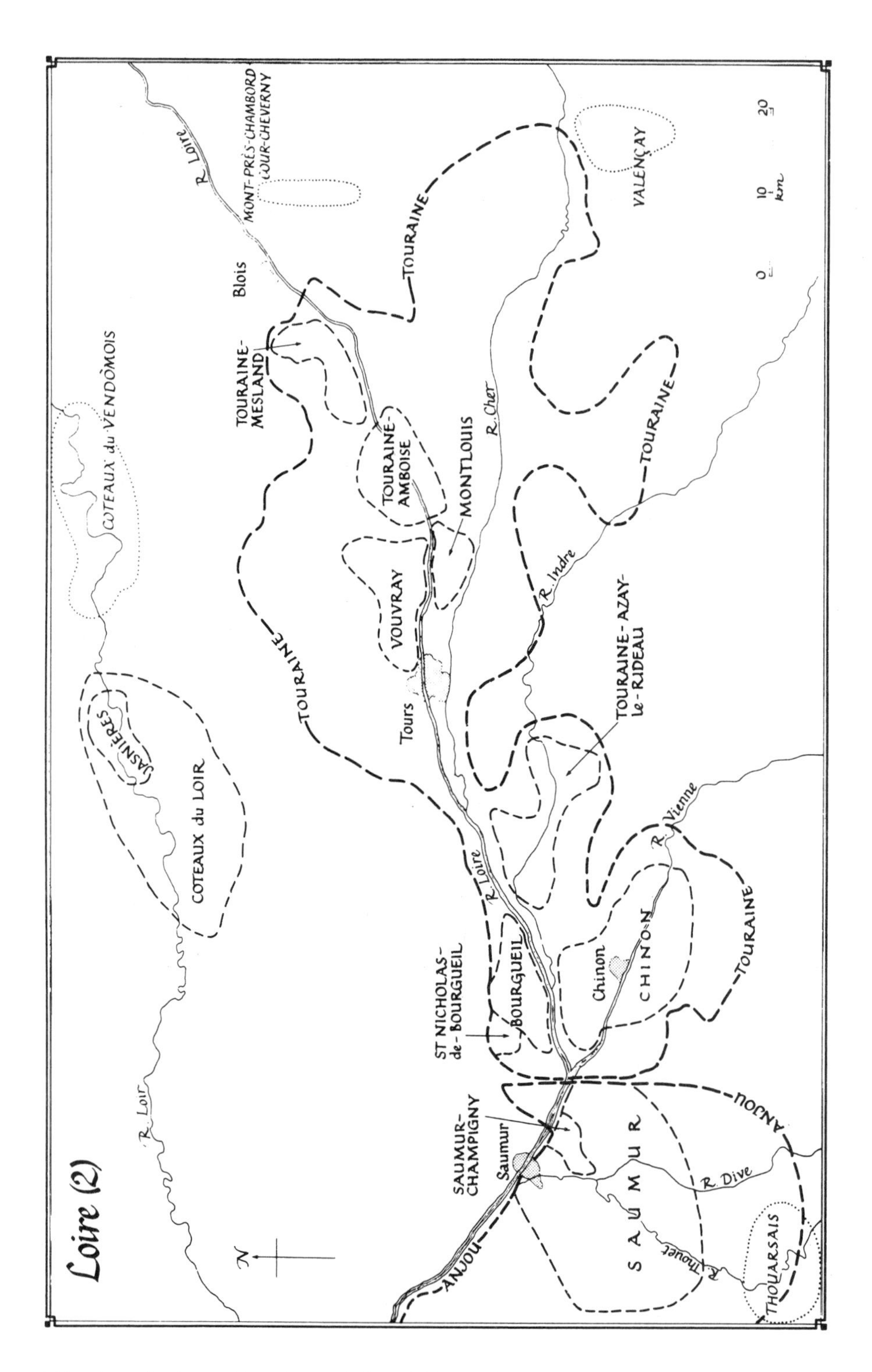

1 The Nantais, Anjou, Touraine, and Central Vineyards.
2 All 59 Appellations Contrôlées of the Loire *may* add them, except AC 'Crémant de la Loire'.
3 Rosé wine.
4 Dry, sharp, white wine, excellent for accompanying fish dishes.
5 Gros-Plant-du-Pays-Nantais, Coteaux d'Ancenis, Fiefs Vendéens.

sub-district of AC Saumur-Champigny, where red wines are made on a limestone soil from Cabernet grapes. These are robust, of a full ruby colour, with a high degree of alcohol and plenty of bouquet.

There are two VDQS districts to the south of the region; Vin de Thouarsais producing light red and rosé wines, and Vin de Haut-Poitou near Poitiers producing white wines from the Sauvignon grape and rosé from the Gamay.

TOURAINE

The vineyard area of Touraine covers some 48,000 ha and produces nine million hl of wine annually, most of ordinary standard, but with a few important wines now respected internationally. The soil in the better districts is of clay and limestone over a tufa chalk subsoil which provides natural caves for cellars. Tufa is chalk which has been boiled by volcanic action and is full of holes like pumice; it is full of minerals and holds water well, and is therefore in great demand by flower-arrangers for making miniature gardens. The climate here is less maritime than that of the Pays Nantais, and the grapes rely on the soil for storing heat. The best known wine is the white Vouvray, from the Chenin Blanc grape; this can be dry or sweet and is said to have a definite taste of quince. Although heavy in alcohol, it is delicate and ages fairly quickly. This wine is often made into AC sparkling wine by the méthode Champenoise. Montlouis on the opposite bank of the Loire produces similar but lighter wines.

The red wine sub-districts adjoining Anjou in western Touraine are AC Bourgueil, producing wines with a distinctive raspberry taste from the Cabernet Franc grape (here called the Breton); its notable commune of AC St Nicolas-de-Bourgueil; and AC Chinon, lying south of the Loire across the Vienne, producing lighter wines with a typical perfume of violets.

Outside the general delimited area of Touraine which itself produces AC red, white, rosé, semi-sparkling, and sparkling AC wines from several grape varieties lie the Coteaux du Loir, 25 km to the north, on soil similar to that of Vouvray. (Le Loir is a tributary of la Loire, running down from the north-east to meet it at Angers.) The best wine is AC Jasnières, a fruity dry white wine made from the

1 Name four grapes typical of AC wine production in Anjou and Saumur.
2 Why are the wines of Savennières superior to most other Anjou wines?
3 Where in Anjou does 'pourriture noble' occur, producing fine Sauternes-type wines?

Chenin and Pineau d'Aunis grapes. Further upstream on this river lies the VDQS district of Coteaux du Vendômois, mainly producing light red wines.

Central Vineyards

This is a term describing a collection of fairly well separated AC and VDQS districts lying around the curve of the Loire between Touraine, Orléans and Nevers.

Quincy and Reuilly

These two AC sub-districts, the former on the Cher and the latter on its tributary the Arnon, make highly presentable wines. Quincy has a poor gravel soil, and its white wines are dry and strong in alcohol with a fine bouquet when young, but they soon lose their style with age. Reuilly, on Kimmeridgian marl, has pleasant white wines of strength equal to Quincy; both come from the Sauvignon grape.

Sancerre, Menetou-Salon, and Pouilly-sur-Loire

These three sub-districts, opposed on either side of the Upper Loire, produce distinctive and characteristic white wines from grapes growing on chalky hills. The wines have been described as slightly smoky, spicy and green, with an aroma of gun-flint. Like the Muscadets from the Nantais, they complement fish admirably. Pouilly-sur-Loire should not be confused with Pouilly in the Mâconnais, where AC Pouilly-Fuissé is made from the Chardonnay grape. AC Pouilly Fumé, also known as AC Blanc Fumé de Pouilly, is made from the Sauvignon grape and gets its name from the white powdery deposit on the grapes which rises like smoke from the vineyards in a breeze. The white wines of Sancerre on the other side of the river are similar, also being made from the Sauvignon grape. The Sancerre AC extends to red and rosé, with light fruity wines being made from the Pinot Noir. The best vineyards in Sancerre are Les Monts Damnés and Clos de Chêne

1 Chenin Blanc, Cabernet Franc, Côt, and Groslot.
2 The commune lies on a bed of volcanic debris washed down from the Massif Central.
3 Coteaux-du-Layon.

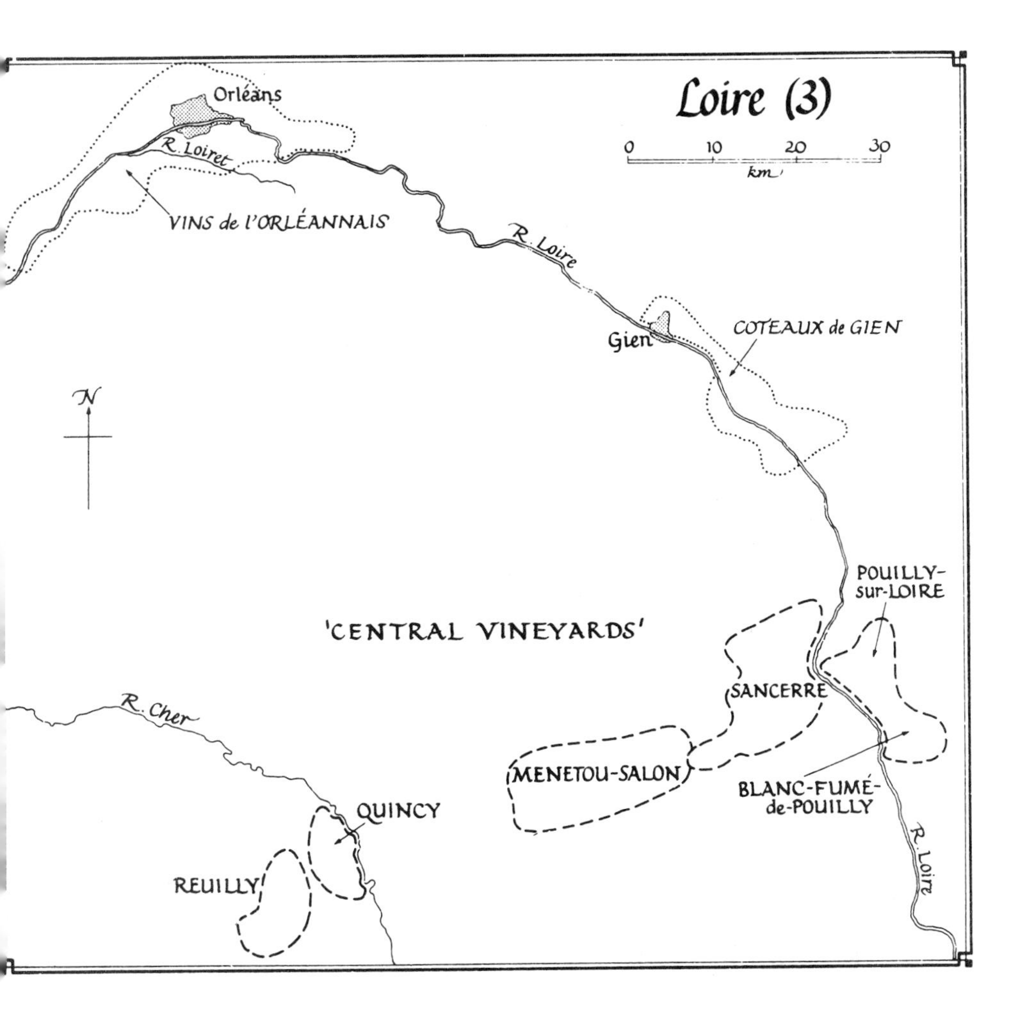

Marchand, and the best in Pouilly Château de Nozet and Château de Tracy. AC Pouilly-sur-Loire covers the same area as the AC Blanc-Fumé-de-Pouilly, but is a lesser wine, being made from the commoner Chasselas grape; it is light, fruity, and attractive in youth, however, and by no means expensive.

VDQS Districts

On the south bank of the Loire adjoining Touraine is the district of Cheverny, embracing the communes of Mont-près-Chambord and

1 What is tufa?
2 (a) In which district is Vouvray? (b) What are the special qualities of its wine?

Cour-Cheverny, producing white wines from the Sauvignon and light red wines from the Gamay grape. Further south on the Cher Valençay produces red, white and rosé wines from the same grapes, and further south still on the same river are found the Gamay reds and rosés of VDQS Châteaumeillant. Surrounding the city of Orléans at the northernmost point of the Loire is VDQS Vin de l'Orléanais; these wines are seldom seen outside their district of origin.

Above Nevers, the Loire and the Allier become wilder rivers to their sources in the Massif Central, but nevertheless support four VDQS districts and one local Vin de Pays. In the Département of Allier lies VDQS St Pourçain, where pleasant red, dry white, and rosé wines are produced, and VDQS Côtes d'Auvergne where, as in the neighbouring Département of Loire with its VDQS Côtes Roannaises and VDQS Côtes de Forez, red wines resembling Beaujolais are produced from the Gamay grape – these districts are, after all, less than 80 km from that region on the other side of the hill! The local Vin de Pays, from most of the Loire Département, is Urfé.

Vins de Pays: There is a regional Vins de Pays du Jardin de la France.

ALSACE

Alsace, the long strip of land 32 km wide running up the west side of the Rhine Valley from the Belfort Gap to Wissembourg, was German (with French vineyard owners) up to the late seventeenth century when, after the Thirty Years War, it became part of France. With the Franco-Prussian war, however, it reverted to German nationality in 1870, and remained German until 1918. With the exception of the German occupation from 1940 to 1945 it has remained French ever since. Bounded to the east by the Rhine and to the west by the Vosges Mountains, Alsace has always had a strong political significance, and it is surprising that her beautiful medieval towns have escaped damage throughout several wars. Under German rule the wines of Alsace were largely shipped down the Rhine for blending with German wines to bolster their small annual production. A period of recovery and planning followed the Armistice of 1945, and

1 Chalk boiled by volcanic action, full of holes, and holding water well.
2 (a) Touraine. (b) A taste of quince, heavy in alcohol, ages quickly and often made sparkling by the méthode Champenoise—dry to sweet.

Alsatian wines were then exported for the first time in their own right as crisp dry white wines of individual character; an important international market has since developed. The administrative centre of the Alsatian wine industry is the city of Colmar in the Département of Haut-Rhin, occupying a central position in the vineyards.

Climate and Soil

The vineyards of Alsace lie some 960 km from the sea and have hotter summers and colder winters than the maritime regions of France to the west. The mountains keep the vineyards free from rain for long periods and shelter them from westerly winds, while little tributaries from the Vosges water the vineyards on their way to join the river Ill and eventually the Rhine. The great dangers are frost in the spring, against which the vines are trained high (*palissage*), and hailstorms in late summer which are countered by seeding the clouds with crystals to make them drop their load elsewhere.

The region is divided between the Départements of Haut-Rhin and Bas-Rhin; throughout, the soil is a rich alluvial mixture of limestone and silica overlaid with loess; however, considerable variations occur, and one vineyard may divide into patches of alluvial clay, of chalk, and of loam, each suited to a different grape type.

The Vines and Wine Names

Six noble vine varieties are cultivated in the region. Their names are important because the wines of Alsace, unlike the wines of most other French regions, are marketed under the grape names rather than under the name of the vineyard, or commune. First and foremost is the Riesling (the 'ie' pronounced as in 'field'), which has been described as 'subtle, gentle and strong'. Next is the Gewürztraminer, producing a wine of fruity aroma and spicy flavour, yet remaining clean and dry. Four other vines, the Muscat, Pinot Gris (Tokay d'Alsace), Pinot Blanc, and Sylvaner are also

1 Name the five AC sub-districts of the Central Vineyard region.
2 From which grape are the white wines of AC Sancerre and AC Pouilly Fumé made?

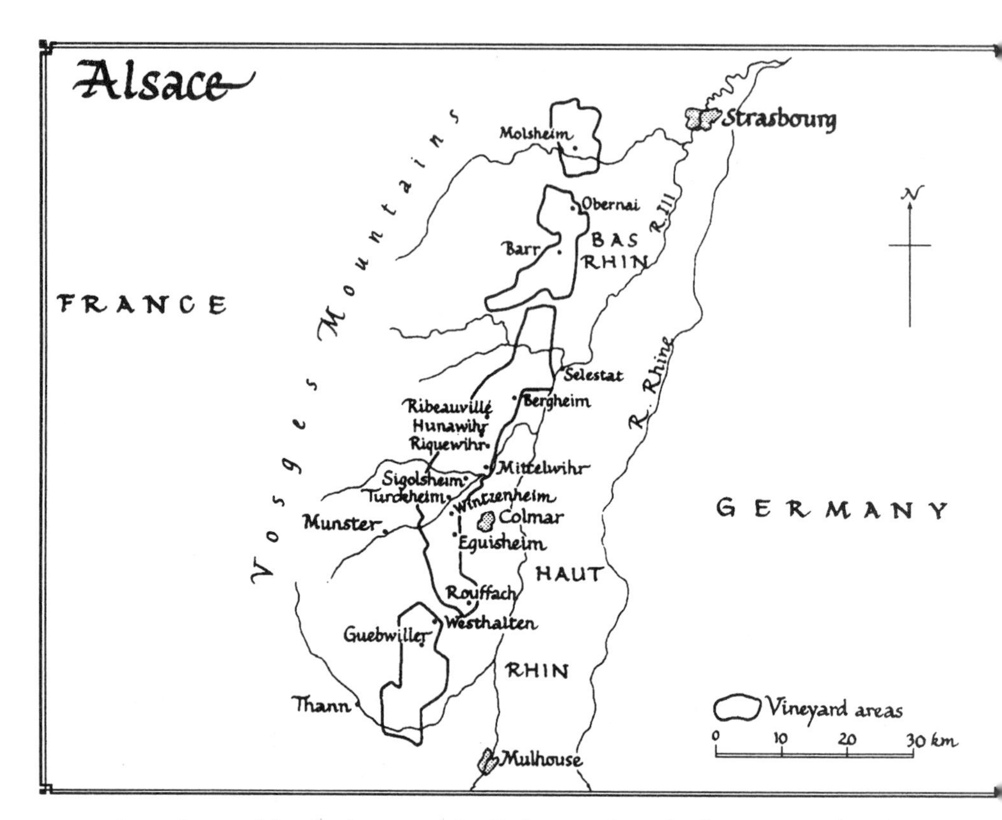

classed as noble; the least noble, Sylvaner, has the largest production.

AC Alsace

AC Vin d'Alsace and AC Alsace may be used to describe wines made from any permitted grape variety, or a blend of them; one of eight grape varieties may be appended to this description, of which all except the Chasselas or Gutedel are classed as 'noble'. The Müller-Thurgau and Knipperlé varieties are no longer permitted. These wines must have a minimum of 8.5% vol. of natural alcohol and the grapes must be harvested according to strict rules laid down each year by the Comité Régionale des Experts. Only the tall slim green bottle, known as the *flûte d'Alsace*, may be used, and all bottling must be done in Alsace.

1 Quincy, Reuilly, Sancerre, Pouilly-sur-Loire, and Menetou-Salon.
2 The Sauvignon grape.

All wines described as Alsace must come from grapes grown 100 per cent in Alsace and if a grape variety is stated, 100 per cent of the grapes must be of that variety. When wines from noble grapes are blended, the blend may be labelled AC Vin d'Alsace Edelzwicker, but the label may not show the name of any of the varieties.

AC Alsace Crémant is a sparkling wine made by the Méthode Champenoise; only the Reisling or Pinot varieties may be used, and the wine must be aged for nine months.

AC Alsace Grand Cru was created in 1975 for selected individual vineyard sites under stricter rules: the permitted yield (1987) was 65 hl/ha; only Gewürztraminer, Pinot Gris, Riesling, and Muscat varieties are permitted, vinified separately, from one year only. Minimum alcohol for wines from the first two varieties is 10°, and for the second two 11°. The name of the vineyard site (*lieu-dit*) may be quoted on the label; as yet there are no more than 50 of these, because the extra price obtainable hardly compensates for the cost of the extra requirements.

Formerly, one might have seen an additional label marked 'Beerenauslese' on especially-fine wines; although Alsace is both French- and German-speaking, this is now forbidden: instead, the wines may be labelled 'Vendange Tardive' if the must has the required Baumé degree (which varies with the grape variety between 13.3° and 14.7°). With another 2.1° Baumé sugar, the wine may be labelled 'Sélection de Grains Nobles'. Only wines from the varieties Gewürztraminer, Pinot Gris, and Riesling are permitted these designations.

Alsace Wine Characteristics

The wines of Alsace serve many purposes. The Chasselas and Sylvaner wines are light, fruity, fresh and elegant, suitable for accompanying fish, white meats, and egg dishes, and may be taken between meals, as thirst quenchers. The Pinot Blanc produces a finer table wine, dry, well balanced, versatile, and pleasantly acid, supreme for all occasions demanding dry wines. The Riesling gives a firm wine, very dry with an expansive bouquet, well suited to complement hors d'oeuvres, oysters, or fish dishes.

1 Name the VDQS of the upper Loire.
2 Name the natural features forming the eastern and western boundaries of Alsace.
3 What is 'palissage'?
4 Describe the soil of the Alsace region.

Tokay d'Alsace is an old Alsatian species of vine – the Pinot Gris – 'grey' (rose coloured) Pinot. Its name derives from a quip of Rousseau, who, on first tasting its wine, said: 'This is the Tokay of Alsace'. It is no relation to the Hungarian Tokay and the name must be qualified – 'Tokay Pinot Gris'. As a wine it is dry, heady and full-bodied, smooth yet slightly tart. Gewürztraminer, similar but with a more pronounced bouquet, will, as Pinot Gris, complement fois gras, roasts and game. Of the two, Gewürztraminer is perhaps the more versatile and may accompany cheese and dessert equally well. Muscat d'Alsace, light-bodied and dry, but fruity with a musky flavour, is good for cheese and dessert as it is as an aperitif.

A certain amount of light red and rosé wine is made from the Pinot Noir in the Bas-Rhin Département, which goes very well with the traditional Alsace dish of knuckle of veal or sausage and sauerkraut.

VDQS of NE France

There are two VDQS appellations of north-east France, lying in Lorraine on the other side of the Vosges Mountains in the Départements of Moselle (VDQS Vins de Moselle) and Meuse (VDQS Côtes de Toul). These are light wines, mostly white, made from the Alsatian grape types. There are as yet no Vins de Pays in this part of France.

1 St Pourçain, Côtes d'Auvergne, Côtes Roannaises, Côtes de Forez.
2 The Rhine and the Vosges mountains.
3 High training of vines – to counter frost.
4 Rich alluvial mixture of limestone and silica overlaid with loess.

Central and Eastern France

BURGUNDY

In the eastern part of France are three wine regions, the first of them, Burgundy, being of major importance, and the other two, Jura and Savoie, of lesser importance. Vineyards were cultivated from time out of mind in the Burgundy region; they were there long before the arrival of the Romans. But from the seventh century onwards, the vineyards were certainly controlled and cultivated by the Church, the lands being given first by the Kings of Burgundy, and later by the Duchy and its nobles. Thus the Church came to found great

1 Name six 'noble' vines of Alsace.
2 What are the percentages of natural alcohol for (a) AC Vin d'Alsace; and (b) AC Alsace Grand Cru?
3 To what is AC Alsace Grand Cru granted, and why are there so few?
4 What is the minimum degree Baumé required for Alsace 'Sélection des Grains Nobles'?

vineyards such as Clos-de-Vougeot, and Clos-de-Tart in the Côte de Nuits.

In the middle of the fifteenth century, Nicholas Rolin built the Hospices de Beaune, creating an institution which ever since has acted as an index, through its annual auctions, of the current price of Burgundy wines. Only after the French Revolution were the great vineyards of Burgundy split up and sold to the citizens – *citoyens* – in small plots, so that now one vigneron may own a part of several vineyards (which incidentally is a good insurance for him against the ravages of hail). Burgundy produces only about half as much wine as Bordeaux in an average year; approximately four-fifths of Burgundy's production is red or rosé wine, and only one-fifth is white.

CLIMATE

The Burgundy region curves in a great 200-km arc from Auxerre in the north to Lyon in the south, and cradles the Charollais whence the beef well complements the wines. The climate is regular, with cold winters and hot summers in a continental-pattern climate. Understandably, the winters are colder in the Chablis district, at the northern tip of the region, than in Beaujolais at the southern extremity, where the summers are hotter, but not very much. Hailstorms are frequent in the early spring and late summer.

SOILS

Soils vary considerably throughout the region. In the Auxerrois, which includes Chablis and St Bris, there is Kimmeridgian clay – this is a stiff clay, well mixed with small chunks of limestone, and takes its name from the little village of Kimmeridge in Dorset, now lying derelict in the Lulworth tank range. A hundred km further along the arc lies the great city of Dijon, from which the limestone hills sweep south on the western side of the Saône valley. The slopes of these hills form the Côte d'Or – the golden hillside of which the golden product is wine. The northern part is the Côte de Nuits; this

1 Riesling, Gewürztraminer, Muscat, Pinot Gris, Sylvaner, and Pinot Blanc.
2 (a) 8.5%. (b) 10 to 11% depending on vine variety.
3 To individual vineyards; the low yield/high alcohol required price it out of the market.
4 15.4 to 16.8, depending on the grape variety.

slope is composed of marl, a hard chalky clay ideal for superb red wines. In the southern part, the Côte de Beaune, the marl is often mixed with limestone fragments, and where these occur, as on the western part of the hill of Corton, superb white wines are produced. Such wines are also produced further south in Meursault, Chassagne, and Puligny where the soil is loess, a granular loam.

South of the Côte d'Or, through the Côte Chalonnaise or Région de Mercurey, and through the Mâconnais, the hills become steeper, and the basic limestone gives way gradually to granite, until in Beaujolais the soil is granite sand. This change is matched by an increase in the mineral content of the subsoil, particularly of iron. The great ironworks of Le Creusot are, after all, only 50 km away.

Grapes

Up to twelve different varieties of grape may be used in Burgundy, but the main varieties for the production of red wine are the Pinot Noir and the Gamay – and for the better white wines, the Chardonnay. The other white grapes, which are of lesser significance, are the Sauvignon (only being found in the VDQS district of St Bris) and the Aligoté which produces a rather acid white wine often rendered sparkling.

Viticulture

The year in Burgundy proceeds much as it does in Bordeaux, but because Burgundy is so much further north and inland, special precautions have to be taken against frost, particularly in the springtime when the leaves are breaking out. Fires are lit in the vineyards in Chablis; great rotating flamethrowers fuelled by propane gas keep the air moving. The vineyards most at risk are those in the valley bottom, where grapes for the lesser wines are grown. Also, because of the cold, the field-grafting system common in Bordeaux is not practised; all stocks are bench-grafted. The vines are generally trained low to the ground to obtain maximum heat. The training systems favoured are the Guyot (Single or Double), the

1 (a) Describe the wines of the Bas-Rhin Département. (b) Which grape is used?
2 Name two VDQS of north-east France. State in the same order the Départements in which they lie.

Cordon de Royat or, in Beaujolais, the Gobelet. Hail has been mentioned as a hazard, but the growers have lessened its impact by employing aircraft to seed the storm clouds far away, as soon as they are detected by radar. Pests and diseases are treated much as they are elsewhere, but when *botrytis* attacks the grapes in a cold wet September, grey rot, not noble rot, will be the result. Growers can only attribute this to the 'luck of the year'. They can pick with care to exclude the worst-rotted bunches, but that is all they can do.

VINIFICATION

While wines of Bordeaux are made slowly and last for a long time; those of Burgundy are matured for a lesser period and have a shorter life. But the real difference between the two is the greater need for blending skills in Burgundy. It has been mentioned that Burgundy vineyards are broken up into smaller parcels than those of Bordeaux. Château Lafite in Bordeaux covers 120 ha and its grapes are all vinified in one chai: Clos-de-Vougeot in the Côte de Nuits, on the other hand, is a walled vineyard of 50 ha, in which the parcels were owned by over seventy-five owners in 1982 and their grapes may be vinified in as many chais. Since all are entitled to the grand cru Appellation 'Clos-de-Vougeot', there can clearly be much variation. The colder climate of Burgundy makes a difference to vinification, in that fermenting musts in Bordeaux usually need cooling to prevent their getting above 'sticking-temperature', whereas the musts of Burgundy often need to be warmed to keep the fermentation going at a temperature which will extract colour from the skins quickly. Although this argument does not apply to white wines, which benefit from a long cool fermentation, the cellars in Chablis may still have to be heated in winter to keep the fermentation going, for below 5°C (40°F) it may stop, to restart in the spring, at the most awkward time. Not all the vineyard owners have cellars and those who do not will sell their grapes to a négociant to vinify; even those who have cellars may run out of space and sell grapes or immature wine to the négociant to make room for the next harvest – or to reduce their overdrafts!

1 (a) White wines but also some light red and rosé wines. (b) Pinot Noir.
2 VDQS Vins de Moselle and VDQS Côtes de Toul. Moselle and Meuse.

Generally speaking red Burgundies receive sixteen to eighteen months in cask; white Burgundies six to nine months. The négociants often have vineyards of their own and also, through brokers, purchase grapes, immature wines and mature wines from several growers within the same AC and year; they will blend these together in a way which is characteristic of their individual establishments. In this way Clos-de-Vougeot of négociant X may be subtly different from that of négociant Y, yet both will be excellent and both grand cru.

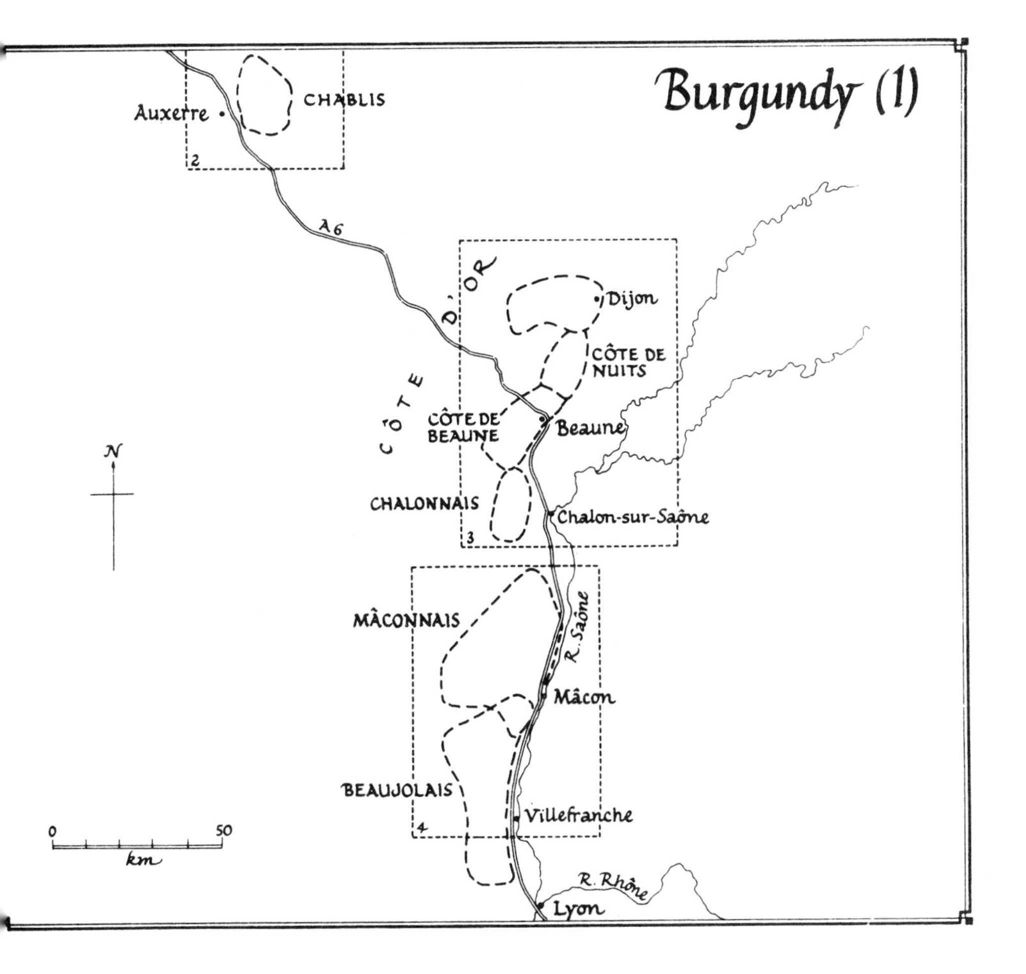

1 (a) When was the Hospice de Beaune founded; (b) by whom; and (c) how is it now maintained?
2 (a)Name a famous AC wine from the Auxerrois. (b) Say what feature of the soil contributes to its characteristics.
3 Name the principal soil constituents in (a) Côte de Nuits, (b) Meursault, (c) Beaujolais.
4 Name the four principal vines cultivated in Burgundy for AC wines and say which are for red or white wines.

DISTRICTS AND QUALITY WINES

There is only one VDQS in the region, and that is Sauvignon de St Bris, coming from the village of St-Bris-le-Vineux and some others, just south-east of the town of Auxerre, south of Chablis. It is a dry light wine, resembling the Sancerre of the Loire Valley 80 km away, and made from the same grape.

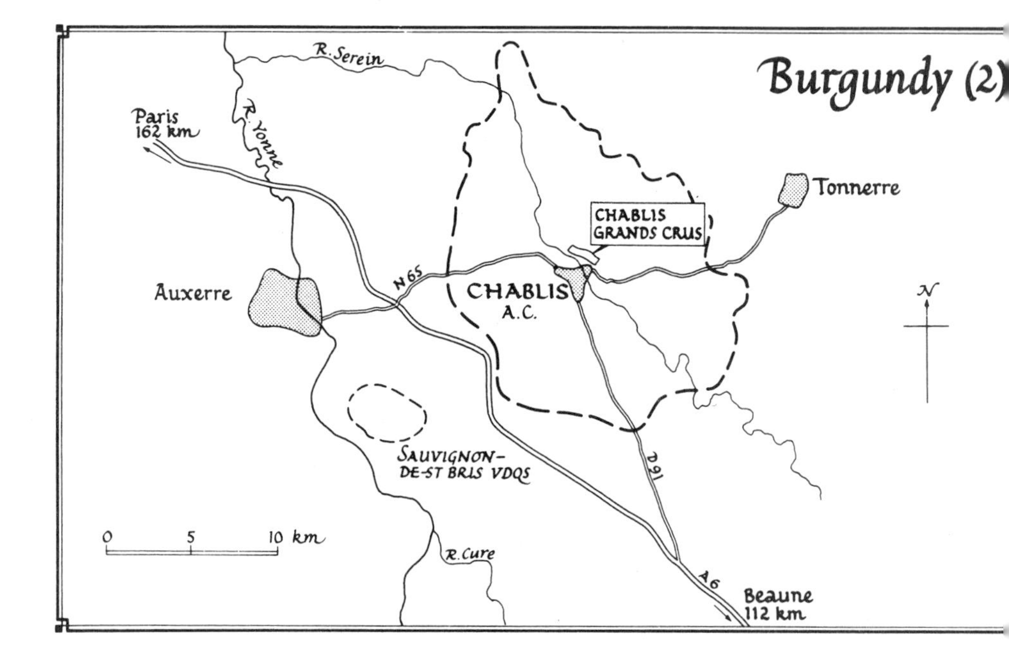

Regional ACs: The general ACs for Burgundy are Bourgogne Grand Ordinaire (the lowest) and Bourgogne, equivalent to the ACs Bordeaux and Bordeaux Supérieur respectively. AC Bourgogne Aligoté applies generally wherever this wine can be made. AC wine made using the Aligoté grape must include the word Aligoté in the name; a new AC – Bourgogne Aligoté Bouzeron was created in 1979 for wines from that village, which lies at the northernmost point of the Région de Mercurey (Côte Chalonnaise).

Bourgogne Passe-tout-grain is the AC applied to a blend made from at least one-third Pinot Noir grapes with up to two-thirds

1 (a) Fifteenth century. (b) Nicholas Rolin. (c) By sale of the produce of vineyards bequeathed to it.
2 (a) AC Chablis. (b) The presence of Kimmeridgian clay.
3 (a) Marl. (b) Loess. (c) Granite.
4 Pinot Noir and Gamay for red wines, and Chardonnay and Aligoté for white wines.

Gamay; it is a fresh, pleasantly light, red wine which does not generally keep long. Grapes for it, like those for AC Bourgogne Aligoté, are grown in the lesser vineyards only. AC Bourgogne clairet or rosé may rarely be encountered. A new regional Appellation for red and rosé wines from the Pinot Noir grape is Bourgogne-Irançy, from the village of that name near Auxerre.

Sparkling Wines: AC Crémant de Bourgogne applies only to red, white, and rosé wines which have been made sparkling by the champagne method; white AC Crémant de Bourgogne may be made from Chardonnay grapes only. The number of different Appellations Contrôlées in Eastern France is greater than in other parts, mainly because selected villages are permitted to add their names to a more general Appellation, but also because there are often separate ACs for *Mousseux, Pétillant, Clairet,* and *Rosé.*

Chablis

The district contains less than 1000 ha of vineyards in rolling country centred on the town of Chablis on the river Serein, tributary of the Seine. It produces only dry white wines, and only from the Chardonnay grape – pale in colour, steely in bouquet, and having great finesse. *Chablis grand cru* comes from a 36-ha slope of seven vineyards, immediately north-east of the town. The AC may include the name of the vineyard: Bougros, Les Preuses, Vaudésirs, Grenouilles, Valmur, Les Clos, or Blanchots. Next down in the scale come the twelve Chablis Premiers Crus wines from named vineyards whose soil and aspect favour production of fine wines; three of these lie in the commune of Chablis and the remainder in the surrounding communes, principally Fyé, Fleys, Fontenay, Maligny, Beines, La Chapelle-Vaupelteigne, Milly, Poinchy, and Chichée. The name of the vineyard (not the commune) may be included on the label in letters the same size as the word Chablis. Further specified vineyards in the Chablis district may produce wine using AC Chablis. Wine from other vineyards within the total delimited district of Chablis may be called AC Petit Chablis; this is the least AC to bear the name Chablis, and the wine will not have matured in

1 Fermenting must in Bordeaux often needs to be cooled. Is this true in Burgundy?
2 For how many months do red and white Burgundies mature in cask?

cask for the eighteen months or more required for finer wines, but will have been bottled in the spring after the harvest or even sent to cafés in barrel to be sold on draught.

Côte D'Or

The 'golden hillside' stretches for about 90 km from Dijon to Santenay along the eastern slope of a low range of hills bordering the rich Saône valley. The lower slopes are of small account, producing AC Bourgogne: the tops of the hills produce AC Bourgogne Hautes Côtes de Nuits or AC Bourgogne Hautes Côtes de Beaune; but the middle section of the slope, deeply indented by minor valleys (*combes*), produces some of the finest wines in the world. The strip is not wide – from only 180 to 540 m; nor is it continuous throughout its length. For these fine wines only the Pinot Noir and the white Chardonnay grape may be grown. The Côte is divided into two districts: The Côte de Nuits in the north, stretching from Fixin to Corgoloin, producing red wines of great fullness, generosity, and bouquet; and the Côte de Beaune to the south, producing soft red wines and astonishingly full fruity white wines. All the red grand crus but one (Corton) lie in the Côte de Nuits, and all the white grand crus save one (Musigny), are in the Côte de Beaune. A full list is given in an appendix to *The New Wine Companion*.

Côte de Nuits

The villages of Marsannay, Fixin and Brochon at the northern end used to be part, with Chenove, of the now-defunct Côte de Dijon; the city of Dijon swallowed up Chenove, but the remaining two still produce deep, slow-maturing wines. The next commune is Gevrey which like many of the villages in the Côte d'Or has added the name of its finest vineyard, le Chambertin, to its own. Gevrey Chambertin has eight grand cru vineyards headed by Chambertin and Chambertin Clos de Bèze; for the six lesser vineyards, the name Chambertin

1 No – the must may even have to be warmed to prevent fermentation stopping.
2 Red Burgundies, 16 to 18 months; white Burgundies, 6 to 9 months.

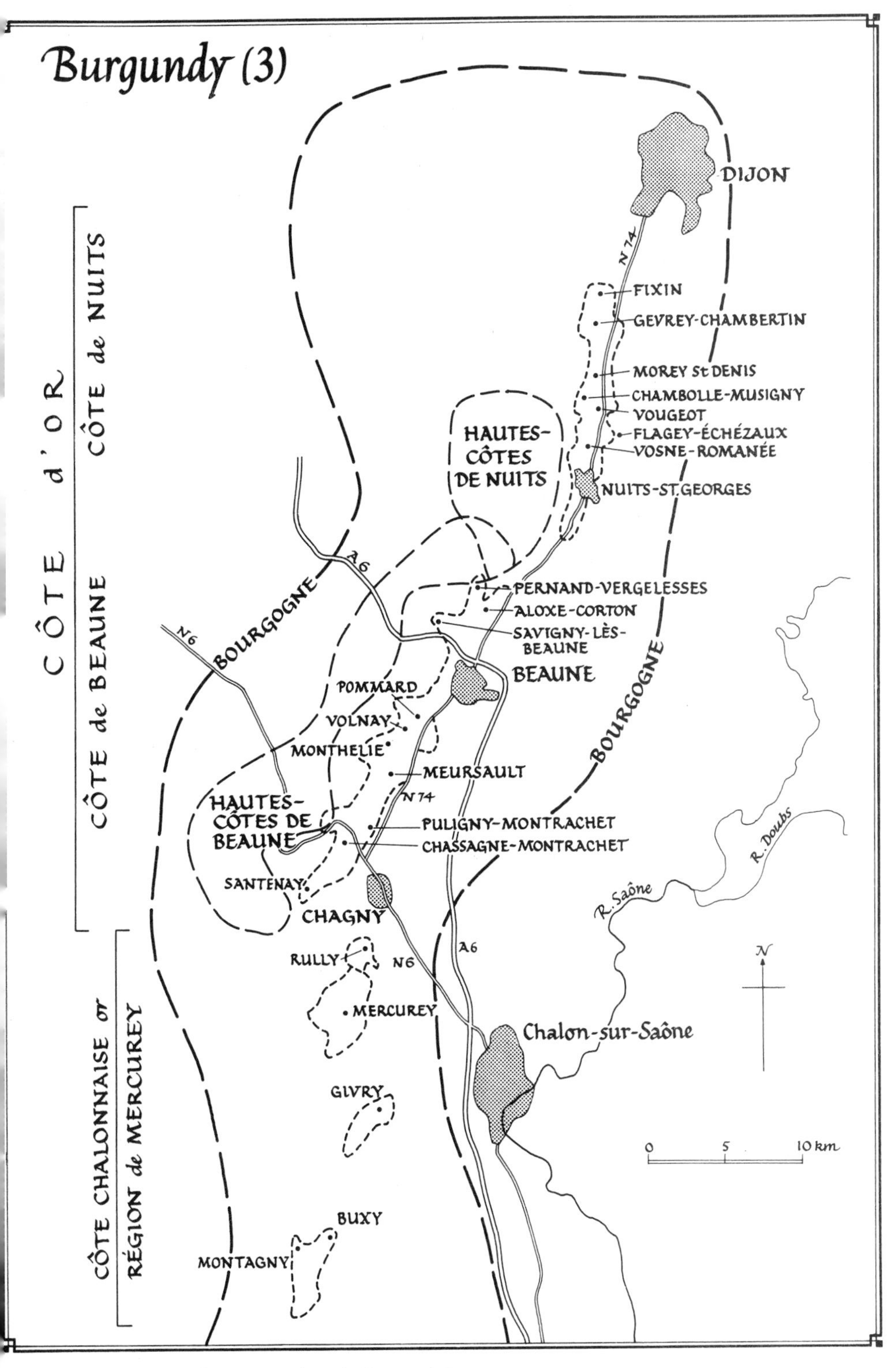

1 What VDQS districts are there in Burgundy?
2 Which two Burgundy ACs *must* include the name of the grape?
3 To what wine does AC Bourgogne Passe-tout-grain apply?
4 (a) How many vineyards are entitled to AC Chablis Grand Cru? (b) How many to Chablis Premier Cru? (c) Which is of higher quality: Chablis or Petit Chablis?

follows the vineyard name. Just south of Gevrey-Chambertin is Morey-Saint-Denis with five grands crus, including Clos-Saint-Denis; two of them Bonnes Mares and Clos-de-Tart, extend into the neighbouring commune of Chambolle-Musigny which takes its name from the grand cru vineyard of Les Musigny. Some white wines are occasionally made here, also entitled to the grand cru AC, and are full-bodied with remarkable bouquet and power. Possibly because of their rarity, they are unforgettable. The red wines are full and are of great finesse, some white grapes occasionally being included in the vintage (the law permits up to 15 per cent of Chardonnay to be included with Pinot Noir throughout the Côte d'Or).

The next village encountered on the southward journey along RN 74 is Vougeot (already mentioned), and beyond that Flagey-Échézaux which, although it includes the grand cru vineyards of Grands Échézaux and Échézaux (pronounced *eshezoh*), is usually included with its larger neighbour Vosne-Romanée. The commune has five more grand cru vineyards, all tiny; the wines they produce are famed for their velvety softness and tremendous bouquet. The bouquet may sometimes owe something to the practice of maturing the wine in barrels of Tronçais oak, which lends its perfume to the wine. It is interesting that all the grand cru vineyards of the Côte de Nuits are in the north-central part of the district; the soil in the south, as in the north, while still favouring production of fine wines lacks the consistency and minerals required for the very best.

The largest commune of the Côte (giving it its name) is Nuits St Georges, adjoining Vosne-Romanée; Les St Georges is the only premier cru vineyard to be so honoured by its commune. On any Burgundy label the premier crus can be distinguished from the grand crus because the former will always have the commune name in the Appellation (as for example, Nuits St Georges Les Cailles), whereas the latter will not, as for example, Corton. More confusing is the habit of including the name of an individual parcel of a grand cru vineyard on the label, as for example, Corton-Bressandes, particularly when the 'parcel' name also happens to be the name of a premier cru vineyard elsewhere (as for example, Beaune-Bressandes).

The wines of the communes of Prissey, Comblanchien, and Corgoloin, to the south of Nuits St Georges, can be sold with those

1 Only one – Sauvignon de St Bris.
2 Bourgogne Aligoté and Bourgogne Aligoté Bouzeron.
3 Wine from a blend of one-third Pinot Noir and up to two-thirds Gamay.
4 (a) Seven. (b) Eleven. (c) Chablis.

of Fixin, Brochon, and Chenove to the north of Gevrey-Chambertin – under AC Côte de Nuits-Villages until recently known as 'vins fins de la Côte de Nuits'. The hills are barer and steeper here, and Corgoloin is more famed for its marble quarry than for its wine.

Côte de Beaune

At the southern end of the Côte de Nuits stands the great tree-crowned buttress of the Montagne de Corton, bearing the most northerly vineyard of the Côte de Beaune. Vineyards surround it from east to west and continue into the deep valleys of Pernand-Vergelesses and Savigny-lès-Beaune, and south towards the city of Beaune through Ladoix-Serrigny and Chorey-lès-Beaune ('lès', meaning 'near' or 'by', should not be confused with 'les' meaning 'the'). The little village of Aloxe, which naturally has added the name of Corton, lies on the south-eastern slope of the hill. Half the hillside – the eastern half in general – is planted with Pinot Noir and produces grand cru red wines, powerful, full, with a pronounced bouquet, from that part of the slope having iron in the subsoil; the other half, towards the summit and west, is planted with Chardonnay, and produces wines of great fruitiness, yet with the steely fineness that many call 'backbone'. Both red and white wines develop well in the bottle. The grand cru ACs are Corton for both red and white wines, and Corton Charlemagne for white only.

AC Beaune has thirty-four premier cru vineyards, but no grand cru; AC Côte de Beaune covers practically the same area – neither should be confused with Côte de Beaune Villages, which applies to *red* wines only, coming from communes less famous for their red wines: Beaune, Aloxe-Corton, Volnay, and Pommard are excluded; just to complicate matters, the 16 communes comprising the 'Villages' Appellation may also label their wines AC Côtes de Beaune preceded by their name! Beaune produces a variety of red wines, which may be identified by the vineyard name whether premier cru or not, and some white wines. But Beaune is chiefly memorable for its walled city, its museum, the Hospices de Beaune, the cellars of its merchants, and its shops selling everyday tools of the vineyard, all combining to make it the capital of Burgundy. There are many

1 What are the two main sub-districts of the Côte d'Or?
2 What and where is le Chambertin?

books about Beaune: some are mentioned in the bibliography (see Appendix 5).

South of Beaune on RN 74 (the road that generally marks the eastern limit of the Côte d'Or fine vineyards) is a fork with RN 73, where the AC commune of Pommard begins. Pommard wines, entirely red, are deep in colour, full, and last well. Further along the RN73 is Volnay, a smaller village with a smaller vineyard – for the hill is steeper here – but including the red wine premier cru vineyard of Les Santenots which is situated in the neighbouring white wine AC commune of Meursault. The vigneron making his own wine in such a village will keep his wines in cask – the 225-litre *pièce* or fût – in a cool dark chai that resembles a half-buried barn. By the light of the only candle the host will draw samples for the visitor from the various casks with a pipette or valinch, a tube dipped into the wine and stopped with the thumb, to pour into a *tastevin*. This is the only receptacle – a shallow bowl with a boldly-patterned base – that can disclose the true colour and clarity of wines by candle-light in a dark cellar. Tastevins vary greatly in depth and patterning, and it is not easy to appreciate the bouquet or even to sip from the shallower ones. It is a good policy for visitors to have their own, and to keep it in a clean handkerchief!

As the RN 73 continues to the south it climbs the hillside separating the red AC communes of Monthélie, St Romain, Auxey-Duresses, and Blagny from the great white wine AC communes of the lower slopes. The wines of Meursault are fresh and fruity in youth, with the long finish that stands for quality; with bottle-age they tend to oxidize, taking on a deeper golden colour and a richer bouquet. But young or old they stand out. So too, to a greater extent, do Meursault's neighbours, the Montrachets, with five AC grand cru white wine vineyards spread over the communes of Puligny and Chassagne; each commune appends the name of Montrachet. Comparing in quality with Champagne, Chablis, and Corton, the wines of Montrachet are the greatest that the Chardonnay grape can produce.

Chassagne-Montrachet also produces sound red wines in its southern vineyards, for the village itself marks the end of the patch of loess soil that favours the white wines. At the southern extremity of the Côte d'Or is Santenay, and the three villages of the Maranges

1 Côte de Nuits in the north and Côtes de Beaune in the south.
2 The finest vineyard of the village of Gevrey.

vineyards, Cheilly-les-Maranges, Sampigny-les-Maranges and, perched on a hill, Dezize-les-Maranges. All produce full-bodied red wines that are soft in texture, often sold as Côtes de Beaune Villages; their white wines may often appear as Bourgogne Blanc.

Côte Chalonnaise

The Côte Chalonnaise, taking its name from the nearby manufacturing town of Chalon-sur-Saône, extends for about 25 km south of the Côte de Beaune. It is not a continuous vineyard but comprises the small sub-districts of Rully, Mercurey, Givry, and Montagny, working from north to south. The largest of these, producing the best wine, is Mercurey, and the district is increasingly called 'La Région de Mercurey'. The northernmost area, Rully, produces both red wines from the Pinot Noir and white wines from the Chardonnay, as well as large quantities of Aligoté for branded sparkling wine. The wines of Mercurey are predominately red, and resemble those of the Côte de Nuits rather than the adjacent Côte de Beaune; they have a deep colour, are somewhat hard in youth and keep well. A few kilometres south, the village of Givry produces heavier red wines with less bouquet and finesse than those of Mercurey. Eight km south again is the area of Montagny, producing white wines around that village and its near neighbour Buxy.

Mâconnais

This district stretches 35 km from just north of Tournus to just south of Mâcon in a range of steep limestone ridges and valleys running north to south. It produces a great deal of rather ordinary red wine in the north and some first-class white wine in the south. The AC system in the Mâconnais is rather confusing and some of its villages are shared with Beaujolais. The ordinary AC reds may be called Mâcon, or Mâcon Supérieur if they have enough alcohol; the best, coming from forty-four villages, may use the AC 'Mâcon-Villages'. The wines are made from Gamay and Pinot Noir grapes. The white wines, on the other hand, are made entirely from Chardonnay, and

1 What commune gives the Côte de Nuits its name?
2 Name the most northerly Grand Cru of the Côte de Beaune.

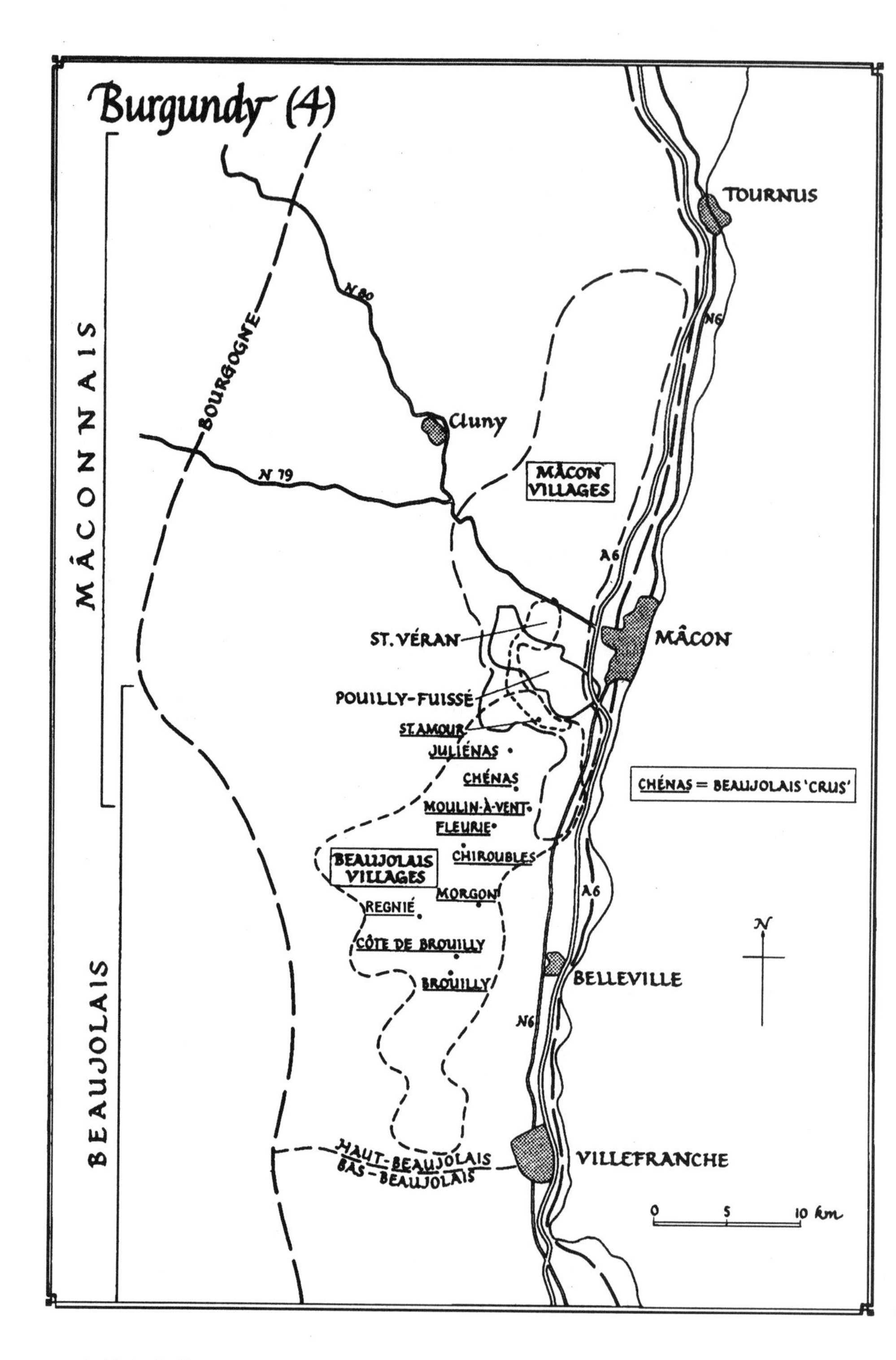

1 Nuits St Georges.
2 Corton, or Corton Charlemagne.

may be called 'Pinot Chardonnay Mâcon', 'Mâcon', 'Mâcon-Supérieur', 'Mâcon-Villages', or 'Mâcon' followed by the name of the village as, for example 'Mâcon Lugny'.

Between Mâcon and Cluny, in the south of the district, is a broad valley studded with immense outcrops like the Rock of Gibraltar, marking the transition between the limestone ridges of the northern Mâconnais and the rounded granitic hills of the Beaujolais. Here lie the best villages for white wine. Pouilly-Fuissé (not to be confused with Pouilly-Fumé from the Loire) takes its name from two of the five villages in the delimited area; this dry wine is neither so delicate as Chablis nor so fine as the white wines of the Côte de Beaune, but is full-bodied with a heaviness of alcohol that can produce a rather coarse effect. Pouilly-Loché, Pouilly-Vinzelles, and St-Véran are lesser ACs, the last coming from a group of eight villages which surround the Pouilly area; some of these villages can also sell their wine as Beaujolais, and one of the villages is confusingly called St Vérand.

Beaujolais

The Beaujolais was not part of the ancient kingdom or Duchy of Burgundy and its differences from the remainder of the province are so great that it is difficult to see why it was ever included in the region. It even has it own *Comité*, or governing body. But then so has the Mâconnais, and ACs Mâcon and Beaujolais Supérieure may be declassified to Bourgogne or Bourgogne Grand Ordinaire; Beaujolais on its own may not be so declassified. The best wines of the Beaujolais come from the part north of Villefranche-sur-Saône where the vineyards cover the rounded hills of granite. The Gamay Noir is the only grape for red wines, and is trained in the bush formation called 'Gobelet'. Ten villages, the 'Beaujolais crus', have their own ACs; working from north to south they are St Amour, Juliénas, Chénas, Moulin-à-Vent, Fleurie, Chiroubles, Morgon, Regnié Brouilly, and Côte de Brouilly. These are Beaujolais at their best, full, fruity, with the unmistakable Gamay 'nose', and will improve with a few years in bottle if they come from a good year. In Beaujolais, the grapes are often put straight into the fermenting vats

1 Describe a tastevin and its use.
2 In which communes do the five grand cru vineyards of Montrachet lie?
3 Which of the following ACs are for red wine only: (a) Corton Charlemagne. (b) Côte-de-Beaune-Villages. (c) Pommard. (d) Meursault. (e) Corton?
4 (a) Name the district lying immediately south of the Côte de Beaune. (b) How long is it? (c) Name its four sub-districts. (d) By what other name is this district sometimes called?

without crushing or de-stemming, enough of the grapes having been broken to allow the yeasts to come into contact with the juice to start fermentation. As the vats are covered by a CO_2 blanket, the carbon dioxide formed by fermentation builds up pressure, forcing yeasts through the skin of the unbroken grapes and causing them to ferment. After two days and two nights, the fermenting must is pressed and the juice finishes its fermentation in vented tanks. Thus Beaujolais retains a light colour with little tannin.

Thirty-nine villages (including the ten 'Beaujolais crus' villages) are entitled to AC Beaujolais Villages or to Beaujolais followed by the name of their village, as for example AC Beaujolais-Beaujeu or AC Beaujolais-Vaux. The former gave its name to the district, while the latter was one of the sites for the film *Clochemerle* – and still displays the sign in its vineyards. Eight of these villages lie in the Mâconnais. Between Villefranche and Lyon, the terrain is flatter, but even more heavily planted in vines. This is the 'Bas Beaujolais' producing on a limestone soil about half the total Beaujolais production of 700,000 hl with little of the finesse of the wines from the north. AC Beaujolais Supérieure requires 1° more alcohol than Beaujolais.

Beaujolais Nouveau: It is a matter of history that the growers of the Bas-Beaujolais obtained permission to sell their new wine before the date set hitherto by law. The wine was rushed to Paris on 15 November to be sold while still fresh. This 'race' enthused the English and created a useful promotional exercise now followed by several other regions of the world.

JURA

40 km across the wide marshy plain of the rivers Saône and Doubs from Burgundy lies the ancient province of Franche-Comté which stretches for 210 km south-south-west from Besançon, between the plain and the Swiss border. Running through the province like a backbone is the limestone range of the Jura mountains which give their name to the Jurassic age of geology, when they were formed. On the western foothills of this range is a strip of vineyards 16 km wide and 72 km long called the Jura region, paralleling Burgundy roughly from Beaune to Mâcon. The wine-making centre of this region is the town of Arbois, where Pasteur lived and wrote his treatises on wine. The wines produced in this region are very varied – red, white, grey, straw, yellow, and 'mad'!

1 A shallow indented silver bowl for disclosing the true colour and clarity of wine by candlelight.
2 The communes of Puligny and Chassagne.
3 (b) and (c).
4 (a) Côte Chalonnaise. (b) 25 km. (c) Rully, Mercurey, Givry, and Montagny. (d) Région de Mercurey.

The climate of this area is more continental than that of Burgundy; that is to say it has colder winters and hotter summers, with a fair amount of rain. The soil is calcareous, and the lower slopes contain much heavy clay so that there is a subsoil of marl.

There are five ACs in the Département of the Jura itself, with a sixth (AC Coteaux du Lyonnais) about 100 km to the south-west. In the Jura, the lowest-ranking AC – AC Côte de Jura – covers the area from Arbois in the north to Lons-le-Saulnier in the south. AC Arbois, which includes the best red and rosé (*vins gris*) wines, covers Arbois and a few surrounding villages, one of which has its own AC, Arbois-Pupillin. They are made from the local black Poulsard and Trousseau grapes, mixed with the white Savagnin. This last is descended from the Traminer grape and should not be confused with Sauvignon. The wines are light and pleasant when young, but age rapidly; rosés are dry and astringent with an onion-skin colour.

While white wines are made around Arbois from the Savagnin grape and are entitled to its AC, the best come from the southern village of l'Étoile which has its own AC – for white wines only – from the Savagnin, Chardonnay, and Pinot Blanc grapes. Much of this light dry wine is made sparkling by the méthode Champenoise if it is to be AC Côte de Jura Mousseux, Arbois Mousseux, or Etoile Mousseux. The largest local producer markets a non-AC sparkling wine under the name '*vin fou*' – hence the 'mad'.

Straw wines (*vins de paille*) are made by leaving the grapes on straw mats for two to three months to dry out; these produce sweet heavy wines resembling white port. The *vins jaunes*, or 'yellow' wines, are specialities of the region and are made from the Savagnin grape alone. The best come from the little village of Château Chalon which has an AC of its own for this type of wine only. This is not the name of a property, as in Bordeaux, but the name of a village around which are many vineyards. After normal fermentation, the white wine is racked into barrels and left, without bungs in and without topping up, for a minimum of six years. A film-forming yeast called *flor* forms on the surface of the wine. While this flor is the same as that found in Jerez and is the creative feature of fino sherry, vins jaunes hardly taste like sherry. The slow fermentation of this yeast changes the constituents of the wine which, being left open to the air, also oxidizes and turns yellow.

1 Distinguish between Pouilly-Fuissé and Pouilly Blanc Fumé.
2 (a) Where are the best Beaujolais wines made? (b) What is the only permitted grape for red wine? (c) What soil predominates? (d) What method of training is used for the vines?
3 Are the Mâconnais and Beaujolais part of Burgundy and, if so, why?
4 Name nine villages in the Beaujolais having their own ACs.

The process is not always successful; much of the wine goes bad and has to be poured down the drain; so vin jaune is rare and expensive. It has a somewhat rancid nose, and is an acquired taste; so powerful is its bouquet that it would spoil the taste of wines coming after. So it usually accompanies nuts and the local Comté cheese, which is full of flavour and like a Gruyère without holes. The wine is also used in sauces and imparts a wonderful flavour to chickens cooked in it, especially those from the town of Bourg, around 40 km to the south, where the Poulets de Bresse have their own AC – for the AC system applies to some foods as well as wine.

Vins de Pays

Vins de Table from the Départements of Haute-Saône and Jura may, if they pass the tests, be given the geographical description 'Vins de Pays de Franche-Comté'.

SAVOIE

This region is spread over the mountainous Départements of Savoie, Haut-Savoie, and parts of Ain and Isère. Because of the nature of the country, the vineyards are widely scattered and lead to a multiplicity of AC and VDQS names, village names being appended to the main name.

Bugey and Seyssel

These two districts are found in the valley where the Rhône flows south from Lake Geneva (Lac Léman). The first is Bugey, a VDQS district producing red, white, and rosé wines, which may be still, pétillant, or sparkling. The red and rosé wines are ordinary, the white better, particularly if made from the Altesse grape, here named the Roussette. There are thirteen different VDQS names, based on 'Vin de Bugey' elaborated by 'Pétillant' or 'Mousseux' to which may be added the name of the best village, Cerdon. For still wines, 'Vin de

1 Pouilly-Fuissé is a full-bodied dry white wine from the southern Mâconnais. Pouilly Blanc Fumé is a white wine from the Central Vineyards of the Loire.
2 (a) North of Villefranche-sur-Saône. (b) Gamay Noir. (c) Hills of granite. (d) Gobelet or bush formation.
3 Yes: ACs Mâcon and Beaujolais Supérieure (but not Beaujolais) may be declassified to Bourgogne.
4 St Amour, Juliénas, Chénas, Moulin-à-Vent, Fleurie, Chiroubles, Morgon, Brouilly, and Côte de Brouilly.

Bugey', to which may be added the name of one of five villages, or 'Roussette de Bugey' for wines made from that grape, to which may be added the name of one of six villages.

In the Rhône valley itself lies the tiny (60 ha) district of Seyssel, producing still and sparkling white wines. The wines are almost water-white in colour and are of a flinty dryness. The vineyards, on soil of sandy clay with flints, lie on the Savoie side of the river as well as in Franche-Comté. The grape varieties Roussette, Molette, and Chasselas (here called Bon Blanc or Fendant) are high-trained on wires, as in Alsace. Being AC, the sparkling wines are made by the méthode Champenoise.

Most of the remaining vineyards in this large and mountainous region are concentrated upon the flat alluvial plain on the southern shore of Lac Léman. One AC, for white only, is Crépy; this wine is made from the Chasselas grape, and is light and fruity. The vineyards face north-west across the lake to the hillsides on the Swiss side of the lake which produce similar, but better, wines.

The AC Vins de Savoie may be red, white, or rosé and come from anywhere in the region. Most of them are white, and the best adjoin Crépy at Thonon-les-Bains (Château Ripaille); but some pleasant white wines are made in the valley south of Chambéry from the Roussette grape. AC Roussette de Savoie or Vin de Savoie Roussette may have the name of one of five villages appended, and AC Vin de Savoie may add the name of one of sixteen other villages. Much of the white wine of the region is elaborated with alpine herbs and fruits to make the famous very dry Chambéry vermouths.

Vins de Pays

In the Départements of Haute-Savoie and Savoie are Vins de Pays d'Allobrogie; and in Isère and partly in Savoie are Vins de Pays des Balmes Dauphinoises, and Vins de Pays des Coteaux de Grésivaudan.

1 How many villages are entitled to AC Beaujolais followed by the village name?
2 Is 'vin fou' of the Jura entitled to AC Vin de Jura Mousseux?
3 What is the exclusive AC for vins jaunes, and from what vine are they produced?

Mediterranean France

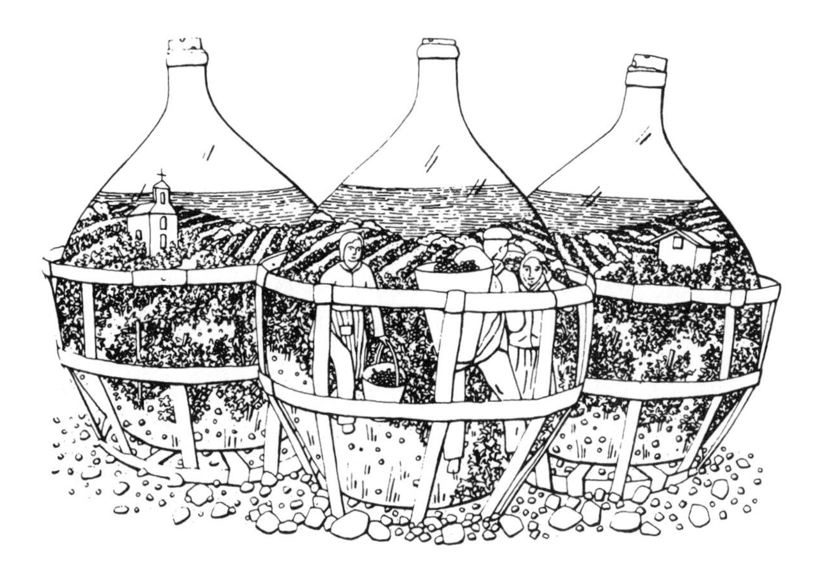

Mediterranean France has three wine regions: first, the important Côtes du Rhône; then Provence; and finally Languedoc & Roussillon – the area stretching east from the Mediterranean as far as Carcassone, watered by rivers that flow to the Mediterranean.

CÔTES DU RHÔNE

The general AC covers the valley from Lyon to Avignon, which is just over 200 km in length. The valley is steep-sided in the north so that the band of vineyards is narrow and precipitously terraced; the

1 Thirty-one.
2 No.
3 Château-Chalon, made only from the Savagnin vine.

southern half of the valley is broader and, to the east, is flatter, giving more room for vineyards. There is a vineless strip 40 km long in the middle, from the point where the Drôme flows into the Rhône, just south of Montélimar. The wines of the north differ greatly from the wines of the south; there is a difference in each of the factors which affect production, and even in the pattern of trade itself.

The region produces about 3 million hl of wine every year, this quantity being midway between the production of Bordeaux and Burgundy.

The wines, of which less than 10 per cent are white, cover a broad spectrum of quality, from ordinary table wines through table wines with indication of origin (vins-de-pays) to qwpsr – both VDQS and AC. About 25 per cent of the total production is AC and 10 per cent VDQS. Here the ratio of quality wines to table wines is lower than in Burgundy or in Bordeaux, yet nearly twice the national average. The region has no grand crus and, with one exception, no AC for an individual vineyard.

Climate

The north of the region has a temperate continental climate, with cool to cold winters and warm summers; the south has a mediterranean climate, with cool winters and hot summers. Professor Wagner divides the viticultural climates of Europe into northern areas with four seasons and southern with only two – winter and summer. His dividing line crosses the Rhône region at Montélimar. This region is also subject to the katabatic wind called *mistral* – cold air from the glaciers in the Alps which starts to flow downwards, as cold air will, into the plains. In so doing it gathers momentum, and arrives in the plains as a wind-force of some 48 to 64 km per hour. Only the sturdiest vines and the sturdiest people can withstand this icy wind.

Soil

The characteristic soils of the north of the region are granite and schist, with lime and iron mixed; the soils of the south are more

1 Does the AC system apply only to wine and spirits?
2 Why are there so many VDQS in Savoie?
3 Name an AC for white wine in the Savoie and the grape from which it is made.
4 With what do you associate Chambéry in the Savoie?
5 Whence comes the Vin du Pays d'Allobrogie?

varied, consisting mainly of clay and sand with outcrops of limestone. In places the glaciers have left beds of large round reddish stones.

Grapes

In the north, wines are made from one grape variety or possibly from a blend of two. In the south, however, as many as five or six varieties and occasionally more may be blended together. The climate is responsible for this because, in a temperate climate, one grape can give all the desired qualities of colour, acidity, bouquet, finesse, aroma, and vinosity; but, in the hot mediterranean climate, different varieties are needed to give these qualities as no one variety can provide them all.

Viticulture

On the steep terraced slopes of the north everything has to be done by hand, including carrying the soil which has been washed down by winter rains back up to the terraces. The vines are generally trained on wigwams of three posts, although some bush-trained vines may be found.

In the south, all are bush-trained and planted at an angle towards the north, so that as they grow the mistral gradually blows them upright. The flatter terrain and larger vineyards make cultivation by machine practicable.

Vinification

In the north wines have more tannin and generally take longer to mature than in the south. They tend to be matured in cask by individual proprietors. In the south, with its higher production, there are many co-operatives providing a central service for vinification under the most modern conditions. As a result, the wines can be made to mature earlier, and bring a return to the co-operatives sooner.

1 No: AC Poulet de Bresse applies to chickens, for example.
2 Vineyards are widely scattered, and village names may be appended to the main names.
3 AC Crépy, made from the Chasselas grape.
4 Very dry vermouth, from wine elaborated with alpine herbs and fruits.
5 The Départements of Savoie and Haute-Savoie.

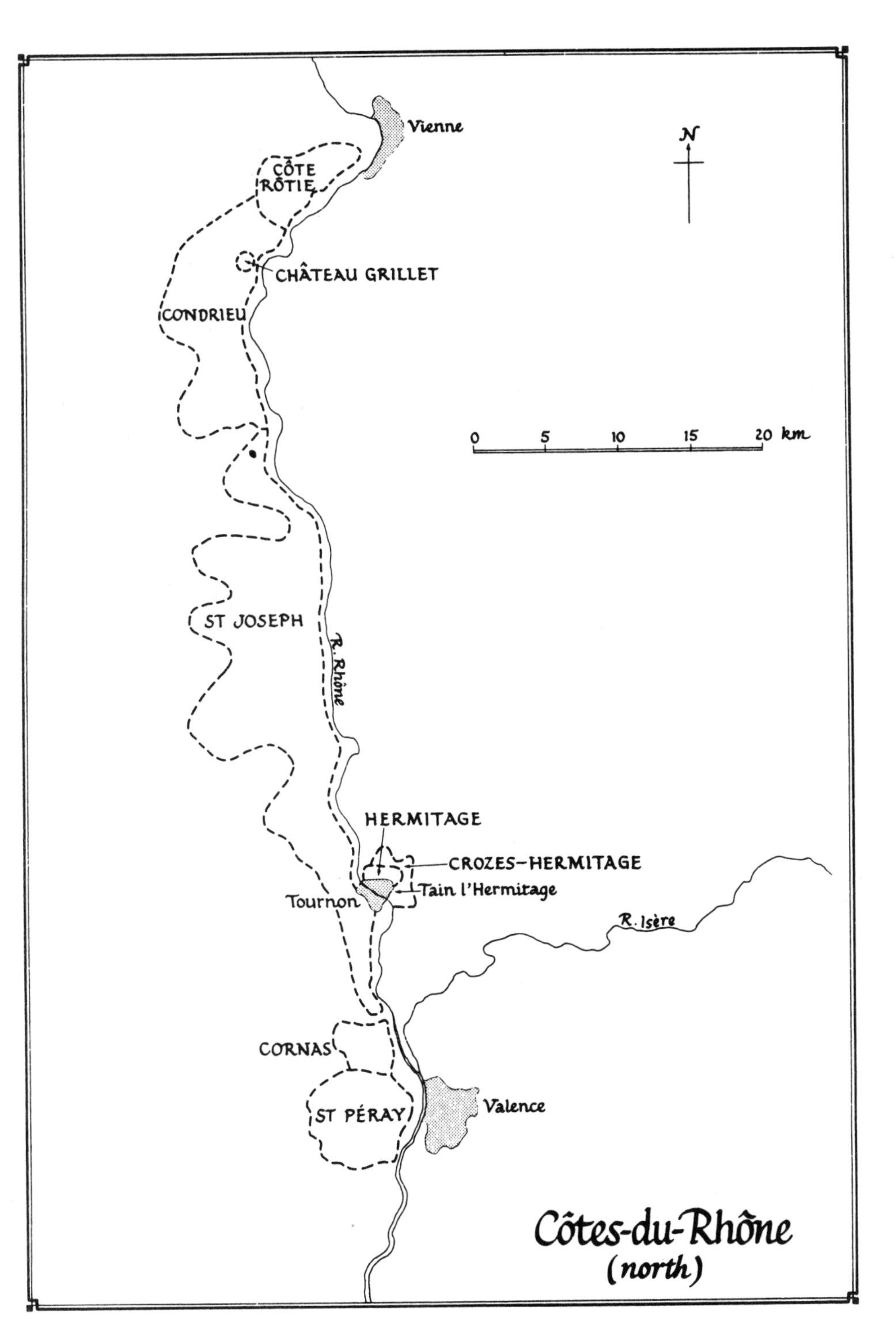

1 Name the three principal wine regions of Southern France.
2 Describe in one word each the climate of (a) the north and (b) the south of the Côtes du Rhône region.

Côtes du Rhône - North

AC Côte Rôtie

This district clings to the precipitous right bank of the Rhône for 3.5 km, south of the little town of Ampuis. It is the northernmost vineyard of the region, and it may be argued that its 60 ha produce the finest red wines: the Syrah or Sérine grape (which is said to have descended directly from the grapes of Shiraz in Persia) is used, and gives a heady tannic wine with great bouquet, improving with long maturation. To moderate the astringent power of this variety, up to 20 per cent of the white Viognier grape is grown in the vineyard to add finesse and delicacy to the wine, which is deep crimson in youth, fading to a deep bronze. The vineyard is divided into two parts, the Côte Brune and the Côte Blonde; legend has it that a medieval owner had two daughters – one brunette and one blonde – and that the vineyards he gave to each took on their characteristics. Certainly the soil in one vineyard is red with iron oxide, and in the other pale with lime.

AC Condrieu

The vineyard continues southward through Condrieu, getting even steeper and terraced like a staircase, and is entirely planted with Viognier, producing a white wine of great finesse, yet full-bodied and very long lasting.

AC Château Grillet

This is a property of no more than 3 ha on a favoured slope within the confines of Condrieu, and has its own AC. In youth the wine is full-bodied and sweet, but loses its sweetness with age, and is noted for its rarity.

AC St Joseph

Some 30 km south of Condrieu, the valley broadens out somewhat; on the right bank, around Tournon, the vineyards of St Joseph produce red wines from the Syrah grape. Although the AC includes white wines, few are produced. The reds are fruity with plenty of backbone.

1 Côtes de Rhône, Provence, Languedoc and Roussillon.
2 (a) Continental. (b) Mediterranean.

AC Cornas

Also on the right bank, just south of St Joseph and opposite the point where the Isère flows into the Rhône, Cornas produces very tannic, heady wines from the Syrah grape. The soil is of granite sand with some beds of large stones.

AC Hermitage

On the left bank, opposite Tournon, stands the town of Tain-l'Hermitage, with the famed Hermitage vineyards lying above it. They take their name from a crusader knight, the Chevalier de Sterimbourg who retired in the vineyards. Both red and white wines are produced, the red from unblended Syrah grapes planted in granite soil, and the white from the Marsanne and Roussanne grapes in more sandy and schistous soil. The best wines come from the top of the hill. The wines have a deep colour, a distinctive bouquet and are full-bodied and soft to the palate. The white wines are dry and delicate, with a full bouquet. Three vineyards, les Bessards, le Méal, and les Greffieux, are recognized as first growths, and ten more as second growths; the most often found is Chante Alouette, whose white wine is better than the red.

Surrounding the Hermitage vineyards are those of AC Crozes-Hermitage, which are lighter and more ordinary, yet fine wines.

AC St Péray

On the right bank, south of Cornas and across the river from the considerable town of Valence, lies St Péray. This district produces white wines from the Roussanne and Marsanne grapes, which are often made sparkling – by the méthode Champenoise to gain the AC St Péray Mousseux. The wines have plenty of flavour and are quite strong.

This completes the catalogue of the north, except for two districts which lie 50 km to the east up the valley of the Drôme, nearer Savoie than the Rhône. AC Clairette de Die is a sparkling wine, made from the Clairette grape which is grown extensively in southern France. The district of Châtillon-en-Diois nearby has AC for red, white, and

1 Why are wines made from only one or two grape varieties in the north of the Côte du Rhône region but from up to thirty in the south?
2 In the south of the Côtes du Rhône, how are the vines trained, and why are they slanted at an angle towards the north?

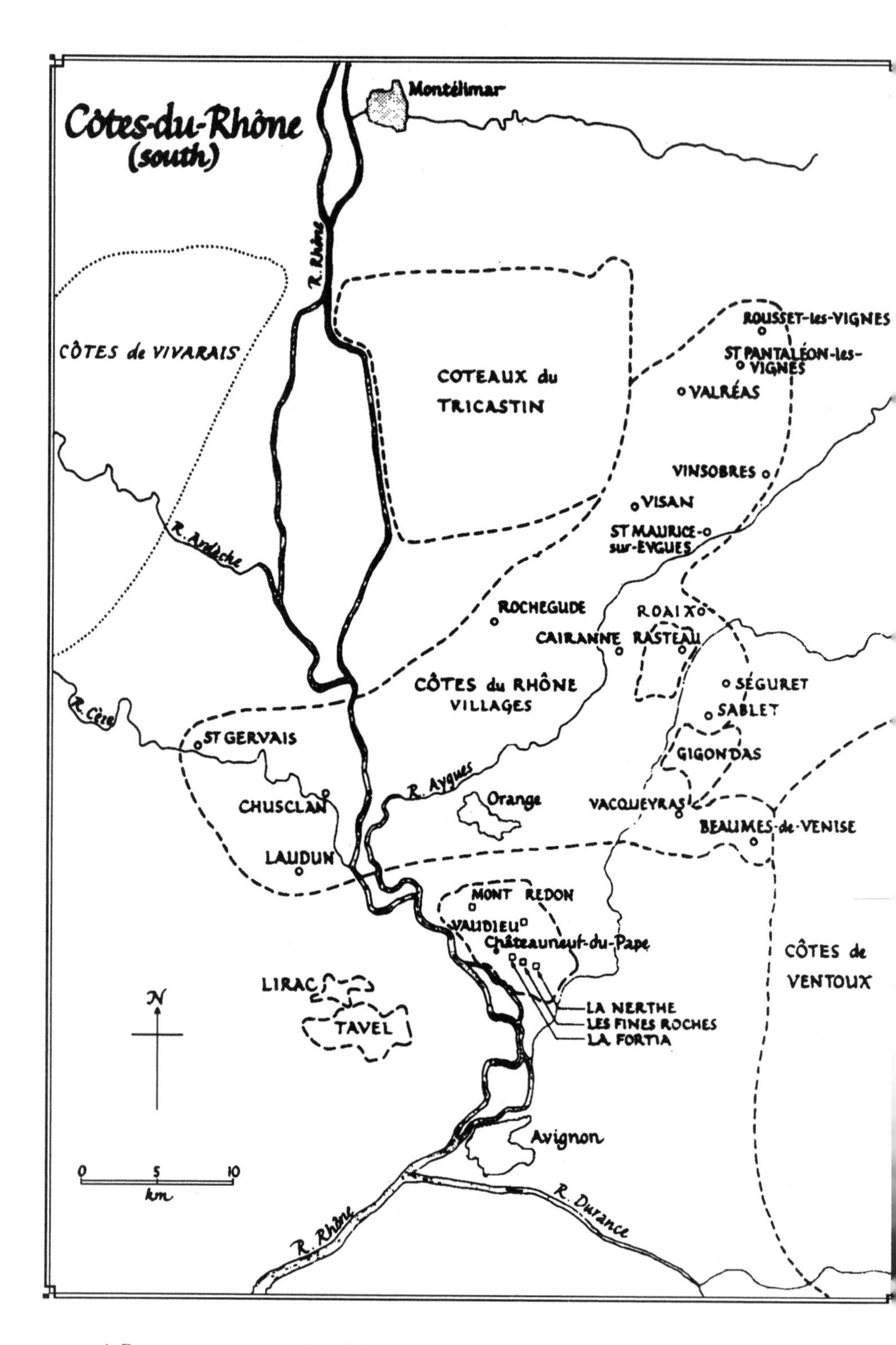

1 Because one grape can give all the desired qualities (colour, acidity, bouquet, finesse, aroma, and vinosity) in a temperate climate, but not in a hot one.

2 Bush trained, and planted so that the mistral will gradually blow them upright.

rosé wines, but is chiefly known for its red wines made from the Pinot Noir and Gamay grapes – both northern varieties, clearly marking the border line.

Côtes du Rhône – South

Côtes de Vivarais VDQS: This district comprises twenty-two villages in the hills west of Montélimar, of which three or four can claim VDQS for red, white, and rosé wines, from five white and seven black grape varieties. If they have 11% vol. of alcohol (rather than 10%) three of them can add the name of their village to the description. This practice is common in the south, and can lead to a confusing multiplicity of names.

AC Coteaux de Tricastin

On the left bank of the Rhône, twelve villages produce mostly red wines from the Grenache, Carignan, Cinsault, Mourvèdre, and Picpoul grapes, the most important black grape varieties of the southern region.

AC Côtes-du-Rhône-Villages

In an area some 20 km wide stretching 60 km south-west from the town of Nyons (east of Tricastin) to the other side of the Rhône lie seventeen villages which, in exchange for stricter levels of yield and minimum alcoholic strength and a tasting test, may add their names to AC Côtes du Rhône, or claim AC Côtes du Rhône-Villages. They are, from north-east to south-west: Rousset-les-Vignes; St Pantaléon-les-Vignes; Valréas; Vinsobres; St Maurice-sur-Eygues; Visan; Roaix; Cairanne; Rochegude; Rasteau; Séguret; Sablet; Vacqueyras; Beaumes-de-Venise; and on the west side of the Rhône, Chusclan; Laudun; and St Gervais. One of the original villages, Gigondas, now has its own AC for red and rosé wines from the Grenache (maximum 65 per cent), and other Côtes-du-Rhône grapes.

1 What colour is AC Côte Rôtie and what grape is used in its production?
2 Is AC Condrieu for red or white wine? What grape is cultivated?
3 How did the Hermitage vineyards acquire their name? Name the vines used for (a) red wine and (b) white wine.
4 By what process is AC St Péray Mousseux made?

Mont Ventoux

Rasteau and Muscat de Beaumes de Venise are AC Vins Doux Naturels, coming from two villages on the slopes of Mont Ventoux. Vins-Doux-Naturels (VDN) are wines to which spirit has been added to arrest fermentation to leave a naturally sweet wine. They are always AC. AC Rasteau is made chiefly from the Grenache grape, and sometimes acquires a curious flavour called *rancio* – which is also an acquired taste! AC Muscat de Beaumes de Venise is made from the Muscat grape and tastes best when young, before the aroma of the Muscat fades. Some sixty villages on the slopes of Mont Ventoux make light red and rosé wines from the Grenache and Carignan grapes. Côtes du Ventoux, previously VDQS, obtained AC status in 1973.

Châteauneuf-du-Pape

The Côtes du Rhône end at the confluence of the Durance and the Rhône, just south of Avignon where in 1309 Clément V was crowned as Pope. Pape Clément is well remembered as Bishop of Bordeaux, for he left to it his vineyard; he should be remembered better as the first occupant of Châteauneuf-du-Pape, a castle lying in a bend of the Rhône halfway between Avignon and Orange. The village lies in a vineyard which had been established before the Romans came, and its wine, now undoubtedly the best known wine of the Rhône, has enjoyed a fine reputation since Roman times. The red wines are full-bodied and deceptively soft, for they are very strong (minimum 12.5% vol.); white Châteauneuf is also made. The soil is very varied, but noteworthy for the large stones which cover the vineyards. Eight grape varieties are commonly used in combination:

40–60 % Grenache, giving finesse and mellowness,
30–10 % Syrah, giving bouquet, body, colour and strength,
20% Cinsaut, Mourvèdre, or Vaccarèse, giving vinosity, and
10 % Clairette, Terret, or Picpoul, giving volume and lightness.

1 Red wine produced from Syrah grapes.
2 White from the Viognier vine.
3 From a crusader knight who retired there. (a) Syrah (unblended). (b) Marsanne and Roussanne.
4 Méthode Champenoise.

Tavel

Just across the Rhône from Châteauneuf lies the village of Tavel, renowned for its dry AC rosé wines. These are made from the Grenache grape, grown on a soil nearly as stony as that of Châteauneuf, although frequently the stones of Tavel are limestone chips. The Grenache grape is vinified together with white Clairette and Bourboulenc to produce a dry, almost austere, rosé wine of not less than 11% vol. alcohol (but usually nearer 13%); the wine has an onion-skin colour, and improves with age. This may be contrasted with the sweet, fruity, pink Rosé d'Anjou of little more than 9% vol. alcohol, which is best in its youth. AC Tavel is bottled in tall slim bottles – differing slightly from the flûte d'Alsace. Adjoining AC Tavel on the north lies AC Lirac, where the soils are similar – red stones, sand, and limestone. Lirac, however, has AC for red and white as well as rosé wines, and of these the red is best known.

Vins de Pays: Collines Rhodaniennes cover the whole of the northern part of the region from Lyon to Montélimar, and up the Drôme to Die. Also in that Département lie the Coteaux de Grignan and, east of Tricastin, the Coteaux des Baronnies – the remains of the erstwhile VDQS Haut-Comtat. Further south towards Châteauneuf-du-Pape, in Vaucluse, Vin de Pays de la Principauté d'Orange is found. On the right bank of the Rhône, several of the twelve Vins de Pays of the Département of Gard come within the region.

PROVENCE

This region extends from the Côtes-du-Rhône southwards from Avignon and east to the Italian border. Its wines are in many ways similar to those of the southern Côtes-du-Rhône; less than 10 per cent are white, but some of these are notable. The soil varies considerably, with sand and granite the principal features; limestone outcrops occur, and sometimes determine the extent of the best areas.

1 (a) How many villages share AC Coteaux de Tricastin? (b) Give the wine colour and the main grape varieties used. (c) Is it part of the Côtes-du-Rhône Villages?

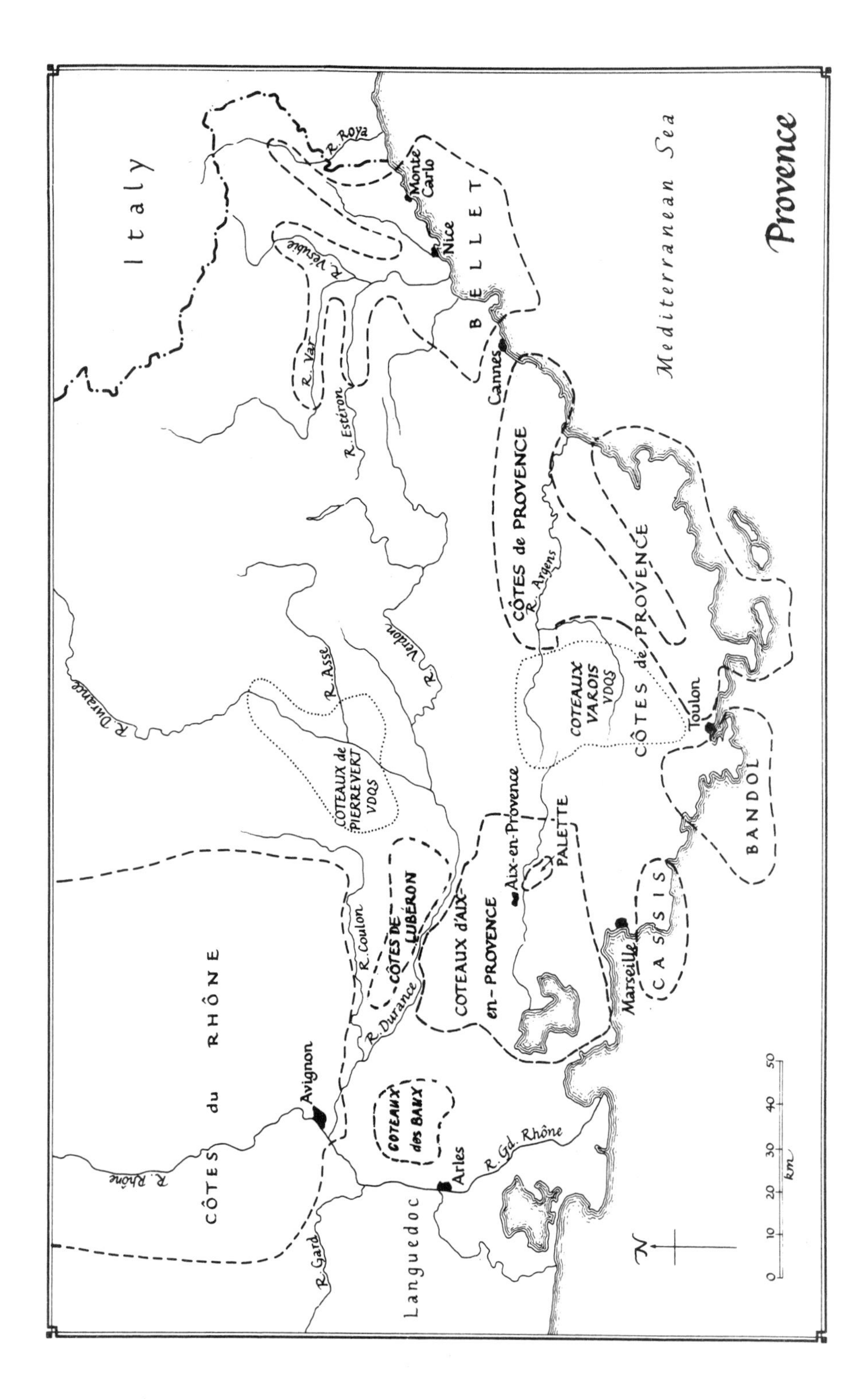

1 (a) Twelve. (b) Red wine: Grenache, Carignan, Cinsault, Mourvèdre, and Picpoul. (c) No.

Viticulture

The grape varieties used are similar to the southern Rhône varieties, but the commercially useful Carignan is increasingly being blended with the Cabernets – Franc and Sauvignon – giving greater finesse to the wine. Because of the mistral, bush cultivation has been standard, but some proprietors wishing to avail themselves of mechanical harvesting machines have converted to Guyot training on wires: it is then necessary to align the rows into the prevailing wind, and to train the shoots between two wires; if tied to one, the shoots are too often broken by the wind.

AC Districts

Provence has eight AC districts, of which four are recent promotions from VDQS as follows:

AC Côtes de Provence: This district surrounding the schistous Massif des Maures between Cannes and Toulon produces large quantities of very drinkable but expensive rosé wine.

AC Coteaux d'Aix-en-Provence: Here sandy soil over limestone extends from the river Durance in the north to the coast west of Marseille giving some sound white, and good quality red and rosé, wines.

AC Coteaux des Baux-de-Provence: This small district lies south of Avignon on the craggy slopes of Les Alpilles, whose aluminium-rich soil gave the name to bauxite ore. The best wine is red, made from Carignan with Cabernet Sauvignon.

AC Côtes de Lubéron: These Côtes lie between the rivers Coulon (Calavon) and Durance on the northern slopes of the Montagne de Lubéron. The red and white wines are excellent value for money.

1 Why is Châteauneuf-du-Pape so called?
2 (a) What is the minimum alcohol by volume in AC Châteauneuf-du-Pape. (b) Describe an unusual feature of its soil. (c) Mention the two most important vines used in its production and the qualities they give to the wine.
3 (a) For what wine is Tavel famous? (b) How does its soil differ from that of Châteauneuf-du-Pape?
4 What is the principal Vin du Pays of the northern part of the Rhône valley?

AC Palette

This tiny district on the southern outskirts of Aix has now only one vineyard – Château Simone – where they make a superb full-bodied red wine by entirely traditional methods. It is fashionable in some quarters to decry *la méthode ancienne* or even to deny that it ever existed, so it may be pertinent to describe that used at Château Simone. Firstly, the vineyard is planted with traditional vine-types, over twenty of them, in their traditional proportions, and replanting adheres to this. Horses, whose emanations are better for the vines than those of tractors, are used in the vineyard. Only wooden implements are used in the cellars – no metal ever comes into contact with the must or wine. After de-stemming and crushing, the grapes are fermented on their skins for two to three weeks according to the temperature, which is kept as near 20°c as possible. Then the wine is run off to mature in hogsheads in the cool cellars beneath the Château; the press-wine is matured similarly, the wines of each pressing being kept in separate hogsheads.

After one year, the new wine is tasted, and blended with as much of the press-wine as the master needs to produce the degree of robustness and keeping quality required: this process is repeated after a further year, though a final adjustment is not often needed (the standard treatments of topping-up and racking have of course been applied, although the wine may only be racked three times in the second year and twice in the third). Fining is carried out before bottling; the fining agent used is casein. The result is a full-bodied generous wine of great depth of colour that will last for several decades; it has a full fruity bouquet and a long finish. Such wines are not cheap to make, nor to bring to maturity, and so they will always be expensive to buy; but they are worth saving up for.

Not far to the south of Aix lie the city-port of Marseille and the naval port of Toulon; between these on the coast two more AC districts are situated.

AC Cassis

This is to the west nearer Marseille, and is noted for its white wine from Ugni Blanc, Oeillade, and Clairette grapes, which is heavy in

1 It is the vineyard surrounding the Summer Palace of the Popes, when they were established at Avignon.
2 (a) 12.5% vol. (b) Large stones cover the vineyards. (c) Grenache, for finesse and mellowness, and Syrah for bouquet, body, colour, and strength.
3 (a) A dry rosé wine of onion-skin colour. (b) Less stony but with large limestone chips.
4 Collines Rhodaniennes.

alcohol with a touch of sweetness; the AC also applies to red and rosé wines. Students should be careful not to confuse this AC Cassis with the blackcurrant liqueur Crème de Cassis, which is often added to the rather acid white Aligoté wine of Burgundy to make 'Kir'.

AC Bandol or Vin de Bandol

This district lies to the east of Cassis nearer Toulon, and is noted for its red wines (although the AC extends to white). Fairly light in colour, sometimes with a pronounced bouquet of violets, these wines are nevertheless heavy in alcohol. They resemble the wines of Bordeaux rather than those of the Rhône.

AC Bellet or Vin de Bellet

The vineyards of this AC lie in the narrow sub-Alpine valleys behind the Côte d'Azur from Cannes to the Italian border beyond Monte Carlo. Here red, white, and rosé wines are made from local grape varieties.

VDQS Districts

There now remains only one in Provence – Coteaux de Pierrevert – producing white wines from the Ugni Blanc grape. The district is situated near the confluence of the Verdon and the Durance, a little to the north-east of Côtes de Lubéron.

Vins de Pays

Vins de Pays de la Petite Crau are found around les Baux in Bouches-du-Rhône; Var has three – Argen west of Toulon, and les Maures and Mont Caume to the east.

LANGUEDOC AND ROUSSILLON

This region, or pair of regions, extends from the Cévennes mountains in the north to the Pyrenees in the south, and from the

1 (a) Between which towns do the Côtes de Provence stretch? (b) What kind of wine is produced there?

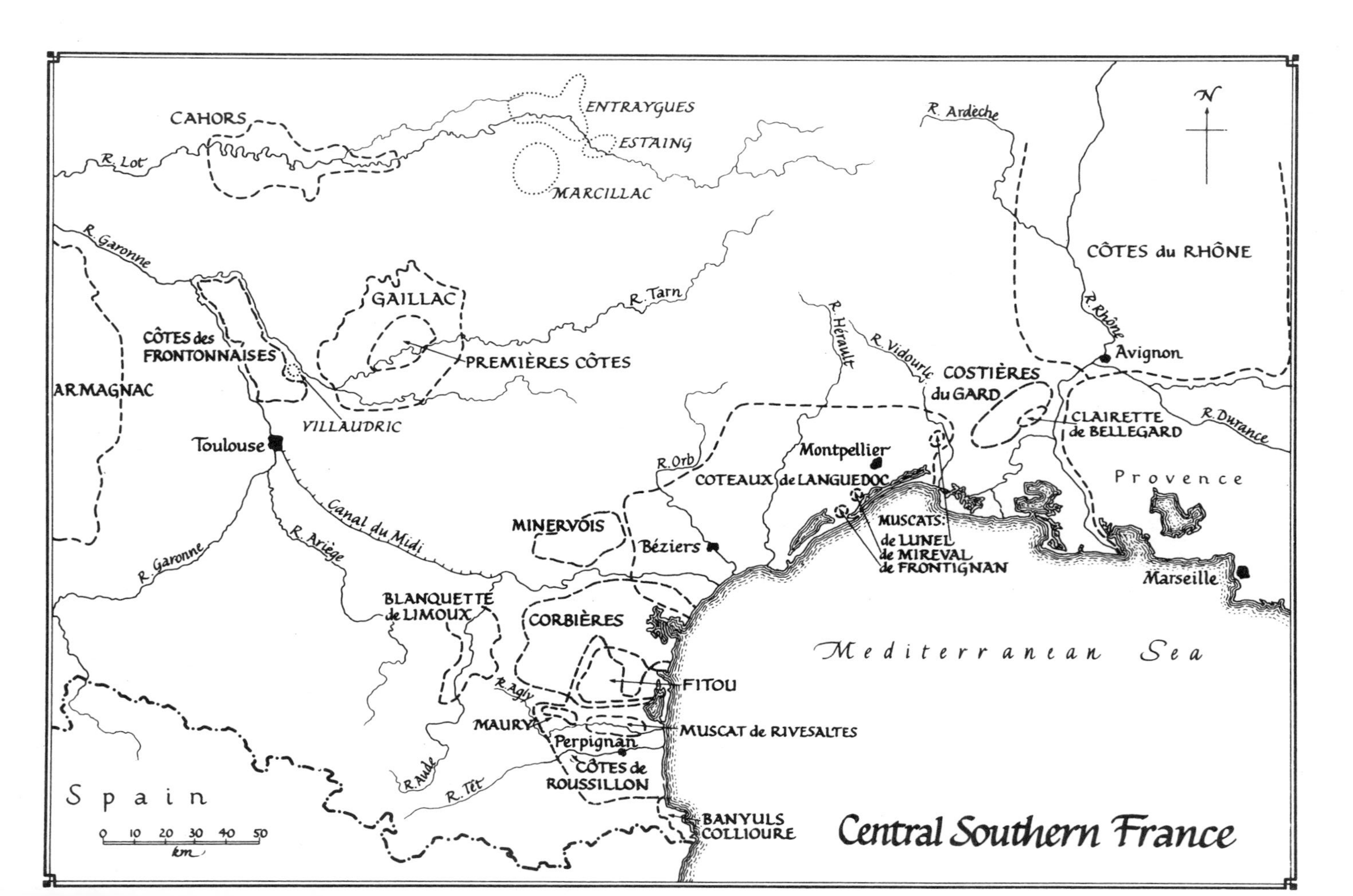

1 (a) Toulon and Cannes. (b) Good but expensive rosé wine.

Mediterranean beaches in the east as far as the city of Toulouse. It is a region of moderate rainfall, great heat in summer and, except in the higher mountain vineyards, is frost-free. Most of it is a vast plain with fairly rich soil, producing vast quantities of very ordinary wines, formerly known as 'les vins du Midi', generally sold under brand names or sophisticated to make Vermouth.

The four Départements of Gard, Hérault, Aude, and Pyrénées-Orientales comprising this region contain between them nearly 40 per cent of France's vineyards. Formerly, they were planted overwhelmingly with the high-yielding fleshy Aramon grape, whose wine needed to be improved by blending with high-strength Algerian or Italian wine in order to reach minimum standards. However, in recent years the Aramon has been replaced with the Carignan, and great strides have been made in persuading growers to plant better varieties such as Cabernet Sauvignon, Grenache, and Syrah, so that much finer wines with distinct local characteristics are being produced.

In fact, it is fair to say that the region produces a large quantity of quality wines, white, red, and rosé, and some very fine fortified 'Vins-Doux-Naturels'. The quality light wines are mostly VDQS, but there are also a number of AC wines; often the districts overlap, as the wines of one village may be AC, of the next Table Wine, and the next VDQS. The reason for this may best be sought in the soils and the grapes, for the climate is equable across the region.

Soil

The River Valleys: Here the soil is rich, and the vines stretch from horizon to horizon. The Aramon used to be the principal vine, which is a large watery grape; it gives large quantities of thin wine which needs to be blended with strong Italian wine to be saleable; previously, Algerian wine did this duty. It is now gradually being replaced with Carignan.

Plains: Between the rivers the plains are more or less clayey, with lime or gravel subsoils. In some places, the gravels, in the form of large pebbles, come to the surface, and it is here that quality wines

1 (a) What vines are used for AC Cassis? (b) Distinguish between AC Cassis and Crème de Cassis.
2 Describe AC Bandol wines.
3 Name the AC districts of Provence.
4 Name three VDQS districts of Provence.

are found. Grapes suitable to these soils are Carignan, Grenache, and Picpoul, found in the districts of Coteaux de Languedoc, Minervois, and Corbières.

Hillsides and Garrigues (moorland): Here the subsoil is of limestone, and is closer to the surface; or of schist also covered by a thin topsoil; the finer wines are found here – and little else. The soil is ideal for the cultivation of grapes for wine.

The Sands: The Rhône over the centuries has formed a delta at its mouth; its finer glacier-ground sands have spread beyond that delta to the south-west and form a coastline of dunes and lagoons. Besides providing some of the most attractive bathing beaches in the world, these sands provide a soil which *phylloxera vastatrix* cannot penetrate, but in which *vitis vinifera* grows readily; the sand has to be carefully fertilized, for it is essentially sterile, yet will produce sound wine from many grape varieties.

Languedoc

Costières du Gard

A few kilometres south-west of Avignon, on stony hillsides, this district is best known for its tough red wines, though some white and rosé are produced from Mourvèdre (50 per cent), Terret Noir, Grenache and other grapes.

AC Clairette de Bellegarde

Enveloped by the Costières du Gard on the north, on calcareous slopes, this tiny district produces good perfumed wines from the Clairette grape, golden in colour, which are much in demand as base wines for vermouth.

AC Clairette du Languedoc

To the west of the preceding two districts, on better limestone soil, the Clairette grape gives sweet white wines; some, with a lighter

1 (a) Ugni Blanc, Oeillade, and Clairette. (b) AC Cassis is Provençal wine; Crème de Cassis is a blackcurrant liqueur.
2 Notably red wines; bouquet of violets; heavy in alcohol; resemble Bordeaux rather than Rhône wines.
3 Bandol, Bellet, Cassis, Coteaux d'Aix-en-Provence and Baux, and Palette.
4 Côtes-de-Lubéron, Coteaux de Pierrevert, and Coteaux d'Aix-en-Provence.

degree of alcohol, develop the acquired taste called rancio already mentioned in connection with Rasteau. The origin of the word 'Languedoc' is interesting: it means 'the tongue, or speech, of Oc', and refers to the local accent: where the rest of France said 'Oui' for 'Yes', the people from this region said 'Oc'.

AC Minervois and Corbières

These two districts, north and south respectively of the main Carcassone-Narbonne road, are centred on the little town of Lézignan-Corbières. These wines are primarily red; although the classification extends to white and rosé, these are little heard of. The wines of Minervois tend to be lighter in colour and more delicate than the Corbières, yet stronger in alcohol; the Corbières give an impression of bitterness when young, which disappears with a little age. The Carignan is the principal grape for both, although the Grenache is gaining ground in Corbières: the difference lies in the soil, which in Minervois is very mixed with high mineral content.

AC Fitou

This district, which used to be known as Corbières-Maritimes, and is entirely within AC Corbières, consists of nine villages near the sea on barren stony soil. It produces strong red wines with good colour and bouquet.

Coteaux du Languedoc VDQS: This general description covers the wines of seventy-four villages stretched over a wide area between Nîmes and Narbonne, on the southern slopes of the Cévennes. They surround the Clairette du Languedoc district, and one is inside it; twelve of them are entitled to add their own village name to Coteaux de Languedoc, or to use it on its own, as separate VDQSs. A few are 'licensed' for red, white, and rosé wines, but most are restricted to only one or two of these categories; they are sound, drinkable wines, usually at reasonable prices. The combination of grape varieties differs from village to village. Two of the original principal sub-districts on the western end of the Coteaux – Faugères and St Chinian – were granted their own ACs in 1982. Both produce

1 Where are the regions of Languedoc and Roussillon situated?
2 By which variety has the Aramon been replaced in Langeudoc?

fragrant red wines like those of nearby Minervois, itself elevated to AC status in 1985.

Picpoul de Pinet VDQS: This small district lying near the coast south of the Clairette de Languedoc district produces dry white wines from the acidic Picpoul grape (the Folle Blanche of Cognac and Armagnac).

The Upper Aude: Here two districts produce good red wine: they are *Cabardès* and *Côtes de Cabardès et de l'Orbiel VDQS*, and *Côtes de la Malepère VDQS*.

'Les Vins des Sables': Although neither AC nor VDQS, these wines deserve mention for the ingenuity of their production. One vineyard in particular between Sète and Agde is separated from the sea only by the beach, road, and railway; it is 1 km wide, between the railway and a salt lagoon, and is 17 km long. Immediately after the harvest, barley is grown to hold the sand together in the winter storms, and to be ploughed in as green manure in the spring. The harvest, for both red and white grapes, is mechanized. In the vinification of red wines, the thermotic maceration process described in Chapter 2 is used. Under nitrogen, the must can be kept almost indefinitely to await the addition of cultured yeasts to become wine, or to be canned and exported for home winemakers, or to be marketed as grape juice, sometimes mixed with other fruit juices. The wine, like the fruit juice, may be put up in one-litre square cardboard cylinders with aluminium ends, lined with impervious material; these save bulk and do not break if dropped; their contents are palatable and cheap, which is all they pretend to be.

Roussillon

The eastern part of the Languedoc merges imperceptibly from Corbières and Fitou south into Roussillon towards the border with Spain. This region is devoted to fortified wines, but some good light wines are made, reputed to have a particular flavour derived from the schistous and marly soils.

1 Between the Cévennes mountains (north), the Pyrenees (south), the Mediterranean (east), and Toulouse (west).
2 Carignan.

The first light wine AC of the region, Collioure, is a tiny area close to the coast on gravelly schistous soil at the end of the Pyrenean chain, and produces red wines from the Cabernet, Grenache, Picpoul, and Terret Noir grapes. AC Côtes du Roussillon created in 1977 covers most of the remaining area south of Perpignan and the river Têt, while Côtes du Roussillon-Villages, created at the same time, covers the area north of the Têt to the Département boundary; the best wines come from the valley of the Agly in which two villages, Caramany and Latour de France, stand out and are allowed to add their names to the general Appellation as a reward for planting better grape varieties and restricting their yield.

Fortified Wines

But the fame of Roussillon lies in its Vins-Doux-Naturels, made either from the Grenache or the Muscatel vines. As yet, they are little known in Britain, where Sherry and Port reign supreme, but they deserve to be.

The different names given to fortified wines in France can be confusing. *Mistelles* are not wines, but unfermented grape juice to which brandy has been added in sufficient quantity to prevent fermentation, and are extensively made in Languedoc and Roussillon for sweetening vermouths. *Vins-Doux-Naturels* (VDN) and *Vins-de-Liqueur* (VdL) are wines which have, like Port, had their fermentation curtailed by addition of brandy, leaving the wine sweet. The difference between them is in the minimum degree of sugar in the original must, and the permitted grape varieties. VDN are always AC: VdL sometimes.

Another difference from Port is that the wines are often matured in casks left in the open to summer heat and winter cold, which imparts to wines made from the Grenache a flavour called *rancio*, particularly if the casks have been left unbunged: it has been likened by some to old Sherry, or to Madeira. The best ACs are Banyuls and Banyuls Rancio from the steep terraced slopes of the Pyrenees where they fall to the sea at the Franco-Spanish frontier; a third AC, Banyuls Rancio Grand Cru, requires a higher degree of alcohol and more age. It is very expensive.

1 What special advantage do the Rhône delta sands have for viticulture?
2 What is the origin of the word Languedoc?

Similar wines are made further up the coast and inland; ACs Maury, Rivesaltes, and Grand Roussillon each have their Rancio counterparts, but ACs Frontignan and Vin de Frontignan do not. The Muscatel grape is also more favoured here, and ACs Muscat de Frontignan, Muscat de Lunel, Muscat de Mireval, Muscat de Rivesaltes, and Muscat de St Jean de Minervois are highly prized; they vary between deep gold and rosé in colour, with a flavour of fine Sultana raisins – with one notable exception: AC Muscat de Rivesaltes, like Muscat de Beaumes de Venise from the Rhône valley, tastes of fresh Muscatel grapes.

The three wines of Frontignan also have ACs for VdL, as also does Clairette de Languedoc.

Sparkling Wines

In the limestone hills of the upper reaches of the Aude lies the little town of Limoux, where AC still, semi-sparkling and sparkling wines are produced, mainly by one co-operative. The grape previously used for AC Blanquette de Limoux was the Mauzac or Blanquette, giving a soft semi-sweet wine which aged rapidly; however, the AC rules have recently been relaxed to allow the Chardonnay and Chenin grapes to be used, which give a wine of greater character needing about three years to mature. The méthode Champenoise is used, and the wine must mature for a minimum of one year. AC Vin de Blanquette, from the Mauzac grape, is semi-sparkling and made by the *méthode rurale*, whereby the wine is filtered before it has completely finished its normal (first) fermentation, and is then bottled. The remaining fermentation is sufficient to give a slight sparkle without creating enough sediment to require removal. A small quantity of (still) AC Limoux is produced, also from the Mauzac grape.

Vins de Pays

This part of France is the birthplace of the Vins de Pays, and their institution was designed to provoke the local producers to make better wines; it is not surprising, therefore, that they have prolifer-

1 *Phylloxera* cannot penetrate them yet *vitis vinifera* grows readily.
2 The 'tongue of Oc' – for here the people said 'Oc' for 'Yes'.

ated to the extent where enumerating them would be wearisome and of little use, as they are unlikely to be encountered away from their origins. Of the four Départements, Pyrénées-Orientales has 4, Aude 20, and Hérault 26 – but then Hérault accounts for nearly 15 per cent of France's vineyards! Gard has 12, some of which lie in the Rhône valley, and one of these, Vin de Pays de Sables du Golfe de Lion, might be found in this country. In time we may see the elevation of some of these Vins de Pays to VDQS or even AC status. Vin de Pays d'Oc covers the whole region.

CORSICA

This island, which more readily brings to mind 'bandits' or 'the birthplace of Napoléon', was nevertheless known for its wines until *phylloxera* wiped out the vineyards and the growers moved to Algeria. Its wines are now staging a considerable revival, and the first ACs were granted in 1976.

The climate is warm, the soil uncompromising rock, but the local Sciacarello and imported Niellecio (Sangiovese) grapes yield heady red and rosé wine, while the white Vermentino, Malvasia, and Muscat grapes yield alcoholic dry and sweet white wines and VDN. The general Appellation Contrôlée Vin de Corse consists mostly of heavy red, white, and rosé wines grown on the rich alluvial soils of the east coast; much of that which does not reach the AC standard (about 80 per cent) is sent to the mainland for blending.

However, there are now five AC districts adding their names to the general one, as AC Vin de Corse Figari, which produce better wines. At the southern tip of the island, Porto Vecchio produces crisp fruity white wines, and Figari and Sartène strong red, white, and rosé, as does Calvi on the north-west coast. The fifth district, Coteaux de Cap Corse, is situated at the tip of the 'finger' which projects north from Bastia; it is reputed for its sweet white wines. The best wines come from this part of the island, particularly near Patrimonio at the base of the 'finger' where there is chalky soil. The wines with AC Patrimonio, granted in 1984, can be likened more to those of the Rhône than to those of Provence; important also is Ajaccio AC.

1 Where does Cabardès VDQS come from and what is it like?
2 Name two villages entitled to add their name to AC Côtes du Roussillon-Villages.

Vins de Pays

There is one Vin de Pays name covering the whole island – Ile de la Beauté.

1 The upper Aude valley; sound red wine.
2 Caramany, Latour-de-France.

4
Germany – General

Germany lies at the limit of the northern climatic band wherein grapes can be grown for wine, and production of wine is therefore a continual struggle between the resourcefulness of man and the inconstancies of nature. At this latitude, black grapes seldom get enough heat during the summer to ripen fully and, although selected from famous French varieties, they produce red wines that are usually no more than mediocre. But the white varieties can be brought to a fully ripened harvest and are tended with such care that their wines are counted among the world's finest, and are widely imitated. Generally speaking, the wines are low in alcohol and high in acid, as one might expect in such peripheral regions. German wine-makers must therefore take greater care than those in more

1 (a) What method is used to produce sparkling AC Blanquette de Limoux? (b) What grapes are now used?
2 How many Vins de Pays originate in the four Départements of Languedoc/Roussillon?
3 Name the main grapes used for Corsican red wines.
4 Name the main grapes used for Corsican white wines.
5 What districts are allowed to add their names to AC Vin de Corse?

favourable climes, and in doing so must turn the vagaries of nature to their own advantage.

German drinking habits, like their wines, differ from those of their neighbours: the French drink wine *with* their meals, and hence prefer wines with some alcoholic weight; the Germans, on the other hand, have preferred to drink before or after meals, occasions when lighter wines are suitable.

The Romans planted vines when they occupied Germany – and the Emperor Charlemagne, who ruled Europe in the ninth century, encouraged plantings; subsequent medieval vine history followed the pattern of one of his other domains – Burgundy. The châteaux and abbeys of France are reflected in the Schlösser and Klostern of Germany. And when Napoleon overran the German vineyard regions, he split up the vineyards among the peasants as in France, so that German vineyards are split up into even tinier parcels (*Einzellagen*) than the *climats* of Burgundy.

Climate

In general this is continental in character with cold winters, hot summers, and moderate rainfall which in winter often falls as snow, protecting the roots of the vine. But the climate varies considerably from year to year, and it is important to know the quality of the different vintages. However, there are two characteristics which help the vinegrower: Whitsuntide – when vines flower – is nearly always fine in the northern vineyards of the Mosel, the Rheingau, and the Nahe; and the vineyards of Germany enjoy long hot autumns.

Soils

These vary considerably between the various regions and will be described separately, but are generally too poor to grow crops other than grapes economically. As many other crops are even less tolerant of climate, growing grapes for wine makes sense.

1 (a) Méthode Champenoise. (b) Chardonnay and Chenin.
2 Sixty-two.
3 Sciacarello and Niellucio (Sangiovese).
4 Vermentino, Malvasia, and Muscat.
5 Calvi, Coteaux de Cap Corse, Figari, Porto Vecchio, and Sartène.

Grapes

The classic grape of Germany is the Riesling, and in the right vineyard it produces unbeatable wines. It is a low-yielding grape, however, and late-ripening which, in a wet autumn, can bode ill for the harvest. The berries have tough skins, are small, and are borne in small bunches.

The Silvaner grape, on the other hand, is more prolific in yield and ripens earlier, but does not give such fine wine.

A scientist named Müller, from Thurgau in Switzerland, crossed these two grapes in 1882, creating the Müller-Thurgau. This crossed variety has many of the better qualities of both its parents, and the wines made from it have a distinctive flowery bouquet with a muscat background. Nevertheless, the grapes are very liable to rot, and German viticulturalists have been at pains to develop further crosses with the object of breeding the perfect grape for Germany: Morio-Muskat, Scheurebe, Kerner, and Faber are widely planted, but others such as Optima, Reichensteiner, Bacchus, Huxelrebe, and Ehrenfelser may also be encountered. Over the years, the Müller-Thurgau has gained popularity, mostly at the expense of the Silvaner, and in 1986 accounted for 24 per cent of white wine plantings; while Riesling more or less retains its 21 per cent share, Silvaner's share has dropped from 17 per cent in 1973 to just 8 per cent in 1986. Kerner at 7 per cent is also worth noting.

The proportional planting of German vineyards changes from year to year: Table 2 shows the proportion obtaining in 1987.

Viticulture

The care taken by German viticulturalists in producing suitable grape varieties by crossing, and by clone selection to establish firm characteristics of new types, has been mentioned. This care is also shown in every viticultural activity, and great ingenuity is used in devising machines to cope with adverse conditions, particularly on the steep slopes found in several of the regions.

Training methods for vines vary: post-training for steep slopes, high-wire training for districts with spring frost-risk, low-wire where this will benefit the crop. Permitted crop levels are higher than

1 Name the Vin de Pays covering all Corsica.
2 Why is fine German wine mostly white?

Table 2
Planting of grape varieties in West Germany by QbA regions, in percentages

Region	*Rheinhessen*	*Rheinpfalz*	*Mosel-Saar-Ruwer*	*Baden*	*Württem-berg*	*Nahe*	*Franken*	*Rheingau*	*Mittel-rhein*	*Ahr*	*Hessische Berg-strasse*
% of total	25	23	13	15	10	5	5	3	<1	<1	<1
White Grapes											
Weiss-Burgunder				3		1					
Elbling			9								
Gutedel				8							
Kerner	8	11	6	1	9	8	5	2	6	1	3
Bacchus	8	2	2			6	9		1		
Faberrebe	7					3					
Morio-Muscat	4	7									
Huxelrebe	4	3									
Müller-Thurgau	24	24	23	38	10	27	46	6	11	12	18
Riesling	6	15	55	7	24	22	2	80	76	15	51
Ruländer		3		12	1	3					
Scheurebe	10	6				7	3		1		2
Silvaner	13	9		3	5	14	19	1	1		7
Gewürztraminer	1	2		2			1				
Black Grapes											
Spätburgunder				22	3			5		36	
Limberger					6						
Müllerrebe					14						
Portugieser	3	8		1						26	
Trollinger					23						
Other varieties	13	12	5	2	–	9	13	6	4	9	11

1 Ile de la Beauté.
2 Because German wine regions are at the northern limit for commercial wine production.

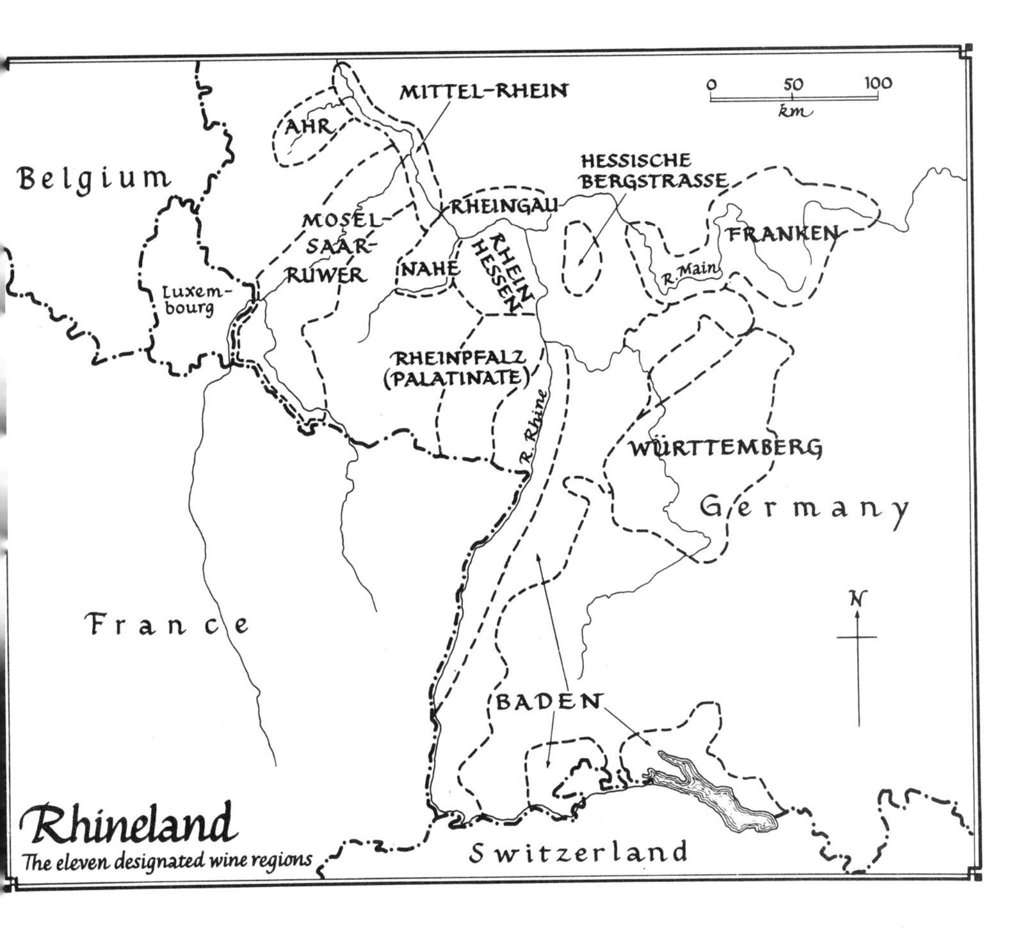

Rhineland

The eleven designated wine regions

in France: 100 hl/ha for fine wines is allowed where the soil (enriched with careful calculation by fertilizers and minerals), the grape variety, and the microclimate permit.

Nor is care only exercised by the owners of large vineyard areas; much wine production in Germany is conducted on an allotment basis. The average (median) holding is less than 1 ha (less than 1 acre even); less than 1 per cent of growers have more than 20 ha (50 acres), and these together own less than 4 per cent of the total vineyard area. Therefore, as it is estimated that a grower needs about 4.5 ha (11 acres) to support himself and his family, almost 98 per cent of the number of growers, owning about 80 per cent of the total vineyard area must have other means of support. In 1984, 60–80 per

1 What are the German equivalents of the French Châteaux and Abbayes?
2 Name two features of the German climate that help the vinegrower.
3 What is the classic grape of Germany?
4 (a) Which vine cross is becoming most heavily planted in Germany? (b) What vines were crossed to make it, when, and by whom?

cent of growers were part-time, according to the region; the co-operative share varied from 80–85 per cent in Württemberg and Baden to 11–15 per cent in Rheinhessen and Mosel-Saar-Ruwer.

VINIFICATION

This is carried out with the same care, even by the proprietors of the smallest vineyards, as with viticulture. Permitted processes are strictly controlled, particularly the enrichment of musts (*Anreichern* and in France, *chaptalisation*) to increase the alcoholic strength. Except in Baden, the must is required to have a specific gravity of 1044 (44° Oechsle) capable of producing 5.0% vol. alcohol before sufficient sugar to raise the alcohol level to 8.5% – the minimum for table wine – may be added. In Baden, the minimum potential alcohol is 6% vol. and only enough sugar to produce 2.5% more may be added. The finest wines, called *Qualitätsweine-mit-Prädikat* (QmP), may not have any sugar added at all.

How is it then that so many German wines are sweet? The late gathering of grapes in a fine autumn enables them to increase their sugar content; it is said that this practice originated in Johannisberg on the Rheingau, when the messenger bearing the permit to start the vintage was waylaid and did not appear for several weeks. Despite gloomy predictions, the harvest turned out to be the best ever, and a tradition was born. In present times, when the weather permits, grapes may be left on the vine even into January of the New Year.

In very recent years, however, the practice of adding *Süssreserve* to fully-fermented wines has arisen. This *Süssreserve* (translatable as 'sweet-reserve'), is grape juice which has been prevented from fermenting normally by filtering out all the yeasts, or by killing them with heat or SO_2 (more usual than CO_2) pressure. Without the Süssreserve, these wines would have attained the requisite alcoholic degree, but would have been unbalanced, harsh, and acid. But when the Süssreserve is blended with the fully-fermented dry, but acid, wine, and kept under sterile conditions so that no stray yeast can re-start fermentation, the balance is restored. This practice is allowed for all German qwpsrs including QMPs, but the Süssreserve must match the wine exactly.

1 Schlösser and Klostern.
2 Fine weather at Whitsuntide and long hot autumns.
3 Riesling.
4 (a) Müller-Thurgau. (b) Riesling and Silvaner, in 1882, by Dr Müller of Thurgau in Switzerland.

WINE REGIONS

The most northerly commercial vineyards of the world are said to lie in Germany on the banks of the river Elbe near Wittenberg at 53°N. These wines are not seen outside East Germany.

In West Germany, vineyards are found in the broad and narrow valleys of the Rhine and its tributaries south of Köln (Cologne), and contain four Tafelwein (table wine) regions – Rhein-Mosel, Oberrhein, Neckar, and Bayern (Bavaria); Rhein-Mosel and Bayern are each divided into three sub-regions, and Oberrhein into two. Tafelwein labels must bear the word 'German' or 'Deutsch', and the wine must be 100 per cent from Germany; the label may bear the name of the Tafelweingebiet or sub-region (provided that at least 85 per cent of the wine comes from there) otherwise, it must bear the description '*Weisswein*' or '*Rotwein*' as appropriate.

The regions (*Tafelweingebiete*) contain fifteen *Landweingebiete.* For classification as a Landwein, the minimum must-weights are raised by 3° Oechsle (equivalent to 0.5% vol. alcohol), and the finished wine must have no more than 18 grams/litre residual sugar (*Halbtrocken* = semi-dry). The labels must bear the words 'Tafelwein', 'Landwein', and the name of the Landwein region; they may give the name of an untergebreve (as ordinary Tafelwein may not), the vintage, and the word 'Halbtrocken' or, if the wine has less than 9 grams/litre residual sugar, the word 'Trocken' (dry).

If the table wine is a blend containing wines from other EC countries, or if the wine was made in a different country from that in which the grapes were grown, these facts must be stated on the label *in the language of the state where the wine is offered for sale*; to avoid misrepresentation, the address of the bottler/blender must be coded, with no mention of a town or country.

For qwpsr, the vineyard area is divided in a different fashion into eleven *Qualitätswein-bestimmte-Anbaugebiete* (QbA) meaning 'designated Quality Regions'. These cover in total the same area as the Tafelweingebiete. The name of the QbA must always be stated on the label, and 85 per cent of the wine must come from the named QbA. The Bereiche are the same as those for Landwein. Grape juice for QbA wines must have a minimum amount of sugar – between 57° and 72° Oechsle depending on the district and the grape variety.

1 Give the German QbA regions in which (a) Müller-Thurgau; (b) Riesling; and (c) Silvaner are planted most heavily.
2 How many designated wine regions are there in Germany?
3 What percentage of German growers have over 20 ha of vineyard area?

Table 3
Elements of German labelling regulations

A *Definitions*

Table Wines: Deutscher Tafelwein, Deutscher Landwein } 100% German Wine
Tafelwein von Ländes des EWG (EEC Wine – including German)
Wein (Third-country Wine)

Quality Wines: QbA (Qualitätswein-bestimmte-Anbaugebiete)
QmP (Qualitätswein-mit-Prädikat)
Kabinett, Spätlese, Auslese, Beerenauslese, Trockenbeerenauslese, Eiswein.

B *Label nomenclature* (O = Optional; E = Essential; X = Forbidden)		*Deutscher Tafelwein*	*Deutscher Landwein*	*QbA*	*QmP*
Type of wine		E	E	E	E
Weinbaugebiete or Untergebiete:	Rhein-Mosel, Oberrhein, Neckar, and Bayern (Rhein-Mosel-Saar)	E	E	–	–
Landweingebiete:	Landwein Region, Main Donan Lindau	X	E	–	–
Anbaugebiete:	Ahr, Hessische Bergstrasse, Mittelrhein, Mosel-Saar-Ruwer, Nahe*, Rheingau*, Rheinhessen*, Rhein--pfalz*, Franken, Württemberg, and Baden.	–	–	E	E
Bereich:	District	X	O	O	O†
Grosslage:	Collective site	X	X	O	O
Gemeinde:	Village	X	X	O	O
Einzellage:	Individual site	X	X	O	O
Vintage:	At least 85% from year stated	X	O	O	O
Vine variety:	At least 85% of stated variety	X	O	O	O
Prädikat:	See A above	X	X	X	E
APNr:	Amtliches Prüfungsnummer - official control number	–	–	E	E
Bottler:	Name and address	E	E	E	E
	'Erzeugerabfüllung (if bottled by Producer)	O	O	O	O
Other Details:	'Trocken (dry) or 'Halbtrocken' (semi-dry)	O	E	O	O
	Seals and other Awards	X	X	O	O
	Brand names	O	O	O	O

* Liebfraumilch must be a Qualitätswein from any *one* of these Regions.
† Prädikatswein must be the product of a single Bereich.

1 (a) Müller-Thurgau 49% in Franken. (b) Riesling 80% in Rheingau. (c) Silvaner 22% in Franken.
2 Eleven.
3 Less than 1%.

Also depending on the sugar-content of the juice at the time of pressing, the type of grape, and the region, some of these wines may be awarded special attributes (*Prädikaten*) to class them as superior QmP (*Qualitätsweine-mit-Prädikat*), which may only come from a single Bereich. The Prädikaten (in increasing degree of quality) are:

Kabinett (67–85° Oechsle). This name was taken from wines reserved by growers for their own use and therefore put in their private cellar, or kabinett.
Spätlese (76–95° Oechsle). These are late-gathered grapes which have had longer advantage of the autumn sun and are therefore sweeter.
Auslese (83–105° Oechsle). Wines with this attribute come from the gathering of whole bunches of grapes which have been allowed to mature longer on the vine, are riper than others, and have not suffered from rot.
Beerenauslese (110–128° Oechsle). This refers to a harvest where individual over-ripe grapes are picked from bunches successively, producing a wine in the style of Sauternes.
Trockenbeerenauslese (150° Oechsle minimum). Wines may be made when the grapes dry out completely without any sign of grey rot.

The attribute *Eiswein* is applied to wine from whole and undamaged grapes that have been left to freeze in the depth of winter, picked when frozen, and taken to the winery and crushed before unfreezing, the ice being thrown away. The small quantity of sticky juice remaining contains much sugar, and little water, yet all the acids are still there, which in a hot summer could have been burnt out of the Beerenauslesen and Trockenbeerenauslesen grapes. Since 1982 the minimum must-weight required has to be the same as for Beerenauslese for a particular region or variety.

All QbA and QmP wines must be tested at an official testing station, and the label must bear the Prüfungsnummer (national certification number). This number comprises between eight and twelve digits, possibly with a letter, depending on the Federal State in which the region lies. The last two digits always indicate the year of application, and the remainder permit identification of the testing

1 What is Anreichern and its French equivalent?
2 When is 'Anreichern' permitted for QmP wines?
3 (a) What is 'Süssreserve'? (b) What is it used for?
4 Which rivers give their names to German table wine regions?
5 Which Tafelweine labels may bear the name of a Bereich?

station, the bottler, and the batch. Wines are submitted with a full analysis, which is checked; the wines are then tasted by a panel. Sealed bottles are kept by the testing station in case of complaints.

Table 3 sets out the essential elements of labelling nomenclature; it does not cover all eventualities, so any divergencies found should be checked with the appropriate regulations.

1 Permitted enrichment of must by addition of sugar; Chaptalisation.
2 Never.
3 (a) Incompletely fermented grape must from which yeasts have been removed. (b) Adding balancing sweetness to fully fermented, acidic, wine.
4 Rhine, Mosel and Neckar.
5 Only those which are Landweine.

German Quality Wine Regions

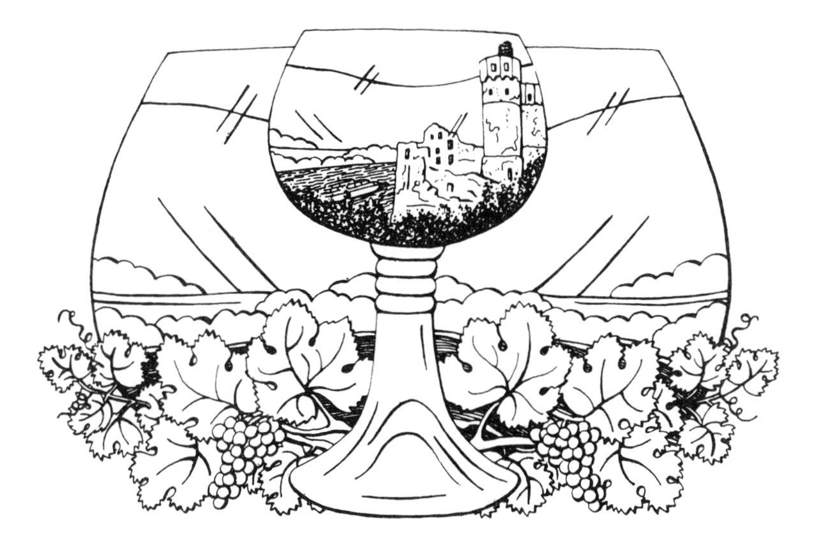

RHEINGAU

Though its vineyard area amounts to no more than 3 per cent of the total in Germany, QbA Rheingau is considered to be the most important on account of the superb quality of its wines, which are derived principally from the Riesling grape. In its journey north from Switzerland to the sea, the Rhine swerves sharply at Mainz to the west to avoid the Taunus mountains and then, some 30 km to the south-west, swings north again into the Rhine Gorge.

The vineyards all lie on the northern side, facing south-east, and extend beyond Wiesbaden along the north side of the Main to Hochheim. These slopes are backed by the wooded slopes of the

1 What do the initials QbA stand for, and what is the English meaning?
2 Translate into English (a) Tafelweingebiet; (b) Bereich; (c) Gemeinde.
3 (a) What are Prädikaten? (b) Name the six Prädikaten.
4 Compare the minimum must-weight of Eiswein and Beerenauslese.

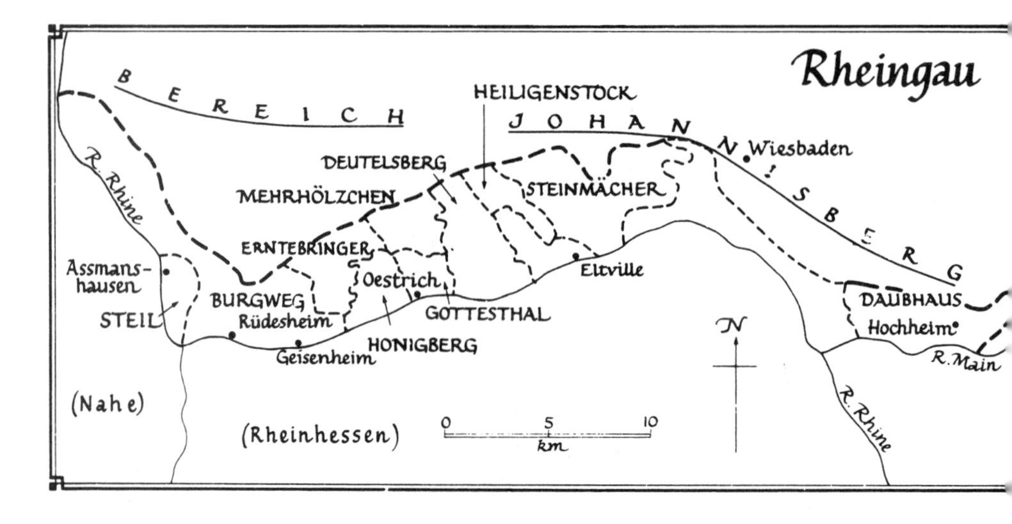

Taunus mountains which protect them from the cold north winds, and the sun's rays are reflected from the surface of the broad Rhine. The river also brings high humidity in the autumn, favouring the development of *botrytis* in its beneficient form, here called *Edelfäule* (noble rot), enabling the great Beerenauslese and Trockenbeerenauslese wines to be made.

The soils vary considerably within the region. At the western end, on the bend opposite Bingen, the village of Assmanshausen lies at the foot of steep slaty slopes planted with the Spätburgunder (Pinot Noir) grape, giving one of the best of Germany's red wines; the most important vineyard, Höllenberg (hellish hill), has been in existence for over 800 years. Moving east, the next town encountered is Rüdesheim, also on slaty soil; its slopes collect so much sun that the grapes ripen early, and its wines sometimes lack acid. Rottland (red land) is one of the best vineyards, its name describing the colour of the schistous soil. Situated on the river-bank upstream from Rüdesheim, Geisenheim is more noted for its Weininstitut (oenological college) than for its wines; but a little inland is Johannisberg, dominated by its vineyard and immense castle of Schloss Johannisberg. This vineyard, on quartzite and loess soil, produces consistently good wines; it belongs to the family of Prince Metternich, who were given it by Napoleon. Their QmP wines are coded with coloured capsules.

1 Qualitätswein-bestimmte-Anbaugebiet, or 'Designated Quality Region'.
2 (a) Table wine region. (b) district. (c) village.
3 (a) special attributes describing quality wines. (b) Kabinett, Spätlese, Auslese, Beerenauslese, Trockenbeerenauslese, and Eiswein.
4 They have been the same since 1982.

From this point until the city of Wiesbaden is reached, the finest wines get progressively better, with such names as Schloss Vollrads, Winkeler Hasensprung (hare-spring or -pool), Oestricher Lenchen (little Magdalen), Hallgartener Schönhell (beautiful light), Hattenheimer Nussbrunnen (nut fountain), and Rauenthaler Rothenberg (red hill). The names of German vineyards are as evocative as the names the Chinese give to their children, but much can be learnt about the vineyards from them – as in the last case, where it describes the red sandy loam of the vineyard. There are 116 of these Einzellagen and ten Grosslagen in the Anbaugebiet, all in one Bereich.

The city of Wiesbaden separates the vineyards of Hochheim from the rest of the Rheingau by 10 km, and their wines are very variable, probably because of the great variation in soils. Nearly every plot differs, and the wines have been likened to Mosel, Rheinhessen, Nahe, and Rheingau wines. The best vineyard is Domdechany (deanery), but the most famous is Viktoriaberg, named after Queen Victoria (who used to drink its produce with soda, created a fashion for German wines, and got them lumped all together as 'Hocks' – a word only the English understand). This practice of diluting wine with soda has re-emerged in the current craze for 'spritzers' – an even lighter party alternative for the driver – although these are now put up ready-mixed.

RHEINHESSEN

This is the next most important QbA region, and has the largest vineyard area – 25 per cent of the total German plantation – spread over the largest area. It is a region of mixed farming as well as viticulture, lying on the rolling plains south of the Rheingau in the bend of the Rhine. It is roughly triangular in shape, bounded by the Rhine on the north and east and a line stretching from Bingen, opposite Rüdesheim in the Rheingau to the cathedral city of Worms.

The vineyards are protected from cold northerly winds by the Taunus mountains, as are those of the Rheingau, and from easterly winds by the Odenwald range across the Rhine.

1 What is a 'Prüfungsnummer' and to what wines does it apply?
2 Compare QbA Rheingau with other QbAs in terms of vineyard area and quality of production.

1 National certification number applying to all QbA and QmP wines.
2 3% of total area: top in quality.

The soils are variable, as they are in the Rheingau. Slate occurs in the northern part near Ingelheim, as it does in Assmanshausen in the Rheingau – and red wines are made in each village. Elsewhere, quartzite and marl are found, but the finest part of Rheinhessen (called the Rheinfront because it fringes the north-flowing Rhine on steep slopes) has outcrops of sandstone.

The grape most heavily planted is the Müller-Thurgau, then the Silvaner, producing soft, fruity, but generally undistinguished wines. The exceptions are the noble wines from the Rheinfront, where the sandstone outcrops are planted with Riesling particularly in the villages of Nackenheim, Nierstein, Oppenheim, and Dienheim.

There are 171 villages (Gemeinden) in Rheinhessen with a total of about 430 individual vineyards (Einzellagen), combined into twenty-four group vineyards (Grosslagen) in three districts (Bereiche) – Bingen, Nierstein, and Wonnegau – the last containing the district of Worms. The average size of holding in Rheinhessen is very small, and many growers operate on a 'weekend' basis – much as the vegetable growers in the UK do on their allotments – and make the wines in their own cellars; nevertheless, if they are to be accepted as good wines for sale, they must pass the required tests.

In the city of Worms is the Church of Our Lady, the Liebfrauenkirche. This church has its own vineyard, the Liebfrauenstifte, and the wines from it came to be known as Liebfrauenmilch or Liebfraumilch. This name became so popular that others used it for their wines, until use of the name had to be controlled; the law concerning Liebfraumilch now decrees that:

- It must be a quality wine.
- It may only come from the regions (QbAs) of Rheingau, Rheinhessen, Nahe, Rheinpfalz.
- Its labels may not bear the name of a grape, a village, a district, or a vineyard (either individual or group).
- It may not have any Prädikat.

So, ironically, the original Wormser Liebfrauenstifte may not now be called 'Liebfraumilch' without losing its original name.

1 What do the Taunus mountains and the Rhine contribute to Rheingau wines?
2 What is Edelfäule?
3 What are the soil types in the Rheingau?

RHEINPFALZ

This region lies immediately to the south of the Rheinhessen, 20–25 km back from the Rhine, and stretches 75 km southwards to the border with French Alsace, called *der Deutscher Weintor* (the German winegate). The region is based on the slopes of the Haardt hills which protect it from rain.

It has a very equable climate, with mild winters, and summers hot enough to ripen almonds, figs, and lemons, generating vineyard-names such as Mandelgarten (almond-grove).

The soils are basaltic in parts, notably around Forst in the heart of the best central district, and by storing heat during the night enable the grapes to attain perfect ripeness. Chalk, and lime with some clay, are found in the rest of the region. The upper Haardt district (higher up the Rhine) produces the most wine, but not the best – it is, after all, a continuation of the lesser-quality Bas-Rhin district of Alsace; nevertheless, the long hot autumns lead to production of many sweet wines to those of Rheinhessen, which it borders.

Nowadays, the Rheinpfalz region is divided into two Bereiche: Mittelhaardt in the north; and Südliche-Weinstrasse in the south. The town of Neustadt marks the boundary between the two. Within the two Bereiche are twenty-six Grosslagen covering about 330 Einzellagen, in 160 Gemeinden.

Previously, the Silvaner was the grape most cultivated for white wine but, as elsewhere, the Müller-Thurgau and other new varieties are taking over. The proportion of Riesling stays at about 14 per cent of the vineyard, producing the finest white wines, where the soil and climate are suitable. The wines range from the fresh, sweetish wines of the south to the full-bodied noble wines of the Mittelhaardt with village names like Forst, Deidesheim, Ungstein, Kallstadt, and Wachenheim.

The region produces more red wine than either Rheingau or Rheinhessen, but not of such good quality; it is all made from the Portugieser grape and the best comes from Bad Dürkheim.

1 Respectively they protect the vines from cold north winds, and reflect the sun's rays, also giving humidity in the autumn.
2 'Noble rot' – beneficent *botrytis*.
3 Schistous on the upper slopes: loam, loess, and clay on the lower slopes.

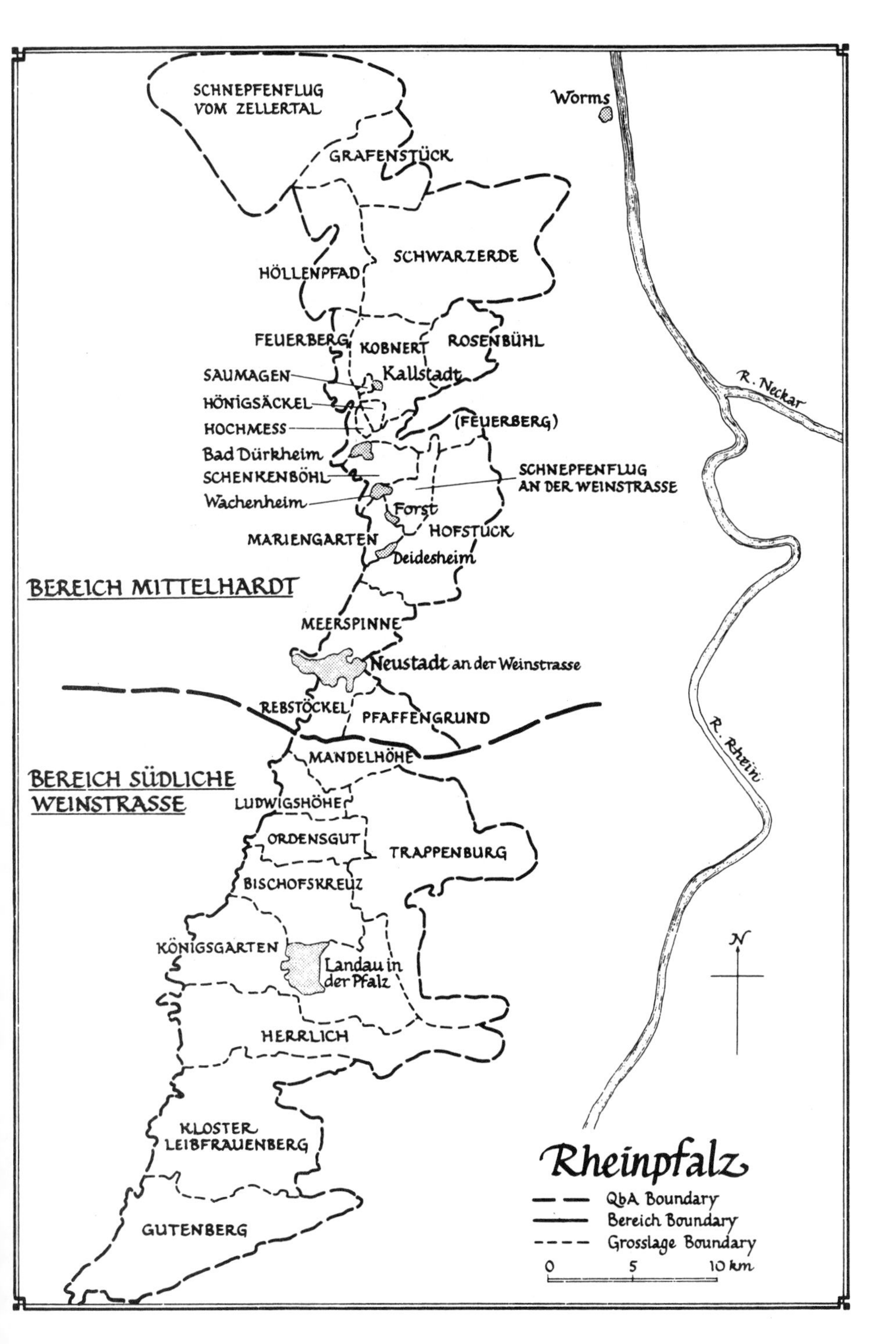

1 What is now the most heavily planted vine in Rheinhessen?
2 (a) Where did Liebfraumilch originate? (b) From which QbA may it now come? (c) What other restrictions are placed on its labelling?

NAHE

This region, large in total area but small in vineyard plantation, adjoins the Rheinhessen region on the west and is generally divided from it by the river Nahe, which flows into the Rhine at Bingen. The Nahe rises in the Hunsrück (dog's-back) range which separates its valley from that of the Mosel and runs north-east parallel to it for

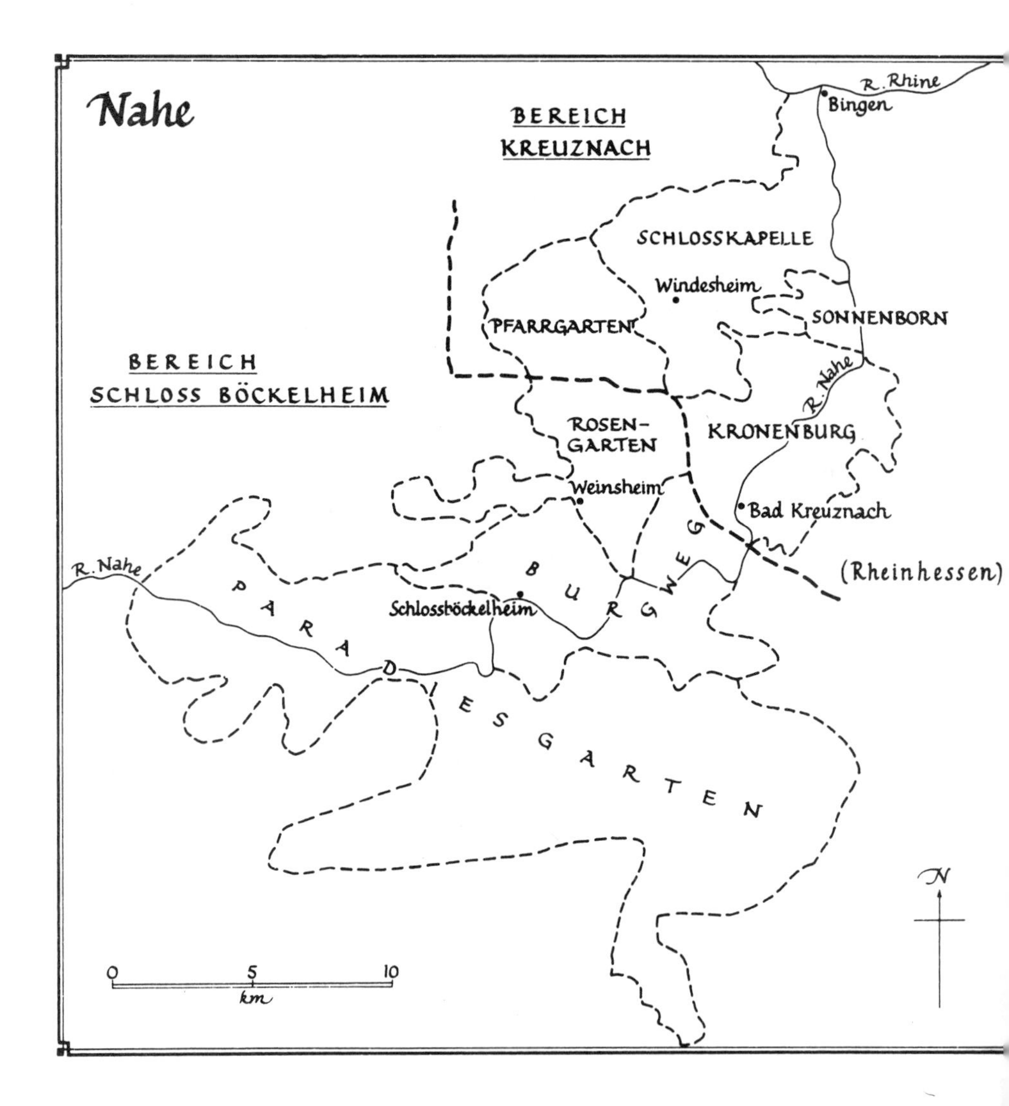

1 Müller-Thurgau.
2 (a) From the vineyard of the Liebfrauenkirche in Worms. (b) Rheingau, Rheinhessen, Rheinpfalz, or Nahe. (c) No Prädikat grape, village or vineyard may be stated.

60 km to Bad Kreuznach; this is the principal town of the region, and the river turns north here for 15 km before joining the Rhine at Bingen.

The vineyards are sheltered by the Hunsrück and Soonwald hills and are generally frost-free.

The soils in the upper Nahe are of sandstone, giving light, fragrant, Mosel-like wines. In the middle, where the best wines are produced, and in the lower Nahe between Bad Kreuznach and Bingen, the soil is slaty like that of the neighbouring parts of the Rheingau and Rheinhessen. These districts produce similar white wines, but practically no red wines. In general, the characteristics of Nahe wines lie between those of the Mosel-Saar-Ruwer and the Rheinhessen regions – understandably, as the Nahe lies between these regions.

The region is divided into two Bereiche (Kreuznach and Schloss-Böckelheim), with a total of seven Grosslagen divided into about 330 Einzellagen, spread over eighty-three Gemeinden. The best known of these are Schloss-Böckelheim, Kreuznach, and Rüdesheim.

The upper Nahe (Bereich Schloss-Böckelheim) used to be planted mainly with Riesling grapes and the lower (Bereich Kreuznach) with Silvaner, but the Müller-Thurgau has encroached on both. The wines have crispness and fruity acidity, with more substance than those of Mosel-Saar-Ruwer, but less than those of Rheingau.

MOSEL-SAAR-RUWER

This long thin region is three times as large as the Nahe, five times as large as the Rheingau, yet only two-thirds the size of Rheinhessen or Rheinpfalz. But, as in the Rheingau, size is no guide to importance; Anbaugebiet Mosel-Saar-Ruwer comprises five Bereiche, twenty Grosslagen and (at the latest count!), 659 Einzellagen. The river Moselle rises in the western slope of the Vosges, and flows in France

1 (a) On the slopes of which hills does QbA Rheinpfalz lie? (b) What is its climate?
2 What are the districts of QbA Rheinpfalz?
3 (a) Compare Rheinpfalz red wine production with that of Rheingau and Rheinhessen. (b) From what grape is it mainly made?

1 (a) The Haardt Hills. (b) Mild winters and warm to hot summers.
2 Bereich Mittelhaardt in the north, and Bereich Südlich-Weinstrasse in the south.
3 (a) Rheinpfalz produces more red wine than either, but of poorer quality. (b) Portugieser.

for half its length before separating Germany and Luxembourg for 50 km, where on its right bank lie the Bereiche of Moseltor and Obermosel up to the point where the Saar flows into the river, now known as the Mosel, from the south; downstream of the old Roman city of Trier, the Ruwer joins, also from the south. These tributaries and the land between them make up Bereich Saar-Ruwer. This Bereich produces very firm austere wines in good years from the Elbling grape, but in less fortunate and regrettably frequent years tends to produce a 'Dreimännerwein' (three-man wine) – 'it takes two men to hold the third one down and make him drink it!'. The soils of the upper Mosel are calcareous, with sandstone outcrops in the Saar valley and slate in the Ruwer.

The middle Mosel, Bereich Bernkastel, is the most beautiful, and produces the most beautiful wines; the wines come principally from the Riesling grape, though cultivation of the Müller-Thurgau and other new varieties is increasing on the richer lower slopes. Here the river winds sinuously in a great gorge, having cut its way through the schistous slate which gives so much nourishment to the vines. In this part, it is said that 'where a plough may go, good wine will not be made'. In this Bereich are all the famous names, one-time Einzellagen whose names were adopted by neighbours and have become Grosslagen – Piesporter Michelsberg, Bernkasteler Badstube, Graacher Münzlay, Kröver Nachtarsch; but Bernkasteler Doktor, the most widely known, is confirmed in its Einzellage status; it could otherwise be part of Badstube, being located within this Grosslage. This vineyard is typical of the Bereich, with a slope as steep as the roofs of the houses below, and as coated in slate.

Here the Riesling is king: while the Elbling may reign in the upper reaches, and the Müller-Thurgau is taking over from the Riesling in the lower reaches, the finest wines still come from the Riesling here.

As the river grows in size downstream from Traben-Trarbach, it winds less and its banks become less steep because the slate is softer. The soils in the valley are richer; here, in Bereich Zell, the wines are fuller but with less firmness: with more bouquet, from the more prevalent Müller-Thurgau grape, but with less nobility. And so the Mosel joins the Rhine at Koblenz.

1 In what range of hills does the river Nahe rise: where does it flow into the Rhine?
2 (a) Name the districts of QbA Nahe, working down-river. (b) Name the principal grape now grown in both.

FRANKEN (FRANCONIA)

This is the most easterly of all the German QbA regions, and lies on both sides of the river Main from Aschaffenburg to Schweinfurt, a distance of 80 km as the crow flies, but considerably more around the windings of the river. The principal centre and the site of some of the most important vineyards is the old cathedral city of Würzburg; the most famous vineyard, Würzburger Stein, has given the name 'Steinweine' to Franconian wines in general, although German law reserves the name for this vineyard only. They are put up for sale traditionally in a flat bottle called a *Bocksbeutel*. The use of the Bocksbeutel is restricted by law to these wines and to some of the wines of Baden. Steinweine are flinty and dry white wines though their best can become rather acid; this is probably due to the climate, which is of an inland continental type with cold winters and late springs. The vineyards are scattered all over the extended region to catch the best of the sun and the soil.

Bereiche and Soils

There are four Bereiche in Franken, comprising seventeen Grosslagen and about 190 Einzellagen, spread over about 158 Gemeinden. The soils differ for each of the Bereiche. Bereich Mainviereck ('Main rectangle' from the shape of the river) has a sandstone soil; none of its vineyards are particularly famous.

Further east in the next Bereich, that of Maindreieck ('Main triangle'), the soil is of a coarse limestone originally formed from the shells of mussels, with some overlay of loess; this Bereich includes the city of Würzburg, and the adjoining village of Randersacker, whose wines are also sought after. In the third Bereich of Steigerwald, the marly soil gives an earthy taste to the wines. The fourth Bereich is Bayerischer Badensee.

The region has until recent times been planted almost entirely with the Silvaner grape but the Müller-Thurgau grape has now overtaken it as Table 2, page 138 shows.

1 The Hunsrück: Bingen.
2 (a) Bereich Schloss-Böckelheim and Bereich Kreuznach. (b) Müller-Thurgau.

BADEN

This is an extremely long and scattered region, stretching along the east bank of the Rhine from Heidelberg and the river Neckar down to Basel on the frontier with France and Switzerland, thence east to the Bodensee better known by the English as Lake Constance; it also stretches north-east from Heidelberg almost to Würzburg, where the wines resemble those of Franconia. The most distinguishing feature in the region is the development of the co-operative system; there are 120 co-operative wineries affiliated to the central wine co-operative of Breisach, which lies just across the Rhine from the best part of French Alsace. This part of the region is influenced by the Vosges mountains just as is Alsace, but is also protected from easterly winds by the Black Forest.

As may be imagined, the soils vary considerably between marl, clay, sandstone, volcanic soil, and glacial moraine over the 400 km of the region. There are seven Bereiche, containing sixteen Grosslagen and about 340 Einzellagen, spread over hundreds of villages.

The most northerly Bereich – Bädische Frankenland – has already been mentioned, and its soil resembles the limestone and clay soil of neighbouring Frankischer Maindreieck. To the south-east of this, towards Heidelberg, lies Bereich Bädische Bergstrasse Kraichgau where, on loess and loam soils, Riesling and Ruländer vines produce wines with good bouquet which have been greatly enjoyed through the ages by students of the University of Heidelberg. At the castle of this beautiful city there is a fountain that used to run with wine on festive occasions. This Bereich runs south to the town of Baden-Baden, a very popular watering-place in Victorian days; no doubt they took the wine as much as the water, which is sulphurous, and enjoyed it more.

Here, in Bereich Ortenau, rich wines are made on the slopes of the Black Forest opposite Strasbourg; a speciality of the district is *Weissherbst*, a rosé wine made from the Spätburgunder (Pinot Noir) grape; the soils are generally tertiary. Bereich Breisgau, the centre of the region, produces flowery wines from a number of grape varieties. To the south of this, between the city of Freiburg and the river Rhine, lie the volcanic hills of Kaiserstuhl and Tuniberg, which give their names to Bereich Kaiserstuhl-Tuniberg. The best wines of

1 What is the principal grape grown in Bereich Saar-Ruwer, and what sort of wines are made from it?
2 (a) Name the districts of QbA Mosel-Saar-Ruwer working down-river. (b) Which produces the best wines and from what grape?

the region are produced here, being heavy and very flowery like the Alsace wines from the same grapes across the river – the Ruländer (Pinot Gris in Alsace) and Gewürztraminer.

Between Freiburg and Basel lies Bereich Markgräferland – literally, the land of the Margrave (Marquis in England) of Baden who was a very important person in late medieval history. Unfortunately, the wines of this region are not as noble as their name, being made from the very ordinary Gutedel table grape, otherwise called Chasselas in France, or Fendant in Switzerland. Nevertheless they are mild, low in acid, and easy to drink.

Lastly, Bereich Bodensee takes its name from the lake through which the Rhine flows after leaving Switzerland; here, gravelly soil gives zest to the white wines made from Müller-Thurgau grapes, and to the rosé Weissherbst made from the Spätburgunder.

WÜRTTEMBERG

This region, comprising two Bereiche, sixteen Grosslagen, and 210 Eizellagen, lies to the east of Baden and south of Franconia, and produces large quantities of red, white, and rosé wines which are seldom seen outside Germany. Formerly, they were of no more than ordinary quality, but have recently acquired recognition for their light red wines from the Spätburgunder grape. The soils are very varied, as in the French region of Alsace.

HESSISCHE BERGSTRASSE

This tiny region, which nevertheless is divided into two Bereiche, three Grosslagen, and twenty-two Einzellagen, is situated between Baden, Rheinhessen, and Rheingau, and seems undecided as to which of these should contain it. Its wines will seldom be seen abroad, but are fine and fruity, with a delicate acidity. They are made

1 Elbling: firm, austere white wines in good years; good for sekt production.
2 (a) Bereiche Moseltor, Obermosel, Saar-Ruwer, Bernkastel, and Zell. (b) Bereich Bernkastel from the Riesling.

principally from the Riesling grape, which suggests that at some future date the region might form part of the Rheingau, particularly as the loess soil resembles that of the eastern part of the Rheingau.

AHR

This small region, the northernmost in West Germany, has but one Bereich, one Grosslage, and forty-three Einzellagen, and extends on their side of the winding river Ahr, which runs parallel to and north of the River Mosel in schistous soil. It is noted for producing red wines, which is extraordinary considering how far north it is. Reflection of heat from the schistous cliffs in the narrow valleys helps to ripen the black Spätburgunder and Portugieser grapes but, except in the hottest years, they have an acidity which is hardly acceptable. The wine is, however, very acceptable in winter in a punch for skiers – called *Glühwein* – and in summer may be mixed with the perfect natural soda water which bubbles up in the spa of Bad Neuenahr – the Apollinaris spring.

MITTELRHEIN

These are the most romantic vineyards of all, stretching down the Rhine from Bingen where the regions of Rheinhessen, Rheingau, and Nahe join at the entrance to the Rhine Gorge, down past Koblenz where the Mosel runs into the Rhine, to the seven hills of Bonn. Its 110 Einzellagen are so strung out that they spread over three Bereiche and eleven Grosslagen. Here the river runs fast through gorges cut in the slate and schist cliffs, on top of which perch castles recalling medieval deeds of valour; in the narrow river the tugs strain against the strong current to take the heavy Rhine barges up to the city-ports of Mainz and Friedrichshaven. Downstream (north) of Koblenz the vineyards are mostly on the right bank, divided into two Bereiche, Siebengebirge and Rheinburgengau. The wines of this region come mostly from the Riesling grape and have

1 Which is the most easterly QbA?
2 What are the 'triangle' and 'rectangle' of Franken?
3 Which is the most northerly Bereich of Baden and what is its soil?
4 What is Weissherbst?

an intriguing after-taste of peach kernels. Upstream (south) of Koblenz the vineyards are principally on the left bank. The region is most spectacular by the Lorelei rock near St Goarshausen, and to be here at dawn or dusk is to conjure up Wagner in the glorious opera-cycle of *The Ring*; but the wines of Germany also conjure up Wagner in other gentler operas such as *Meistersinger*, and it is in that mood that they should be enjoyed. The third Bereich, Bacharach, joins the region to that of Rheingau.

1 Franken (Franconia).
2 Bereiche Maindreieck and Mainviereck.
3 Bereich Bädische Frankenland: limestone and clay.
4 A QbA rosé wine made from the Spätburgunder grape.

5

Italy – General

Italian wine history reaches back beyond the days of the Roman Empire, though it remained for the Romans to control the culture of vines and the quality of wine production in a recognizable industry.

In the Dark Ages that followed the barbarian invasion, Italian wine production became a rural industry with no central control and certainly no chronicle of its experience, and fine wine production seems to have ceased; wine production is next recorded around Florence *circa* AD 1000. In AD 1282 a guild of wine merchants (*vinattieri*) was established in Florence, possibly the earliest record of a controlling body for the production and sale of wine, and certainly the forerunner of the district *consorzi* set up throughout Italy in the twentieth century.

1 What is Glühwein?
2 For what type of wine is Württemberg noted?

As today, most of the wine produced was consumed at home, although there are records of Trebbiano from Campania and Vernage from northern Italy being shipped to London in the late fourteenth and fifteenth centuries, and of a monopoly granted to Southampton by Henry IV for the import of certain Mediterranean wines, including those of Italy. There is little evidence of a co-ordinated Italian wine industry developing, and it was not until the middle of the eighteenth century that agricultural *Academia* were set up to include the study of viticulture, vinification, and wine maturation and storage. But even then the Roman skills were not revived, and in 1850 Cyrus Redding observed 'Italian wines have stood still and remain without improvement'. The plain fact is that

1 A hot sweet punch based on the acid red Spätburgunder of QbA Ahr.
2 Light red wine from the Spätburgunder grape.

policies to control and improve Italian wines were not formulated until Italy acquired a central government in the nineteenth century.

Climate

Italy stretches from the 37th to the 47th parallel of north latitude, and is dominated by mountain ranges – the French and Swiss Alps and the Dolomites, which form a crown along her northern borders, and the Apennines running like a spine from Liguria down to the toe of Italy. The climate to the north of the river Po is one of cold winters and long hot summers: it is tempered by the Alps which give protection from northerly winds, and also by the Italian lakes of glacial origin, Maggiore, Como, and Garda. The climate south of the Po but north of the Apennines is gentler and continental in character, producing a considerable variety of subtle wines. In central and southern Italy, however, the climate is mediterranean, with warm winters and hot humid summers, so the wines tend to be bigger and meatier. Turin is nearer to London than to Sicily and the climatic difference between Turin and Sicily is greater than that between Turin and London.

Italian Vines

Grapes grow practically everywhere in Italy, and viticulture is not restricted to the plains but thrives also in the mountains – in fact wherever the land is inhabited.

The vine varieties of Italy are many, and some have local names which disguise their identity. A few are cultivated over several regions, such as Moscato, Malvasia, Trebbiano, Sangiovese, Verdicchio, and Vernaccia. Others are used exclusively or predominantly in one or two regions, and these include Nebbiolo, Freisa, Cabernet, Merlot, Pinot, and Dolcetto. Although the Cabernet Sauvignon of Bordeaux is not generally authorized as a variety it has proved popular with growers, and blends using this grape are, as *vini da tavole*, sometimes reaching better prices than DOC wines from the same area.

1 Name (a) the principal vine used in QbA Mittelrhein; (b) a characteristic after-taste of its wines; (c) its two Bereiche.
2 What were the vinattieri?

Viticulture and Production

The best grapes are trained on wires, high or low, according to local requirements. The Italians are great individualists and much of the ordinary table wine (the bulk of total production) is produced by small farmers (growers), owning less than a hectare each, for their own and strictly local consumption.

Two million labourers work in the vineyards of Italy, representing nearly one third of the national labour force, though this number is decreasing with modern technology and the attraction of labour away from the villages by higher wages paid in the cities.

In 1984, Italy's wine production was 71 million hl (in most years Italy is the largest producer in the world).

DOC Law

In July 1963, the Italian Government issued Decree 930, setting out the basis for national control of the wine industry according to EEC principles. The *Denominazioni di Origine* were established as protected names for delimited geographical regions, which might be qualified by the name of a vine or vines or other definitive quality of the wine product of a region. A Denominazione di Origine could be extended to include wines from neighbouring areas, providing such wines had been produced for at least ten years prior to 1963, were made from vines typical of the delimited region by traditional methods, and had similar physical and organic characteristics.

Decrees of Denominazione di Origine, issued by the Ministry of Agriculture, were divided into three categories of quality. The lowest, *Denominazioni di Origine Semplici* (DOS), or simple names of origin, applied to wines typical (as to vine, and viticultural and vinification methods) of a delimited production area, but manufactured in normal rather than legally defined conditions. DOS, like the French Appellations d'Origine Simples, are now finished. They have been replaced under DOC regulations by 'table wines' – vini da tavola – which, if they conform to stricter rules of production and

1 (a) Riesling. (b) Taste of peach kernels. (c) Siebengebirge and Rheinburgengau.
2 A guild of wine merchants established in Florence in AD 1282.

are certified by a tasting panel (like French vins de pays and German tafelweine) may add a geographical name, as *vini tipici*. A *Denominazione di Origine Controllata* (DOC), or controlled name of origin, applies only to a wine made under conditions laid down in its particular decree.

A *Denominazione di Origine Controllata e Garantita* (DOCG), or controlled and guaranteed name of origin, applies only to an ex-DOC wine bearing a government seal of guarantee, bottled locally, and sold in containers holding less than 5 litres. DOCG have to be applied for, and by 1988 six had been granted. A National Committee, chosen from governmental and wine industry nominees, considers applications for DOC and DOCG and makes recommendations to the Ministries of Agriculture and of Industry and Commerce for award of the Presidential Decree. The committee has twenty-eight members and a chairman. Application to the committee must be supported by growers representing 30 per cent of the total district volume of sparkling or fortified wine, or 20 per cent of still wine.

Italy reinforces the EEC laws against unfair trading by adding to the normal sanctions (of fines and imprisonment), closure of wineries, cellars, or warehouses for up to one year, and by requiring details of the actions and penalties to be published in local, trade, and national papers.

The Wine Label

According to individual decrees, Italian wine names may be geographical and represent a big area (Chianti), or a district (Asti), or simply a commune (Montalcino). Or wines may be named from the principal or sole grape variety used in their production (Barbera); for quality (qwpsr) wines, the grape name must be qualified by a place-name (Barbera d'Asti, Barbera d'Alba and various other Barbera place-names throughout Italy from Piemonte to Sicily). The word *Classico* after a district name means that the wine comes from a better part of the district.

The producer's name may be prefixed with such words as *Produttori* (producer), *Cantine* (cellars of), *Cantina Sociale* (co-

1 Between what parallels of north latitude does Italy lie?
2 Name three mountain ranges that dominate Italy.
3 Which is nearer Turin: Sicily or London?
4 Which Bordeaux grape is blended in such proportions as to produce wine competing with Italian DOC wines, and how will such wines be legally categorized in Italy?

operative producer), *Casa Vinicola* (wine house), or *tenuta* (estate); and the initials DOC or DOCG (*q.v.*) must also appear, when appropriate, on the main label. Bottling is indicated by *imbottigliato* (bottled), or *infiascato* (put in flasks), *in zona d'origine* (in the growing area), or *nello stabilimento* (at the producer's premises). Vintages, often shown on neck labels, are described by *annata* (year) or *vendemmia* (vintage). If the grower belongs to a consorzio, its paper seals will be gummed over the closure. Permitted wine descriptions that may also appear on the label include *vecchio* (old), *riserva* (reserve), *superiore* (with greater minimum age and alcoholic content), *bianco* (white), *rosso* (red), *spumante* (sparkling), *recioto* (rich), *passito* (wine from dried grapes), *amarone* – only for Valpolicella – (bitter), *secco* (dry), and *dolce* or *abboccato* (sweet).

The Consorzi

It was not until the early 1930s that the consorzio system was introduced, and even then it was somewhat haphazard. Each consorzio had a constitution authorized by Royal Decree, by which it was to control the production of wines in a specified district, but each constitution was different; and as the system was voluntary, producers in many districts had no consorzio. Each consorzio was charged with the general responsibility for controlling wine production, and made its own rules. Usually these included the naming of permitted vines, fixing of geographical limits, specification of minimum alcoholic content, and restrictions on imports of wine into the district; in most cases approval of wines was shown by affixing the consorzio seal to each bottle or shipment. The consorzi were made responsible for implementing governmental regulations, and promoting sales at home and overseas. The Ministry of Agriculture and Forestry approved the delimitation of the production areas. However unsatisfactory the consorzio system may have been, it was a first step from chaos towards relative organization in the Italian wine industry.

1 The 37th and 47th parallels.
2 The French and Swiss Alps, the Dolomites and the Apennines.
3 London.
4 The Cabernet Sauvignon, as *vini da tavola*.

In recent years, the consorzi have been resurgent, reflecting the growers' dissatisfaction with the application of the DOC system. They apply their own, often stricter, rules and only those who agree to abide by them and do so may bear the Consorzio label on their bottles. The Consorzio label is generally the guarantee of a better wine.

1 What are the initials of the Denominazioni di Origine in descending order of wine quality?
2 What is the Italian equivalent to French 'Vin du Pays'?
3 (a) What body considers applications for Denominazioni di Origine' and (b) by what law are they granted?

Northern Italy

About half the Italian wine imported into the UK is quality wine (DOC), bottled in Italy. The other half is table wine, often shipped in casks for bottling in 1- or 2-litre bottles on arrival, for sale under trade names. Veronese and Piemontese wines figure substantially among the DOC exports; they are typical of the subtle wines to be found north of the Apennines, where the climate in the high plains is strongly influenced by the protecting Alpine mountains. The Italian *Reggioni* (comparable with French Provinces) of Piemonte, Valle d'Aosta, Liguria, Lombardia, Veneto, Friuli- and Venezia-Giulia, and Trentino-Alto-Adige combine to form the 'thigh' of the long Italian leg, and between them they enjoy a continental climate, as opposed to the mediterranean climates of central and southern

1 (1) DOCG. (2) DOC.
2 'Vino da tavola con geografica'.
3 (a) A National Committee. (b) Presidential Decree.

Italy. The Compartimenti are divided into *Provinci* (counties), equating to French Départements. For the purposes of the British wine trade, Reggioni are referred to as regions, and collections of wine-producing villages as districts. All the wines mentioned in this chapter have been granted Denominazione di Origine Controllata, unless otherwise stated.

PIEMONTE

The name of this region means 'foot of the mountain', and describes it perfectly; as a producer of fine wines, this region of Italy is possibly unequalled. Two DOCG and more than forty DOC have so far been granted for its wines, at least twice as many as granted to Veneto, the next most honoured. Piemonte is particularly famous for red wines made from the Nebbiolo and Barbera grapes, and for white wines made from the Moscato grape. Three of the most important wines derive from the Nebbiolo. The first is Barolo DOCG, taking its name from the district in which it is made, a big full-bodied wine with excellent bouquet and flavour, which must be aged for three years (of which two must be in wooden barrel), before being offered for sale. It will continue to improve, and can last a century, taking on a brick-red colour with age; it is said to have an aroma of tar.

Barbaresco DOCG (from the same grape and neighbouring district) has been described as softer and less austere – 'Barolo's younger brother'. It takes its name from a picturesque village above the river Tanaro – the district extends to the neighbouring villages of Treiso and Neive. Barbaresco is ruby-red and fully mature after three years. The reader must be careful not to confuse Barbaresco, the village, with Barbera, another grape. Gattinara, the main town of the Vercelli county, gives its name to wines made from a blend of the Nebbiolo grape, known there as Spanna, and other grapes such as the Croatina. Gattinara is not as powerful or full-bodied as Barolo, yet qualifies as an ideal wine to accompany red meat. It is superior to other wines of the district known simply as Nebbiolo – from the grape. Nebbiolo d'Alba is the best known.

1 Give the English equivalents of the following: (a) classico; (b) cantina; (c) imbottigliato.
2 Give the English equivalents of the following: (a) vecchio; (b) rosso; (c) amarone; (d) secco; (e) abboccato.
3 What are the Consorzi?
4 How may a Consorzio wine be recognized?

1 (a) Wine from best part of district (usually the centre). (b) Co-operative producer. (c) Bottled.
2 (a) Old. (b) Red. (c) Bitter. (d) Dry. (e) Sweet.
3 Bodies set up in specified wine districts to control wine production.
4 By the coloured label of the Consorzio on the neck of the bottle.

As already mentioned, Barbera is another noble grape giving its name to wine, the name always being followed by a place name if the wine be DOC. The best known is Barbera d'Asti, a dry fruity wine but not so robust as Barolo or Barbaresco. Wines from the Barbera grape may also be semi-sweet and *frizzante* (slightly sparkling). The Freisa grape, producing a wine less strong than Barbera, is noted for its scent of violets and requires only moderate ageing. The best-known Freisas are those of Chieri, of Monferrato, areas north-west and north-east of Asti; and of Asti itself. The Grignolino grape gives its name to another notable red wine which is bright and clean, dry and light-bodied, with a bouquet of roses; for DOC it adds its name to Asti and is mature after two years but best at three. Without place-name, it has been called 'the table wine par excellence' of Piemonte – an unconscious tribute perhaps to neighbouring France?

The sparkling white wines of Piemonte are the best known of the region, particularly Asti Spumante, which is made from a single fermentation without any addition of sugar, or carbonation. It is made from the Moscato grape, which gives its pungent fruitiness to the wine. Small quantities of Pinot and Riesling grapes may be added to mellow the strong scent of the Muscat. The wine itself is sparkling, sweet, yet low in alcohol: all of which it owes to its particular and traditional method of manufacture.

The wine undergoes but one, natural, fermentation – in a sealed tank. After some time, the fermenting must is filtered under pressure: the yeasts are excluded, but the gaseous carbon dioxide product of fermentation is retained. The yeast spores remaining are regenerated by protein still remaining in the must, multiply, and continue fermentation. Often nowadays, half-fermented filtered juice is held in cold store and processed as needed. The process is repeated until the protein is exhausted, resulting in a stable wine of high sweetness and low alcoholic strength, and sparkling as well. For the same reason the wine is crystal bright, and retains its sparkle when bottled at low temperature under pressure, and sealed with champagne-type corks. Asti Spumante needs to be drunk young and chilled. As it owes its sweetness to natural sugar left unfermented, it must not be allowed to age, as the taste and fragrant bouquet are fugitive. Asti Spumante's low actual alcohol content ranges between 7.5% vol. and 9%.

Even lower in actual alcohol (6% vol. to 8%) is Moscato d'Asti, a sweet dessert wine made from the Moscato Bianco grape.

1 What is the minimum maturation time required for Barolo DOCG?
2 For how long will Barolo improve?
3 Distinguish between Barbaresco and Barbera.

LIGURIA AND VAL D'AOSTA

Liguria, sheltering under the Apennines, is one of Italy's most beautiful districts and has the ancient yet extremely modern and efficient port of Genoa at its centre. Liguria is small in area, and is not an important wine producer. At its southern tip, however, in the region of La Spezia, the vine growers have battled with nature to transform the steep cliffs rising out of the sea into thriving vineyards (some accessible only by boat), and here white Cinqueterre wine is produced, named for the district of 'five lands'. The wine comes from a blend mainly of Bosco vines, with Albarolo and Vermentino added. The wine is dry, with a delicate flavour.

Val d'Aosta, a diminutive enclave among the Alpine peaks, is more famous for its skiing resorts than for its wines. Two red wines are produced: Donnaz from the Nebbiolo grape, and Enfer d'Arvier from the Petit-Rouge grape. The Donnaz has a distinct flavour of almond which increases with age. It is the rougher wine when young and must age for three years, as against only one year for Enfer d'Arvier.

LOMBARDIA

Lombardia is one of the larger Reggioni of Italy, but here the vine competes with other agricultural crops, notably rice, and with industry spreading out from its capital, Milan. To the north, the land rises into the mountains, and here the Valtellina, the valley of the river Adda, high above sea level running east from the northern end of Lake Como, produces the wines of Lombardia best known outside Italy. To the south, in the province of Pavia, the red wines of Oltrepo Pavese have been known for centuries.

Both red and white Valtellina and Valtellina Superiore are made; the red Superiore is the best, for which the grapes must be 95 per cent Chiavennasca (the local name for the Nebbiolo), and the wine must have 12° alcohol. Four sub-districts may add their names for wines emanating solely from grapes grown in the sub-district – Sassella, Grumello, Inferno, and Valgella – of which the first two have the

1 Three years, of which two must have been in wooden barrel.
2 For a century.
3 The first is a DOCG district and a village, the second is a grape.

best reputation. The Valtellina wines are lively red in colour, dry and tannic, with a fine bouquet; the Superiore must age for two years of which one is spent in wooden casks. Valtellina DOC may contain up to 30 per cent of other grapes, and needs only 11° alcohol; it is by no means so fine. A specialty of the region is DOC Valtellina Sforzato, which is made from grapes which have been dried for two months before vinification; like Recioto della Valpolicella, it is richer and heavier (14.5°) and complements the local cheeses well.

On the southern shores of Lake Garda, bordering the region of Veneto, a rather deep-coloured rosé wine is made from Sangiovese grapes, called Chiaretto del Garda ('Chiaretto' meaning 'claret').

VENETO

Among the northern wine regions, Veneto is second only in importance to Piemonte, and the names of its great wines have a world-wide reputation. The region is mountainous in the north and flat in the south-east where its capital, Venice, sinks slowly into the sea. Its wine-producing region is based on Verona, south of Lake Garda. The three most famous wines are Bardolino, Valpolicella, and Soave. DOC Bardolino is a red wine blended from Corvina, Rondinella, and Molinara grapes, possibly with some others. This wine is produced in a group of sixteen communes on the eastern shore of Lake Garda, in the Province of Verona, six of which form the inner Classico district. It has a pale cherry colour darkening with age to garnet, and is dry, with a slight salty-bitter tang; it can be frizzante. It has 11.5% vol. alcohol and, after ageing in the province for a calendar year commencing on the first of January following the vintage, may be described as superiore.

Valpolicella is now well known, and its large production is appreciated on the UK market. The district lies on the hills to the east of the Adige river, and spreads over the river to the south round Verona. It comes from a blend of grapes including Corvina, Rondinella, and Molinara, made in a group of nineteen communes, of which five have classico rating. Wines produced in the Valpantena valley may add this name to the DOC. The yield of must per tonne

1 (a) Where are the principal wine districts of Veneto situated? (b) Name the best-known wines from these and their colour.
2 (a) Name the best-known sparkling wine of Piemonte. (b) From what grape is it made? (c) Is it sweet or dry? (d) When is it ready for drinking?

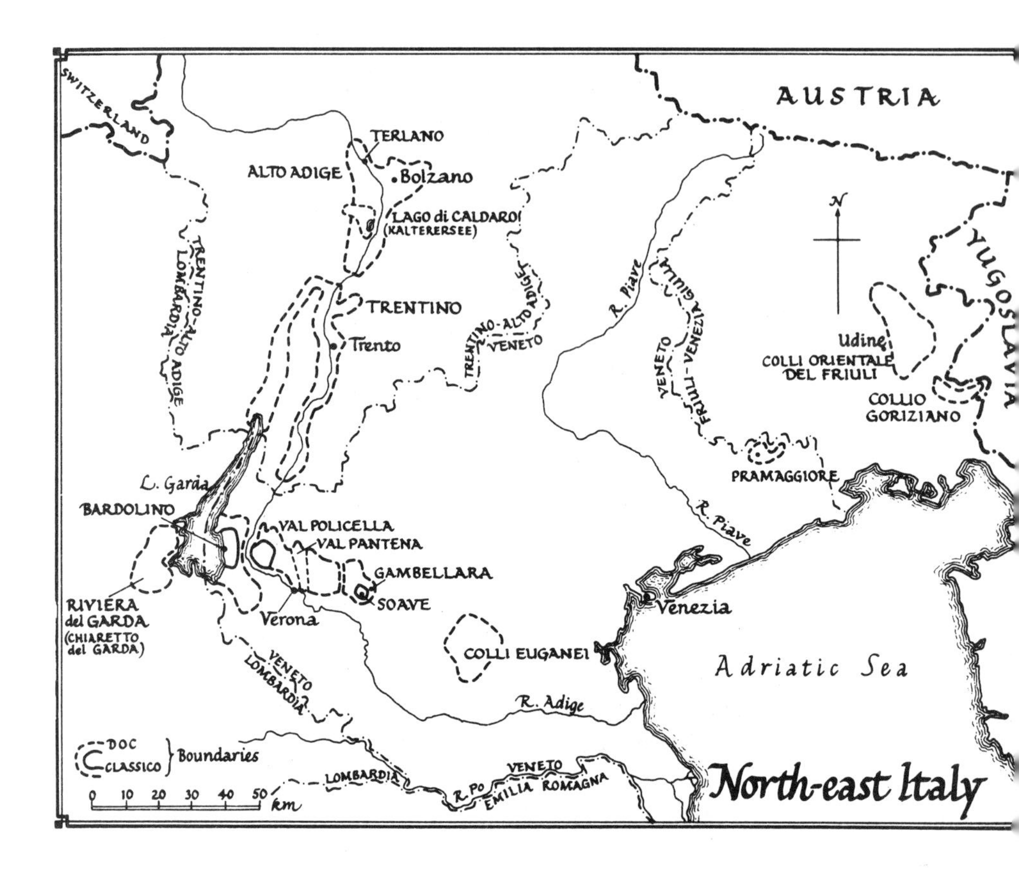

of grapes is restricted more for Recioto del Valpolicella than for Valpolicella. The recioto was made from the ripest grapes in the bunch – usually the 'ears' or '*recie*' characteristic of the Corvina grape – which are then dried on wire racks under cover to concentrate their sugar. These days selected whole bunches are more often used.

The wine is semi-sweet, with a honeyed nose, delicate though full in the mouth, and deep ruby in colour. With ageing, it 'eats its sugar' and dries with a hint of bitterness, becoming the much sought-after Recioto della Valpolicella Amarone. Valpolicella must have an alcoholic strength of not less than 12% vol. and, as with Bardolino, can be rated superiore when aged. It is ruby-red deepening to garnet with age, has a delicate bouquet with a suggestion of bitter almonds; it is semi-dry, velvety, and well balanced.

1 (a) Near Verona, east of Lake Garda. (b) Bardolino (R), Valpolicella (R), and Soave (W).
2 (a) Asti Spumante. (b) Moscato. (c) Sweet. (d) It must be drunk young.

Soave is a small town 19 km east of Verona, giving its name to the balanced, smooth, fresh white wine made there. The grapes used are the Garganega and Trebbiano (Ugni Blanc), grown in a group of thirteen communes having a soil predominantly of clay. The yield is restricted for Soave and Recioto di Soave, as for Valpolicella. Soave is a straw-coloured white wine with a deep and delicate scent, dry and crisp, with an attractive bitterness. The recioto is pale golden yellow and fruitier in the mouth, and is semi-sweet with a slight taste of bitter almonds. Again the descriptions classico (from certain inner communes) and superiore may be used.

The Euganean Hills in the province of Padova were covered with vines over a thousand years ago, and here some seventeen communes produce red and white Colli Euganei wines. Garganega and Serprina are the principal grapes used in the bianco, producing a dry to semi-sweet wine, soft and velvety, with a grapey bouquet. The rosso comes from the Merlot grape, with some Cabernet and Barbera grapes added. The wine is ruby-red with good body, dry to semi-sweet. A Moscato is also made from Muscat grapes, and has all their sweet pungent characteristics. All the Colli Euganei wines may be made sparkling as well as still, and have a minimum total alcoholic strength of 10.5% vol. Superiore wines must have 12% alcohol, and sparkling up to 12%, depending on type.

TRENTINO-ALTO ADIGE

Trentino-Alto Adige – the Italian Tirol – is Italy's most northerly region, and German is spoken as much as Italian; labels may be found in either language. Here, where the winters are cold and hard, the climate remains reasonably temperate, the mountains shielding the long Adige valley from northerly winds. Viticulture in Alto Adige dates back to Roman times; Italian-speaking Trentino was incorporated into the Tirol by the Austrians, and the ill-matched pair were given to Italy after 1918. German-speaking Alto-Adige prefers to be called Süd-Tirol, and their red and white wines have some similarity to those of Germany and Austria (whither they are exported).

1 (a) What is Cinqueterre? (b) Why is it so-called? (c) In which Reggione of Italy is it made?
2 What is the Valtellina?
3 Sassella, Grumello, Inferno, and Valgella are four wines. (a) What is their colour? (b) How were they named? (c) Where are they produced?
4 Describe Valtellina Sforzato.
5 What does chiaretto mean?

The best red wine is Santa Maddalena, deep in colour with good body and a bouquet of violets and almonds, made in the province of Bolzano from different forms of the Schiava vine; classico is produced in an inner delimited area. Lago di Caldaro (also called Kalterersee), is another fruity red wine from the same grape, grown on the shores of the little lake of Caldaro, near Ora. The white wines of the region, often named after their grapes such as Traminer (which originated in this region) and Riesling, tend to be bitter and strong, the former with a vanilla perfume. The white wines of Terlano are made from not less than 50 per cent white Pinot; all the DOC wines of Süd-Tirol are entitled to QbA labelling in addition to DOC.

Trentino starts where the Adige valley broadens out into a flat plain between the mountains. Here the vine is intensively grown on the *pergole trentine* 1.5 m verticals with 1.2 m arms jutting at 45° from the top either side – a larger version of the Geneva Double Curtain. The principal grape is the Teroldego, which produces full-bodied, long-lasting reds and perfumed rosé. Several producers make excellent sparkling wine by second fermentation in bottle from the Pinot Bianco. A great deal of Chardonnay is also now used.

FRIULI-VENEZIA GIULIA

This small region is wedged between Veneto, Yugoslavia, and the Adriatic, in the extreme north-east of Italy. It has a mild climate but is open to the 'bora' – a violent cold north-east wind. Wine has been made here for over two thousand years – and possibly much longer. Collio, a small area west of Gorizia, has marl and sandstone soil, ideal for the production of well-balanced dry white wines, some of them sparkling, entitled to DOC Collio Goriziano. These wines are made from the Ribolla grape with proportions of Malvasia and Tocai added. DOC Collio Goriziano may be followed by the name of a grape from a specified list for wine made solely from that grape. The list includes Tocai, Malvasia, Pinot (Bianco and Nero), Sauvignon, Traminer, Merlot, and Cabernet. Some eight white varieties of Collio Goriziano and three red are beginning to appear in the UK.

1 (a) Dry white wine. (b) Named for the district of 'five lands'. (c) In Liguria.
2 An Alpine valley running east from Lake Como which gives its names to wines.
3 (a) All red. (b) For the sub-districts in which they are made. (c) In the Valtellina.
4 DOC Valtellina wine from grapes dried for two months before vinifying; it is heavier and rich.
5 Deep rosé colour, or 'Claret'.

The district of hills (Colli) east of Udine, named Colli Orientali del Friuli, also has DOC status for its red and white wines, among which may be noted Verduzzo, Pinot Grigio and Ribolla (white), Picolit (white dessert), and Merlot di Latisana, Cabernet di Latisana, and Refosco (red).

1 (a) What is peculiar about recioto wines? (b) What does the word recioto imply?
2 Describe Soave wine.
3 (a) Are the wines of Colli Euganei still, sparkling, or both? (b) What is their colour? (c) Where are they made?

Central and Southern Italy

The length of Italy with the Apennine chain running through it is divided into ten Reggioni, each constituting a wine region of greater or lesser importance; the regions lie either to the east or west of the mountains, never across them. The soils, as may be expected, are predominantly granitic and basaltic, reflecting the eruption of volcanoes over the ages. The climate is one of long hot summers and mild winters, except that in the Apennine mountains winters can be very hard indeed, as veterans of 1944 will remember. These conditions tend to produce big full wines in which strength can overrule character; nevertheless, the wine-makers' skills produce fine wines which enjoy a good export market.

1 (a) Wine made only from grapes from the top and outside of the bunch. (b) The recie are the 'ears' which receive most sun.
2 A straw-coloured white wine dry and crisp, with a deep delicate scent.
3 (a) Both still and sparkling. (b) Red and white. (c) Province of Padova.

EMILIA-ROMAGNA

This is one of the largest wine-producing regions in Italy, though even the Italians will admit that most of its wine is best drunk where it is made, east of the Apennines, to wash down the generous portions of heavy local food. The wine that has world renown is Lambrusco, made from the grape of that name (and not on any account to be confused with *vitis labrusca* of America), where Lambrusco is so popular as a mass-produced light, sweet, ruby-coloured wine with a slight sparkle, deriving from carbonation.

The best Lambrusco is di Sorbara, taking its name from a little town north of Modena, and having a flowery bouquet and a good acid balance. Although the grapes are grown on high trellises, they are harvested late, and do not contain an excess of malic acid, so the pétillance does not come from malolactic fermentation. After being allowed to macerate on the skins for two to three days, about 10 per cent of the must is drawn off and filtered to prevent further fermentation; in the spring, this sweet low-alcohol *filtrado* is returned to the wine which has fully fermented, and bottled as soon as possible with corks tied down with string. Sufficient fermentation takes place to produce a pétillant wine with the necessary 11° alcohol. Its acidity is sometimes marked, and complements the rather oily spaghetti bolognese and also pig's trotters – a local speciality. It comes from a group of ten communes of the county of Modena. Carbonation is forbidden. Other wines of Emilia-Romagna come from the Sangiovese and Trebbiano vines.

TOSCANA

Toscana, the beautiful region on the other (west) side of the Apennines, produces wines both good and famous. The Tuscans themselves hasten to extol the virtues of Chianti, and have recently done much to improve this wine and the conditions in which it is made. But there are also lesser-known wines of even finer quality which merit study here, particularly Brunello di Montalcino and Vino

1 What labelling practice is peculiar to the Quality wines of Alto Adige?
2 What is the principal grape used in Trentino, and what sort of wine does it produce?
3 Name the wine from Friuli-Venezia Giulia now appearing in the UK, both as red and white wine.

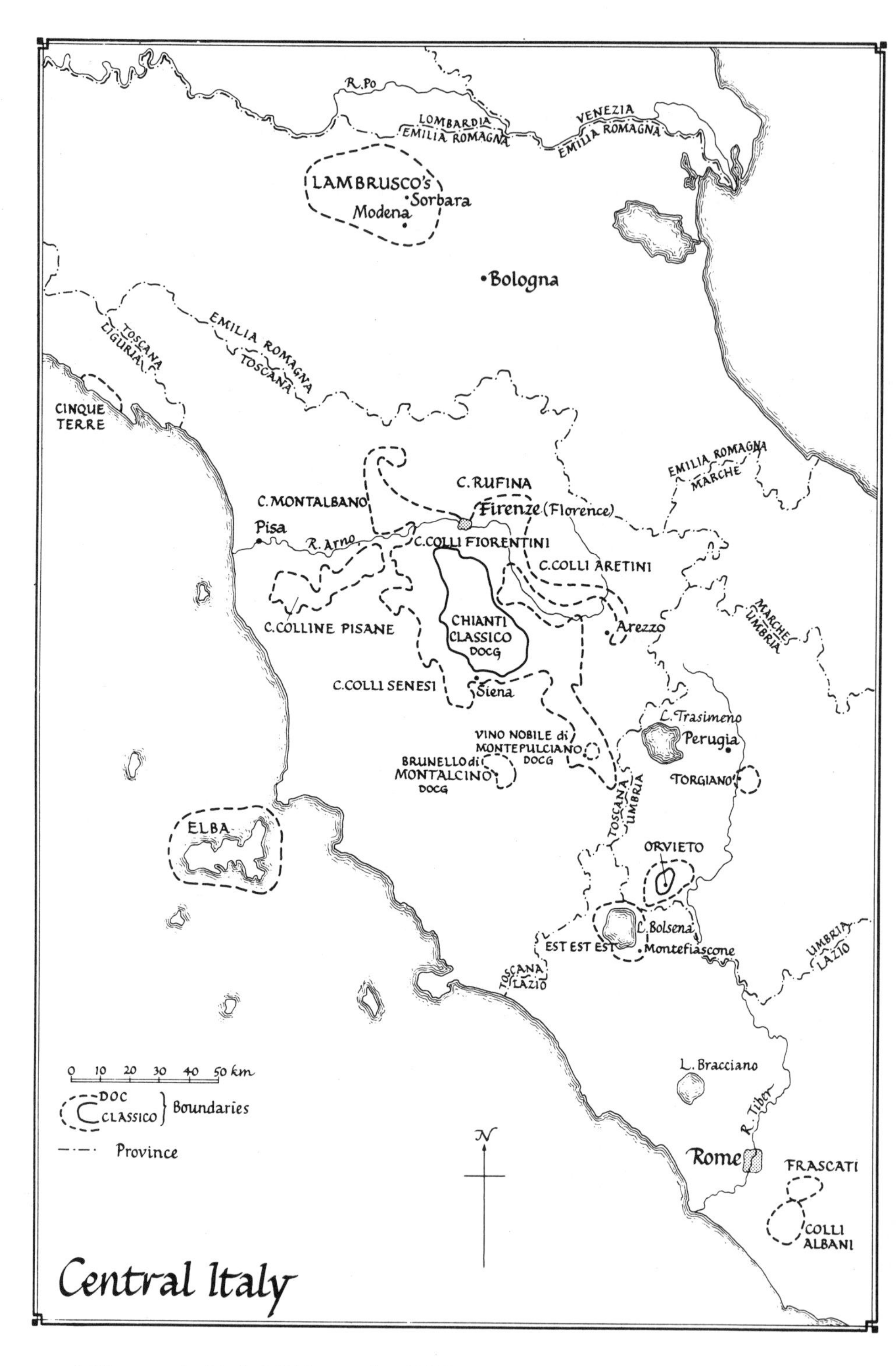

1 They may be labelled QbA as well as DOC.
2 Teroldego. Full-bodied, long-lasting red, and perfumed rosé wines.
3 Collio Goriziano.

Nobile di Montepulciano, both of which, with Chianti, have now achieved DOCG status.

Chianti producers have had their problems, the greatest of which was the protection of their famous name. The true Chianti growers had difficulty in establishing the boundaries of their zone because neighbouring producers of lesser wines pirated the name and tried to extend the boundaries. As Chianti Classico must be vinified *and* bottled in the Classico district, several producers of wine made from grapes grown in the Classico district are excluded from the consorzio, because their wineries lie outside the delimited area at places like Siena. Not all wine from the diminutive original area of Chianti production is therefore entitled to be called Chianti Classico. Six more districts in surrounding areas have wines entitled to the name Chianti associated with their particular district of origin; they jointly subscribe to the Consorzio Putto, and are: Chianti dei Colli Fiorentini, Colli Senesi, Colli Aretini, Montalbano, Colline Pisane, Colli Fi, and Rufina (which must be distinguished from the Tuscan wine-producing firm of Ruffino).

Each Chianti consorzio has its own label and symbol, a valuable recognition point for customers: that of Chianti Classico is a black cock, while that of Chianti Putto is a cherub. Chianti Classico is allowed a yield of only 7.5 tonnes/ha, compared with 10.0 for Putto; and it must have 11.5% vol. alcohol, 0.5% more. This is another reason why Classico is more expensive.

The Chianti vineyards lie between 500 m and 650 m above sea level, and have a soil of marl and lime; the alluvial soil in the valleys is unsuitable for viticulture. Colli Senesi is the largest Chianti district, and has three sub-districts, the first in the Siena hills south and west of the Classico district, the other two surrounding the hill towns of Montepulciano and Montalcino.

A small proportion of Chianti wines are made by the old system – *Governo all'uso del Chianti*. In this method, up to 10 per cent of the grapes – usually the white Malvasia and Trebbiano – are left to dry on wicker frames while the remainder – the 70 per cent-odd of Sangiovese and the 5 to 10 per cent of Canaiolo Nero – are fermenting in vats. When their fermentation is finished, the *passiti* (dried) grapes are added; this addition of sugar restarts the fermentation, resulting in a light fruity wine with a trace of residual sugar, but one which may be bottled early and drunk young.

1 (a) How many Reggioni are there in Italy? (b) Generally what soil do they all share?
2 Distinguish between Lambrusco di Sorbara and *vitis labrusca*.
3 With what wine is the filtrado process associated?

All Chiantis share the common characteristics of being red, fairly dry, and slightly tannic, with a ruby colour deepening to garnet with age; they have a velvety full-bodied texture and bouquet of violets or irises; all are smooth well-balanced wines, excellent with roasts and game. Connoisseurs will serve Chianti carefully at a temperature around 15°C.

Chiantis allowed to age for two years are entitled to the description vecchio, or old – if they have a minimum alcohol of 12% vol. (Chianti) or 12.5% (Chianti Classico) when reaching the consumer. After a third year's maturation the description riserva is allowed. Chiantis will age well and keep for decades.

The Vino Nobile di Montepulciano DOCG has been produced by the aristocracy of Montepulciano as a religious wine since the thirteenth century. This name might seem to contradict the saying that wines are named after either grape varieties or village names, because Montepulciano appears as a grape-name in Montepulciano d'Abbruzzo and as a place-name in this other instance – Nobile di Montepulciano. No connection is traced, except that a rough translation would be 'beautiful mountain'. The Vino Nobile is said to have been a favourite wine of Pope Paul III. In the fourteenth century it was already being exported. It is made from the Prugnolo Gentile grape (a clone of Sangiovese the species of both is *vitis vinifera:* see clones, p. 4), with small additions of Canaiolo, Malvasia, and Trebbiano, and comes from the portion of the hills having tertiary rock formation. The wine is of a deep garnet colour, with a faint violet bouquet, dry and smooth, with 12% vol. alcohol. It may not be sold with less than two years ageing in wood, and may become riserva after three years; it can continue improving for twenty to thirty years.

Brunello di Montalcino DOCG is one of the best red wines of Italy, made from the Brunello grape (another variation of the Sangiovese) grown near the village of Montalcino south of Siena. It is more full-bodied than Chianti, and must be aged in cask for longer – the vinification and ageing must take place in the commune of Montalcino – and when bottled is said to be good for a century. It is scarce, and expensive.

There is no white Chianti – in fact, there never was – but some wines made from the Trebbiano grape, sometimes blended with

1 (a) Ten. (b) Granite and basaltic.
2 The first is a wine of Emilia-Romagna, the second a wild grape.
3 DOC Lambrusco.

Malvasia, were marketed under this name until Chianti as a red wine received international protection. These Toscana Bianco vini da tavola are pleasant golden, and full-bodied, but a much better vino da tavola bianco can be found under the name Galestro; this wine, which has a maximum strength of 10.5°, is made from Trebbiano blended with Sauvignon, Chardonnay and Pinot Bianco. Another, launched by the Chianti Classico consorzio, is Bianco della Lega, the Lega being the wine fraternity of Chianti. It should also be mentioned that there are five DOC white wines of Tuscany, of which the best is Vernaccia de San Gimignano.

The Tuscan wine-makers have long been convinced of the value of the Cabernet Sauvignon grape which the authorities are slow to recognize, and have created vini da tavola which fetch better prices than DOC or even DOCG wines. Sassicaia, which originated in the coastal plain near Livorno, has a greater proportion of this grape than the great wines of the Médoc, and matures superbly; Tignanello, with 80 per cent Sangiovese and 20 percent Cabernet is made in the hills, is softer and matures more quickly. Cabernet Sauvignon is also finding its way into Chianti as one of the 10 per cent 'other grape varieties' permitted, and is allowed up to 10 per cent for DOC Carmignano, west of Florence.

On the Island of Elba, off the Tuscany coast, DOC Elba Bianco and Elba Rosso are made from Trebbiano (locally called Procanico) and Sangiovese grapes respectively. The wines are dry and well balanced.

MARCHE

This region, on the Adriatic side of the Apennines, is noted for its white wines, made from the Verdicchio grape, which have been known from Roman times. They are strong, and nearly demoralized Rome's attacker, Hannibal, as his elephants did Rome.

Verdicchio dei Castelli di Jesi (a village near Ancona) is the best of the Marche wines, and is put up in bottles shaped like ancient amphorae. The wine is very pale in colour and astringent; fruity, it is best drunk young. There is an inner Classico district in the oldest production area south of the river Misa.

1 (a) Where is Chianti made? (b) To claim Chanti Classico, what conditions must be fulfilled?
2 Name the consorzi district and sub-districts of Chianti.
3 What is the symbol of Chianti Classico?
4 (a) What are passiti grapes? (b) How are they used?

A fairly recent arrival is DOC Rosso Conero from Ancona, which is a medium full-bodied red wine made from the Montepulciano and Sangiovese grapes.

UMBRIA

This region lies in the centre of Italy, under the western slopes of the Apennines, from which the Tiber flows down through the region towards Rome. The Trebbiano and Sangiovese grapes reign supreme in central Italy, and Umbria is no exception.

In the upper reaches of the Tiber valley near the medieval town of Assisi, home of St Francis, is the diminutive area of Torgiano, producing DOC red and white wines which have improved dramatically since the war. The full red wine, called Rubesco, is made from the Sangiovese grape with the addition of some Canaiolo and Montepulciano. The resulting wine is heady, full-bodied and fruity, with a fairly long finish; it is gaining recognition in the UK market.

Downstream the old Etruscan town of Orvieto, just north of Lake Bolsena, gives its name to an internationally famous white wine produced from the Trebbiano grape vineyards on the terraced slopes which surround it. Orvieto was once a Holy See, and early laws imposed heavy penalties for damaging the vineyards; considerable quantities of wine were shipped down the Tiber to Rome. The DOC extends to fifteen communes surrounding the old city. Orvieto can be a sweet and slightly frothy wine or a refreshing dry one, and matures in cellars cut out of the massive rock on which the city is perched.

This wine is straw-coloured, and 'dances with golden lights'; it has 12% vol. alcohol. Besides the 50 per cent minimum Trebbiano, it may have Malvasia added to give smoothness to the wine, with minor proportions of Verdello (no relation of Madeiran Verdelho) and Grecchetto. The resulting wine has great character, and the semi-sweet abboccato was the pride of Orvieto for many years, being light and delicate without cloying. In some years, the sunny humidity of autumn finds beneficent *botrytis* – *'il muffa nobile'* – affecting

1 (a) In Toscana. (b) The wine must be made from grapes grown in the Classico district and must also be bottled there.
2 Classico, Colli Fiorentini, Colli Senesi, Colli Aretini, Montalbano, Colli Pisane, Colli Pistoiese, and Rufina.
3 A black cock.
4 (a) Grapes left on wicker frames to dry in the sun. (b) 10% of such grapes are incorporated in Chianti wines made by the method 'Governo all'uso Chianti'.

the grapes to give a special lusciousness to the wine. In recent years, the drier secco, which is crisp and refreshing, has gained in popularity.

LAZIO

This province contains Rome and the lower reaches of the Tiber. In the north of the province, separated from Umbrian Orvieto by Lake Bolsena, lies the village of Montefiascone, where a sweet white wine – DOC Est! Est!! Est!!! di Montefiascone – is produced. There are many versions of the origin of the name 'Est! Est!! Est!!!', but generally it is acknowledged that about the end of the eleventh century, a certain Bishop Fugger was travelling from his diocese of Fulda in Germany to Rome for the coronation of Emperor Henry V. It seems that he sent his manservant forward to seek out inns with good wine, which would be identified by the word '*est*', marked upon the door; '*est*' being an abbreviation of '*vinum bonum est*' – 'the wine is good'. Apparently the wine at Montefiascone was so good that the manservant wrote 'Est! Est!! Est!!!' on the door, as a result of which the good bishop abandoned his journey to Rome and remained at Montefiascone to enjoy the wine to his life's end.

Originally a sweet dessert wine made from Muscat grapes, this wine is today made from Trebbiano grapes with a minority of Malvasia and some others. It is produced in the area around Lake Bolsena, some 80 km north of Rome. The drier version complements stewed eels (a speciality of Viterbo) admirably.

From the earliest days, wines were made in the volcanic hills (*castelli*) rising to the south of Rome, where the wine industry still flourishes today. The best known of the 'vini dei Castelli Romani' is Frascati, a soft velvety white wine – dry, semi-sweet or sweet – with a refined scent. Malvasia grapes predominate in its production, with Greco and/or Trebbiano added. A sparkling variety of Frascati is permitted within the DOC. Some DOC red wines are made in Lazio from Merlot or Cesanese grapes, but they are not yet commercially important.

1 Name two wines using the name Montepulciano.
2 What are Galestro and Bianco della Lega?
3 What are Sassicaia and Tignanello?
4 Describe Verdicchio dei Castelli di Jesi.

ABRUZZO AND MOLISE

This remote and mountainous region of Italy, east of the Apennines, previously noteworthy only for its wolves, has been transformed since the war by government redevelopment. Although Abruzzo has not hitherto been noted for fine wines, they are getting better as a result of this redevelopment. DOC Trebbiano d'Abruzzo, the white wine of the province, is already very good. It is a dry velvety wine which goes well with the fish dishes of the Adriatic coast. DOC Montepulciano d'Abruzzo is a slightly heavier red wine (12°) than DOC Rosso Conero of Marche and DOC Biferno de Molise which need only 11.5° – although the latter has a Superiore version with three years' minimum age and 13° actual alcohol.

CAMPANIA

The region of Campania stretches down the west coast of Italy from the boundary with Lazio and forms the gateway to southern Italy. The region is steeped in history, and is more renowned for its beauty than its wines, for here is the Bay of Naples with the Isles of Capri and Ischia, the volcanic cone of Vesuvius overhanging all. In the black volcanic soil around Vesuvius and along the coast of Campania much of the early wine for Rome was made, and today Vesuvio DOC (previously known as vino da tavola Lacryma Christi del Vesuvio) is produced in both red and white versions from local grape varieties on the seaward slopes of Vesuvius. There are various legendary origins of the name – that Christ wept to see the destruction of innocents in Pompeii, and that his tears fell on the slopes of Vesuvius and fertilized vines. That the wine comes from the slopes of Vesuvius is fact, but Our Lord's tears were more likely for the sins of Naples.

Aglianico is a true grape of the Mezzogiorno, and is the only constituent of DOC Taurasi, a deep-coloured rich red wine; it also appears with Primitivo in non-DOC Falerno, the modern equivalent of the Romans' beloved Falernian.

While Ischia produces red and white wines, which are DOC and pleasant, they are not often seen outside the region. Greco di Tufo

1 Montepulciano d'Abruzzo and Vino Nobile di Montepulciano (of Tuscany).
2 Fine white *vini da tavola* from Toscana.
3 Extremely fine (and expensive) red vini da tavola from Toscana.
4 A dry white wine from Jesi, near Ancona, pale in colour, astringent, and fruity. The best wine of Marche.

DOC merits mention as drawing attention to its soil, the 'boiled limestone' called tufa; the wine has some similarity to Vouvray, where the soil is similar.

PUGLIA, BASILICATA, AND CALABRIA

These three regions make up the 'heel', 'instep', and 'toe' of Italy; none of them is distinguished for fine wine, but jointly they play an important part in the Italian wine scene. Puglia produces more wine than any other region of Italy, including Sicilia; its wine, like that of the other two regions, is not generally noble, but is very strong, and is in great demand as base-wine for Italian Vermouth, or for blending with the weaker wines of southern France or Germany to produce EEC Table Wine. That the French preferred it neat – as opposed to the blend – led to riots in the Midi during 1976.

Of the finer wines from Puglia, one or two deserve mention. DOC San Severo Rosso comes from Foggia, near the border with Molise, and is made from the Montepulciano grape; the Bianco version, also DOC, is made from the Bombino Bianco. DOC Aleatico di Puglia is a dark red sweet strong dessert wine; a third red, DOC Primitivo di Manduria, improves with age – this vine has been genetically identified as the ancestor of Californian Zinfandel. DOC Locorotondo is a dry white wine made from the Verdeca and Bianco d'Alessano grapes.

Basilicata has the least rewarding soil of all Italy, and the vine is the only crop that can be sustained on the slopes of Mount Vulture, a burned-out volcano. The only wine of note here is Aglianico del Vulture, a garnet-red astringent wine, which after three years may be described as vecchio, and after five as riserva.

Calabria must always be renowned for its rugged beauty rather than its wines; but one golden amber dessert DOC wine called Greco di Bianco has an exquisite perfume of orange blossom and a high alcoholic content of 17% vol. It is made in the hills around Gerace from the Greco grape, and the small amount available for export is much in demand. Calabria also produces several 'roast meat reds' from the Gaglioppo grape, for example DOC Pollino, DOC Savuto,

1 Where does Rosso Conero DOC come from?
2 (a) Where is DOC Torgiano made? (b) Which Saint is connected with the region?
3 Give the colour, degree of alcohol, and the main grape of Orvieto wine.
4 What is 'il muffa nobile'?

and DOC Ciro Rosso; DOC Ciro Bianco is made from the Greco grape which, like Aglianico and Gaglioppo is a relic of the Grecian occupation of this part of Italy in the far distant past.

SICILIA

Across the straits of Messina from Calabria is Sicily, with its ancient civilization and history; today only Puglia produces more wine than Sicily.

Although much of Sicily's wine is shipped elsewhere for strengthening blends, the Istituto della Vite e del Vino is promoting Sicilian wines in their own right. Many Cantini Sociali (co-operatives), equipped with the most modern machinery, are raising the standard of wine, both for local consumption and also for export. As a sign of quality, a large letter Q enclosing the words 'Regione Siciliano' is now granted in a system reinforcing the DOC system. The best light wines are white, particularly vini de tavola Regaleali (based on the Sauvignon grape) and Corvo (based on the local Inzolia grape), both from the north-west of the island near Palermo. The red Corvo is also excellent and DOCs exist for Faro and Etna Rosso, both made from a mixture of local grapes.

DOC Etna Bianco wines complement fish and hors d'oeuvres well, and are made predominantly from the Catarratto grape in the Province of Catania. Sweet fortified wines from the Muscat and Malvasia grapes are locally renowed and very pleasant.

First and foremost among Sicily's wines, however, is Marsala, the fortified wine created by the Engishmen Woodhouse and Smith in 1773. In its production three vines, the Catarratto, Grillo, and Inzolia are used; *vino cotto* (cooked wine), produced by boiling down grape must, is added with brandy to the normally-fermented wine to achieve the dark golden sweetness that is Marsala. Marsala with four months of age and 17% vol. alcohol is known as *Fine* and may be dry or sweet. Marsala kept in cask for two years, with 18% alcohol, may be called superiore. Without vino cotto, but kept five years in *solera*, it emerges lighter and drier, and may be called *Vergine. Marsala*

1 Near Ancona, in Marche.
2 (a) Between Perugia and Assisi. (b) St Francis.
3 Straw-coloured, 12% vol., Trebbiano.
4 Noble rot – beneficent *botrytis*.

Speciali are sophisticated with egg-yolk – 'Marsala all' Uovo' – or with bananas, coffee, or almonds.

A strong golden dessert wine which goes well with fruit, called Moscato Passito di Pantelleria, is made on the island of that name from the Zibibbo grape.

SARDEGNA

Little is heard of export of Sardinian wines, but tourists have found a fair range of red and white wines, with a few speciality wines. DOC Nuragus di Cagliari, from a local grape, is straw-coloured wine of 11% vol. alcohol, dry with a pleasant bouquet. DOC Vermentino di Gallura, and DOC Cannonau di Sardegna are among the better wines. The fortified Vernaccia di Oristano, resembling Sherry, has been described as 'the most Sardinian of all Sardinia's wines'.

1 In what Province and district is Lacryma Christi made?
2 What is the principal red wine-producing grape variety of the Mezzogiorno?
3 Name the three regions of the 'heel', 'instep' and 'toe' of Italy. What important function in the wine world do their products serve?

6

Spain – General

Spain and Portugal, the subject of this and the next chapter, are the latest countries to join the EEC and also have in common a joint land mass; a few lines about this Iberian Peninsula may help the reader to understand the climatic differences between the countries better.

The Peninsula is a huge square of land 1300 km in breadth and depth; the western sixth is Portugal, and the rest Spain. South-westerly winds from the Atlantic prevail, bringing with them moisture which falls as rain as it is driven up by the mountains into the colder upper air. There is therefore a maritime climate in the western and northern coastal regions, with rainfall about twice that of Britain. To the east of the mountains, the climate varies from temperate with about 635 mm of rain a year – as in Britain – to

1 Campania, on the seaward slopes of Vesuvius.
2 Aglianico.
3 Puglia, Basilicata, and Calabria. Often these strong wines are used as base wines for Italian Vermouth, and for strengthening EEC table wines.

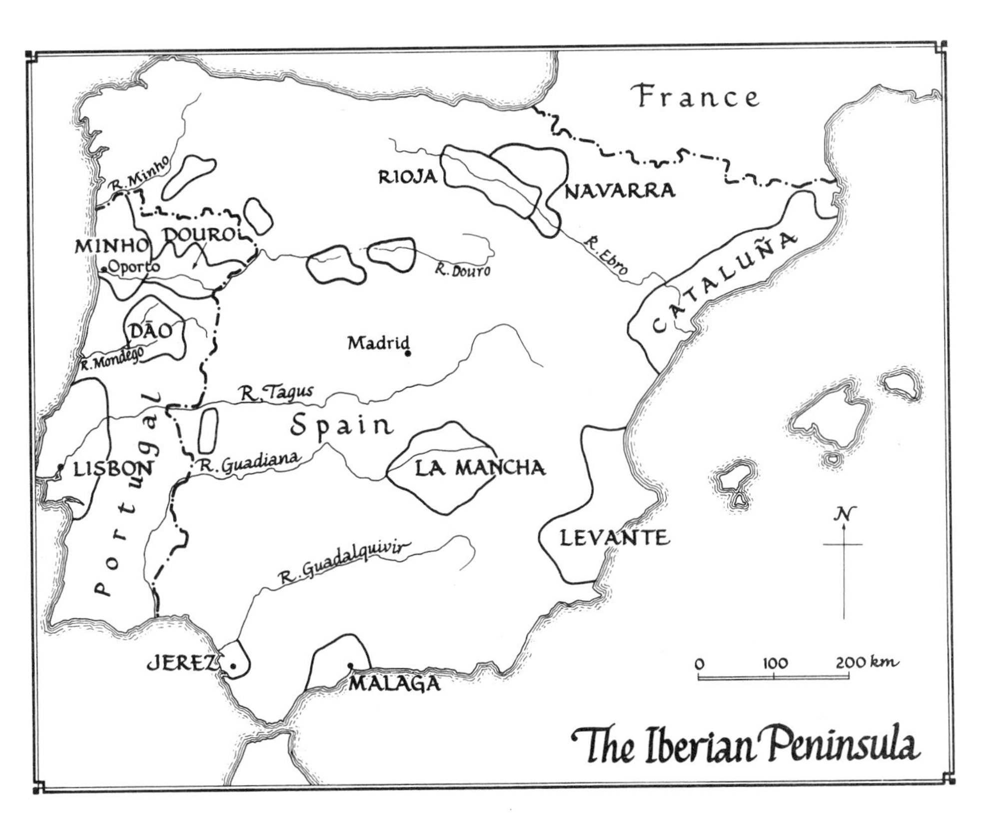

mediterranean. The central plateau has an extreme temperature variation, and is dry. Rainfall generally is greater in the north, falling off very considerably in the south. Reputedly, Spain has more soil under vines than any other country, but this includes areas of mixed planting. What is significant, however, is that the mountains of central Spain give rise to four important rivers – which with their tributaries drain the land and make vine-growing possible on a vast scale.

The Duero (Douro), Tajo (Tejo), Guadiana, and Guadalquivir run down to the Atlantic, the first two flowing through Portugal to Oporto and Lisbon. A fifth river, the Ebro, flows east through northern Spain to the Mediterranean, rising in the Cantabrian mountains west of the Pyrenees.

1 Describe the labelling feature of non-DOC Sicilian wines that attributes quality.
2 Who created Marsala, and when?
3 (a) What is the most famous wine of Sicily? (b) Describe this wine. (c) What is the driest form of this wine?
4 Where in Italy are the grape varieties Nuragus and Cannonau grown?
5 What is the 'most Sardinian of all Sardinia's wines'? What wine does it resemble?

Table 4

Spanish Denominaciónes de Origen		
GALICIA	Ribeiro	Valdeorras
OLD CASTILE	Campo de Borja Cariñena Navarra Ribero del Duero	Rioja Rueda Somontano Toro
CATALUÑA	Alella Ampurdán Conca de Barberá Penedés	Priorato Tarragona Terra Alta
CENTRAL SPAIN	La Mancha Manchuela	Mentrida Valdepeñas
LEVANTE	Alicante Almansa Jumilla	Utiel-Requena Valencia Yecla
ANDALUCIA	Condado de Huelva Jerez/Xeres/Sherry	Malaga Montilla y Morilés

HISTORY

The history of Spanish vineyards stretches back to Phoenician times, and continues into the Christian era through Greek and Roman occupations. The greatest medieval influence was that of the Muslim Moors who, strangely enough, improved the quality of the vines, being by no means strict teetotallers. Historically, however, apart from the wines of Xerez, the wines were never very good, and neither Wellington's nor Napoleon's commissaries had much good to say of them.

Spanish wine had the reputation of being an impure product stored in unhygienic conditions. Students of Cervantes will recall with what zest the *cueros* (pigskins) filled with wine were attacked by Don Quixote – perhaps he had a higher purpose? For a cuero was a whole

1 A large letter 'Q' enclosing the words 'Regione Siciliano'.
2 The Englishmen Woodhouse and Smith, in 1773.
3 (a) Marsala. (b) A white fortified wine, coloured with vino cotto and often sweetened and flavoured. (c) Marsala Vergine.
4 Sardegna.
5 The fortified Vernaccia di Oristano: Sherry.

pigskin, minus a foot (for access), turned inside out and daubed with pitch to render it waterproof. In this grisly container, Spain safely secured most of her wine for home consumption! With the coming of the English, the quality of the wines reaching the ports gradually improved. Even in the nineteenth century, however, Spain's critics found much to decry and little to praise in her wines, though Redding was then able to aver that Spain had excellent red and white wines to be discovered by the *patient* traveller.

This Spanish story is by no means unique. As in other countries, it is only in the present century that government control has improved the quality of wines to the extent that some of them are certificated 'fine' wines, when hitherto they had been merely *blanco* (white) or *tinto* (red), the latter corrupted in Shakespeare's day by the English soldier to 'Tent'. This control enables the purchaser to be sure of what he is buying. Various trade organizations in the industry had already upgraded the better wines somewhat, but not until 1970 did the Spanish Ministerio de Agricultura introduce its 'Statute of Vines, Wines and Spirits'. By this statute, power was delegated to regional *Consejos* or Committees (of which there are now about twenty) to regulate and administer the new system of Denominaciónes de Origen (DO), parallel to the French AC and Italian DOC systems.

1 Where in the Iberian peninsula is rainfall heaviest, and why?
2 Name the five principal rivers draining the Iberian Peninsula.

Spanish Quality Wine Regions

The six main regions and the twenty-nine Denominaciónes de Origen registered with the EEC in 1986 are listed in Table 4 on page 192. The best-known still and sparkling light wines of Spain are produced in Old Castile, Cataluña, Central Spain, and Levante, while Andalucia is the home of fortified wines. The wines of Galicia closely resemble those of the adjoining Vinhos Verdes region of Portugal, and will be considered with them.

The wines of Old Castille and Cataluña are noted for their superior quality, those of Levante for their strength, and those of Central Spain for their immense quantity.

GALICIA

The wines of Ribeiro and Valdeorras in this region, which lies in the far north-eastern corner of Spain, are light and fruity, resembling the Vinhos Verdes wines of Portugal, immediately to the south.

OLD CASTILE

Rioja

The Rioja sub-region lies in the upper reaches of the Ebro river, and takes its name from the Rio (river) Oja, a tributary which runs into the Ebro at Haro. The sub-region has three distinct districts, Rioja Alta (upper), Rioja Baja (lower), and Rioja Alavesa on the north side

1 In the western and northern coastal regions: moisture in prevalent south-westerly winds falls as rain on the mountains.
2 Duero (Douro), Tagus (Tajo), Guadiana, Guadalquivir, and Ebro.

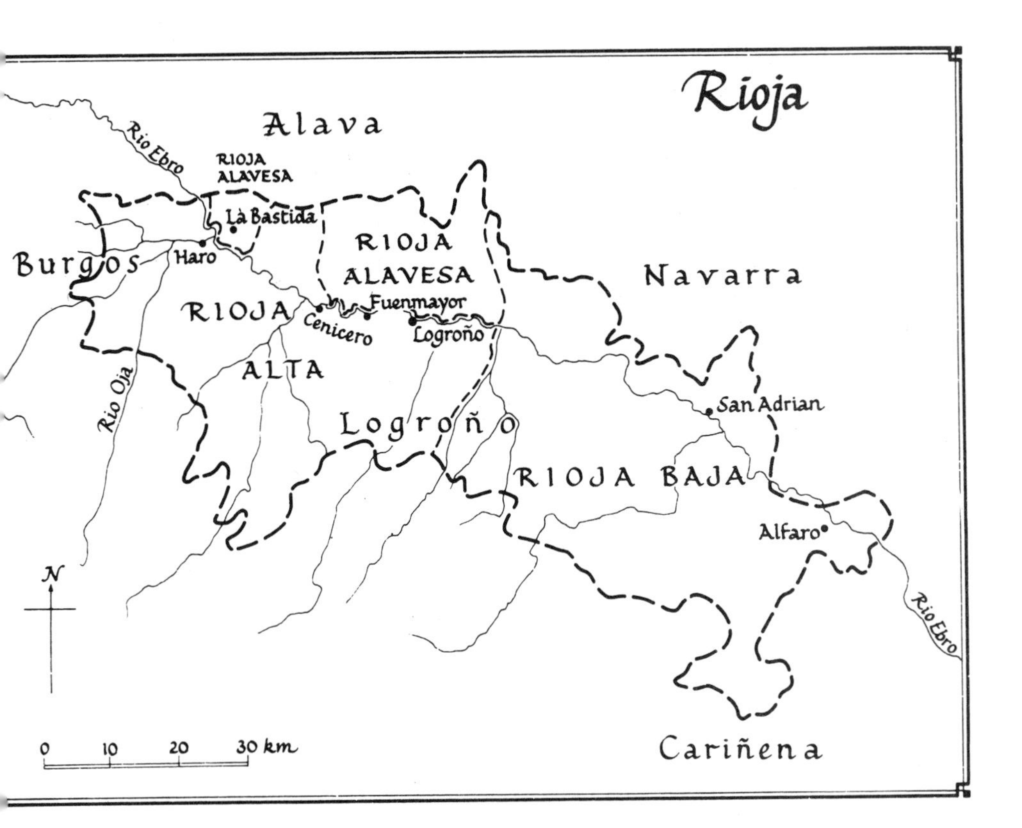

of the Ebro in the Province of Alava. Logroño is the principal centre, and is situated on the Ebro where Alta and Baja meet. Apart from Haro in the Alta district, other important centres are Fuenmayor and Cenicero (also meaning 'ashtray'!).

When in the 1880s *phylloxera* struck the Bordeaux region, many of its vignerons crossed the border into Spain, bringing their wine-making skills, and probably some of their sound vines, with them. They set out to serve the needs of the French market with wine made in the Rioja according to French methods; the *barrica*, identical with the Bordeaux barrique of 225 litres, was introduced to mature the wines for two to three years, and is still used in Rioja to this day.

The western parts of Rioja – Rioja Alta and Alavesa – lie in the Atlantic airstream in the lee of the Cantabrian mountains and enjoy a temperate climate with moderate rainfall (600–700 mm a year). This

1 What was a cuero and how was it prepared?
2 What is the Spanish equivalent of French AC and Italian DOC.

gives their wines greater character than the heavy wines produced under the hot sun. The climate changes rapidly between Rioja Alta and Baja from sub-humid to semi-arid, and even before the Baja begins, the average rainfall lessens and the climate becomes hot and dry, which makes a great difference to the grapes and the resulting wines. In the Baja the acidity is burned out of the grapes, but their sugar content is much higher, producing heavier, flabbier wines, which contrast with more acidic full-bodied wines of lower alcohol from the Alta. Frequently the two are blended together.

Viticulture and Wine Types

Rioja produces both red and white wines, the principal grapes for the red being the Tempranillo and the Garnacha (the Grenache of southern France), and the principal grape for white wines, the Macabeo, here known by the name of Viura.

The vines in Rioja are trained in the bush style, without support, on a soil of clay with much gravel and limestone.

The Rioja wine-types range from deep ruby-red tintos which are full-bodied, through lighter *claretes* with a marked bouquet, and rosado wines of varying strength and dryness, to white wines, which vary between green and gold in colour, and are best when dry. The tintos and blancos have been likened to Burgundies and the clarets to the clarets of Bordeaux.

Navarra

This developing sub-region is divided into three districts: Baja Montaña to the east of the town of Pamplona; Ribera Alta between Baja Montaña and Rioja Baja to the south; and Ribera Baja, lying across the Ebro downstream of Rioja Baja. The wines are lighter

1 A wine container of a whole pigskin (minus a foot), turned inside out and daubed with pitch.
2 Denominación de Origen.

than those of Rioja but full-flavoured, being made from the Garnacha and Graciano grapes. Some claretes, of unusual colour and bouquet, are called *ojo de gallo* (partridge eye).

Cariñena, Campo de Borja, and Somontana

The first of these three sub-regions in the ancient kingdom of Aragon gave its name to a grape, also known in Spain as Mazuelo and now widely planted in France as Carignan. This variety produces rather thin, acid wine and now forms less than half of the blend with Garnacha and Bobal making up the balance, which has considerably improved the wines.

Rueda and Ribero del Duero

The Duero river rises in the mountains to the south of Rioja and flows west to the Portuguese border; halfway along this 300 km journey it traverses a cool upland plateau of chalky clay where these two small sub-regions are located. In Rueda, the western sub-region, excellent white wines are produced from a local grape variety Verdejo Bianco, somewhat like the white wines of Rioja but drier.

Peñafiel is the main centre of Ribero del Duero, producing quantities of sound red wine from the Tempranillo grape; but the main interest of this sub-region lies in the smaller town of Valbuena, where the Vega Sicilia vineyard produces the best red wine of Spain – from a blend including Bordeaux grape varieties. It is said that refugees from Bordeaux during the persecution of Protestants following the Revocation of the Edict of Nantes in 1685 brought their vines with them; certainly, more were brought – as a result of phylloxera – to the district in the 1860s. The wine is aged for eleven years in cask and far from being exhausted by this is full of colour and bouquet: it is, of course, very expensive, and the vineyard produces wines with only five to nine years' ageing under the name Valbuena – these also are very fine wines.

1 Describe the wines of Ribeiro and Valdeorras.
2 From what does the region of Rioja take its name?
3 Which is the best-known DO in the region of Old Castile?
4 Why could Rioja wines be said to resemble those of Bordeaux?

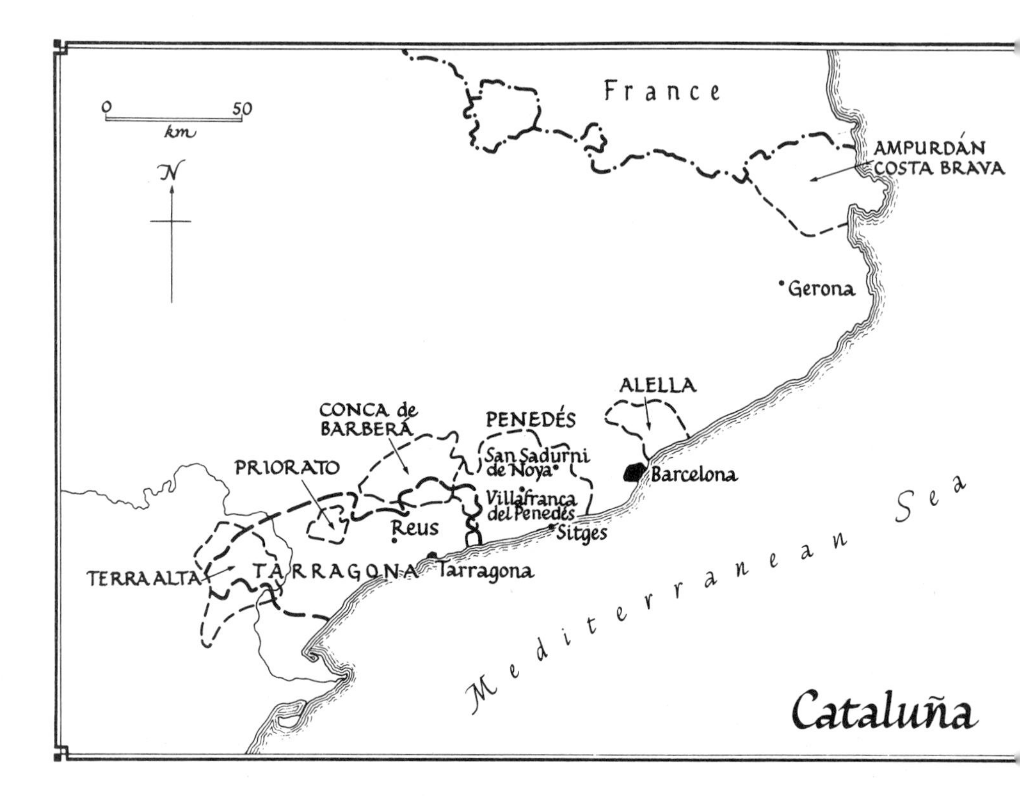

CATALUÑA

In the north-eastern region of Cataluña a vigorous wine industry exists and six Denominaciónes de Origen have been granted to its wines.

Ampurdán

This sub-region lies on the slopes of the Pyrenees at the northern end of the Costa Brava, and produces red, white, but mainly *rosado* wines from the Garnacha, Piquepoule (Picpoul), and Macabeo grapes.

1 Light, fruity, and slightly pétillant like the Vinhos Verdes of Portugal.
2 The Rio (river) Oja, a tributary of the Ebro.
3 Rioja.
4 Because Bordeaux vignerons came to Rioja as refugees from *phylloxera*, in the 1860's. (Others had come as refugees from religious persecution: see page 197.)

Alella

This sub-region lies to the north of Barcelona on the coastal range of hills; its wines are white, from the Garnacha Blanca and Malvasia grapes, and of about 11% vol. alcohol. The wines are matured in oak for one or two years and sold after three. Blanco Seco is a fresh wine suitable for fish dishes in the tradition of the Loire, the grapes from north- and south-facing slopes being carefully blended to produce an acceptable acidity – Blanco Abbocado is sweeter. The annual production is modest.

Penedés

The sub-region lies to the south of Barcelona, centred on the town of Villafranca del Penedés on the hilly plateau at a height of about 750 m, but stretches to the coast also. The town boasts one of the best wine museums of Europe, and a group of statuary, composed of men standing on each other's shoulders, with boys likewise above, commemorates an occasion in the middle ages, when such a 'circular pillar' was formed in order to get warning of the approach of enemies, by which the town was saved. The occasion is celebrated each year with a feast, at which villages from the region compete against each other to contrive the highest pillar.

The sandy coastal part of the sub-region, around the little port of Sitges, is noted for sweet white Malvasia de Sitges; Malvasia and Sumoll grapes, dried in the sun, produce this fortified wine, which is similar to the French VDN across the border.

The upland vineyards, on limestone soil, are noted for their dry white wines, excellent for accompanying shellfish. The grapes most commonly used are Parellada, Xarel-lo, and Macabeo, although growers have been successful in growing Gewürztraminer and Muscat d'Alsace, and experiments are taking place with Chardonnay and Pinot Blanca.

Penedés also produces large amounts of sparkling wine. About 1000 hl of *vinos espumosos* are produced annually by a second fermentation in bottle; where this is done, the bodega may be known as a *cava*, to distinguish the quality of its product; CAVA is itself a

1 Name the principal vines of Rioja used for (a) Red wine. (b) White wine.
2 What are the three districts of Navarra?
3 What does 'ojo de gallo' mean, and what does it describe?
4 Describe the wine of Rueda.
5 Where is the Vega Sicilia vineyard, and for what is it famous?

DO, denoting sparkling wines made by this method. About the same quantity of sparkling wine is produced in bodegas by the tank or impregnation methods. The best sparkling wines are made from white grapes, recalling the Blanc de Blancs from Champagne. There are now some quality reds being made in middle Penedés these days too, from Cabernet Sauvignon and Merlot in addition to traditional varieties. The grapes are harvested with great care and broken grapes are discarded. Spanish sparkling wines are labelled, according to their sweetness, *bruto* (extra-dry), *seco* (dry), *semiseco* (semi-dry), *semidulce* (semi-sweet), and *dulce* (sweet), corresponding generally to their French equivalents.

Tarragona

This is the largest sub-region, and extends for 80 km along the coast either side of the port of Tarragona, and for some 40 km into the hills. It used to be known for its high-strength sweet red wines, still popular in certain parts of the UK as 'poor man's Port', but now concentrates more on the production of drier table wines, for which the principal production centre is the town of Reus. Included in its area is the sub-region of Priorato.

Priorato

This lies to the west of the port of Tarragona and has its own Denominación de Origen for its production of some 50,000 hl a year of dry red wines, nearly black in colour. These are made from Garnacho Negro and Cariñena grapes and are full-bodied and very high in alcohol (up to 18% vol.). Also coming from Priorato are sweet white dessert wines from the Garnacho Blanco, Macabeo, and Pedro Ximénez grapes, and substantial quantities of *mistela*, a grape-must whose fermentation has been prevented by the addition of grape spirit; it is used for blending into dessert wines and vermouths. Priorato takes its name from the ruined Priory of Scala Dei in the Sierra de Montsant, where the soil is volcanic slate, ideal for the growth of good vines.

1 (a) Tempranillo and Garnacho. (b) Viura.
2 Baja Montaña, Ribera Alta, and Ribera Baja.
3 'Partridge eye', describing the colour of the clarete wines of Navarra.
4 Full-bodied white wines from the Verdejo Bianco grape.
5 In the village of Valbuena in Ribero del Duero district; famed for fine red wines of great maturity.

Terra Alta

This small sub-region lies in the hills to the south of the Ebro river, and produces wines similar to those of Priorato, but less fine.

CENTRAL SPAIN

La Mancha

This consists of a vast high plateau lying to the south of Madrid and producing a third of the table wine of Spain. The principal distinguishing feature of this region is its microclimate which, due to its altitude (615 m above sea level), and its location in the centre of a large land mass, has considerable temperature variations, both seasonal and daily. This does not advance the production of fine wines and the accent is therefore on quantity rather than quality. La Mancha wines tend to be light, low in acidity, and dry. Recent improvements from earlier harvesting are producing fruitier, less alcoholic wines of various styles. But the mass of the region's produce is distributed to other regions for blending for export in bulk, or is converted into brandy or industrial alcohol.

Valdepeñas

This sub-district lies in the middle of La Mancha on rich stony red soil with a chalky subsoil, and has a justifiable reputation for red wines made from the Cencibel and Garnacha grapes and white wines from the Airén, Pardillo, and Cirial. Even now some wines are still being fermented and matured in earthenware vessels called *tinajas;* as these contain no bunghole at the base, the wine has to be drawn from the top, and since this might not all be drawn off before the new vintage is added, a form of solera results. Valdepeñas has its own Denominación de Origen, as does La Mancha generally and the neighbouring districts of Manchuela and Mentrida. The wines appear ubiquitously in carafe, or are drawn directly from the wood

1 Which is the northernmost DO of Cataluña?
2 Which Spanish wine region surrounds Barcelona?
3 Describe Malvasia de Sitges.
4 What are vinos espumosos and where are they made in Cataluña?

in cafés. Manzanares and Daimiel also are important centres for wine production, and Tomelloso is the centre of production for brandy and industrial alcohol.

LEVANTE

This region, on the eastern slopes of the La Mancha plateau, comprises the districts of Valencia and Utiel-Requena in the north, Almansa in the centre, and Alicante, Jumilla, and Yecla in the south. All are situated on the coastal Mediterranean hillsides, and have individual Denominaciónes de Origen for a total annual production

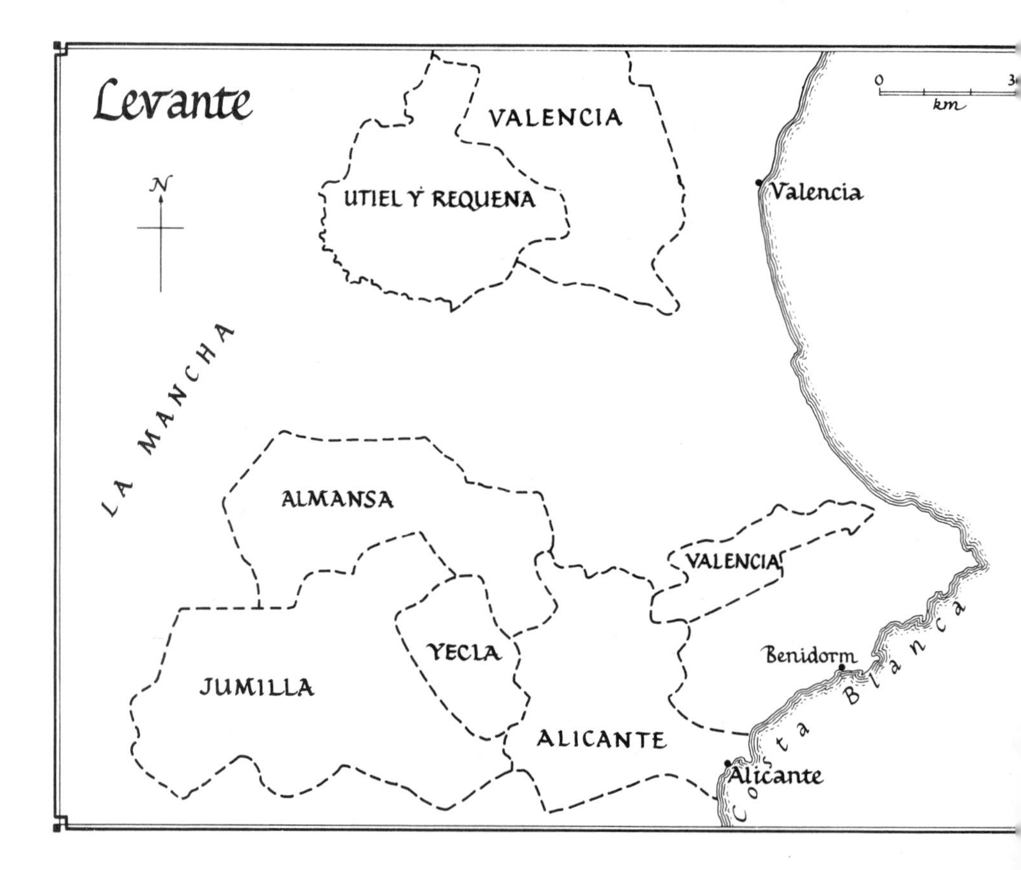

1 Ampurdán Costa Brava.
2 Cataluña.
3 A sweet white fortified wine, made from Malvasia and Sumoll grapes.
4 Sparkling wines: principally in the Penedés district.

approaching 1.5 million hl. On the hilly border with La Mancha the winters are severe, and light red and rosé wines, of up to 12% vol. alcohol, are produced from the Bobal grape. In Jumilla and Yecla, lower down where the climate is drier and hotter, stronger wines of 13% to 17% vol. alcohol are produced from the Monastrel grape. As the plateau drops down to the coastal plain the climate and soil lend themselves admirably to general agriculture production, and the grape has to compete with oranges, lemons, rice, sugar cane, and almonds. The soil is alluvial and often too rich for the production of fine wines.

ANDALUCIA

Jerez/Xeres/Sherry

In the southernmost corner of Spain, reaching out almost to touch Africa, lies a small vineyard region less than 70 kilometres in diameter, an undulating treeless plain of blinding white, whose wine has been renowned throughout the world for more than a millennium. This wine has been, and is, widely imitated, yet remains unique; nevertheless, it has moved with the times, and in its infinite variety has both led and followed fashion. It is Sherry.

Many people have settled this land – possibly first to bring the vine were the Phoenicians about 3000 years ago. They were followed by the Greeks who named the town of the region Xera; the Romans came and stayed 500 years, until driven out by the barbarian Visigoth and Vandal tribes from Central Europe, the latter giving their name to the province of Andalucia. They only stayed 300 years, and were in turn driven out by the Muhammadan Moors from North Africa. In the 400 years that followed, Sheris (as the Moors called it) was peaceful, as the Moors had driven the battle north through Spain into France. Although abstainers – in theory anyway – the Moors encouraged viticulture, and boiled down the grape must to give a sweet drink called *arrope*, still used for sweetening. It was possibly at this time that the normally-fermented dry wine became known as Sheris Sack. In the 100 years before 1264 the town was again in the battle-

1 What is DO Cava?
2 Name a wine of La Mancha having Denominación de Origen.
3 What two districts adjoin La Mancha?

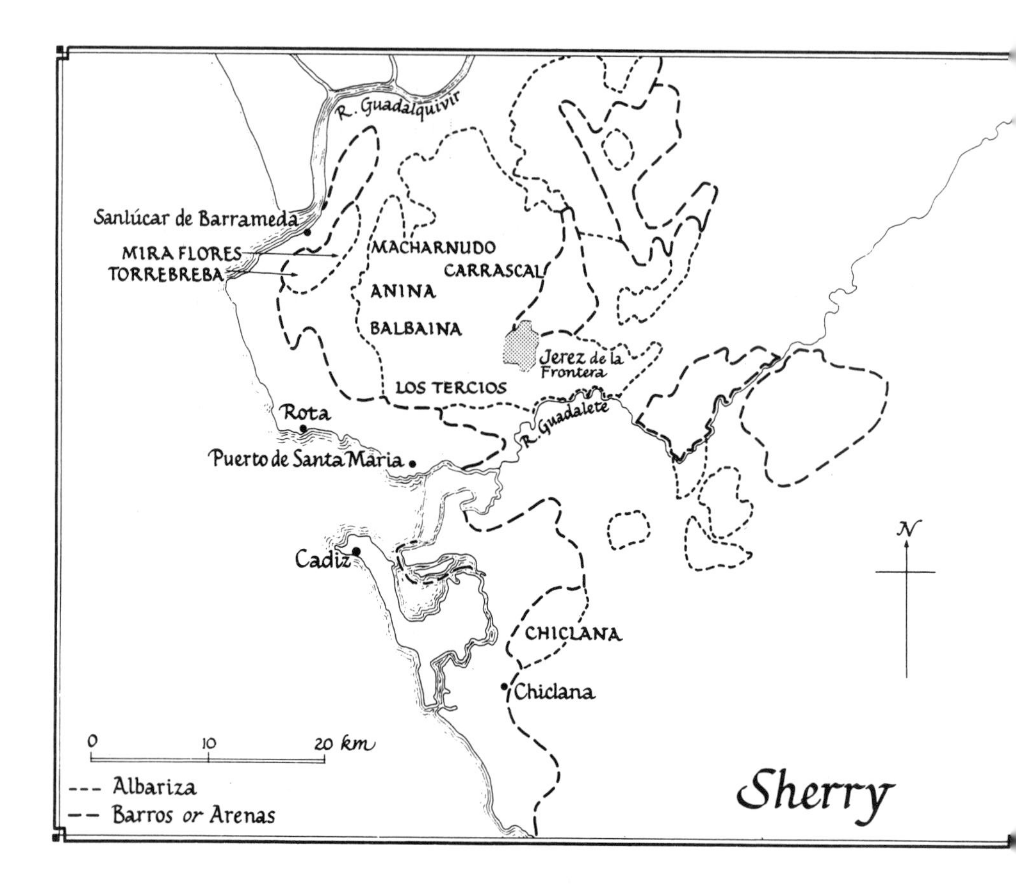

zone, and gained its present-day name of Jerez de la Frontera – 'the frontier'.

The wine, as we know from Shakespeare, has been popular in England since the days of Elizabeth I; when Drake 'singed the King of Spain's beard' by burning his fleet in Cadiz harbour, he took care to remove the wine first! About a hundred years earlier, Columbus embarked on his second voyage to America from Puerto de Santa Maria, carrying Jerez wine as ballast. Today, Sherry is seldom exported in bulk (barrels or containers) and may only be so from these two ports.

Climate

The rainfall is, surprisingly, much the same as in England – 560 mm a year; much of it falls in October and November, and the land is dry

1 The Denominación for Spanish wines made sparkling by second fermentation in bottle and sediment removal by riddling.
2 Valdepeñas.
3 Manchuela and Mentrida.

and hot between mid-May and mid-October. The winters, though without frosts, are quite cold enough to rest the vine, so grafting can be done in the field. The prevailing winds are south-west from the Atlantic, bringing moisture; occasionally the Levante, an easterly wind from the Mediterranean, blows too strongly to be welcome but dries out the soil.

Soil and Grapes

The soil is all-important in the Sherry region; the more chalk it contains, the better it is. *Albariza* has about 40 per cent chalk, is very water-absorbent, but dries out from a gluey consistency to a hard pan which has to be broken up. When this has been done, the resultant brilliant white tilth reflects the sun and acts as a mulch. The map opposite shows the principal districts of this soil – Macharnudo, Balbaina, Añina, and Carrascal.

The grape grown on this soil is almost exclusively the white Palomino, and produces the finest finos; the other two grapes (also white), the Pedro Ximénes (PX) and the Moscatel, are grown on lesser soils and are only used for sweetening and colouring.

The other two soils, *Barro* and *Arena*, only contain about 10 per cent chalk. Barros are clayey, grow many weeds, and produce coarser wines, while Arenas are sandy, reasonably weed-free, but only good for the sweet wine grapes. There is an in-between soil, a barro with 25 per cent chalk, known as *Albarizona*, which is found near Trebujena in the north and Chiclana in the south, producing 'in-between-quality' wines.

Viticulture

The nature of the climate and soil make this a hard labour. The autumn rains must not be lost, so long V-shaped troughs have to be dug between the rows of vines to collect the water and prevent erosion. In the less mechanized vineyards a circular cup-trench may be hoed out round each vine. The vines are grown as single bushes and trained in a style similar to the Guyot Simple, with one main branch and a reserve shoot which will form the next year's main branch. Training on wires is increasingly practised, with the rows 2 m apart, enabling machinery to be used. The mechanical auger, as

1 Name three of the wine districts of Levante.
2 (a) Where are Levante's strongest wines made? (b) What is their vol. alcohol? (c) What grape is used?

used for digging holes for telephone poles, is one of the most useful machines, for it not only makes the holes for replanting, but also those for burying fertilizer, applied every four to five years.

Weeding and spraying (for which helicopters are now much used) are constant labours from March onwards, as they are in every vineyard in the northern hemisphere; an additional labour in the Jerez region is to break up the top crust of the earth after the rains, and in May. The vines bear a heavy load of grapes, and those not trained on wires have to be supported with Y-shaped props. The vintage takes place during the first three weeks of September; the yield varies considerably with the year, and has been quoted over the range of 24–120 hl/ha: the normal appears to be about 70 hl/ha.

Vinification

Where previously the grapes were trodden in the vineyard *lagares*, and the process was considered the culmination of viticulture, nowadays the grapes are pressed (after light crushing but not destalking) in cylinder-screw presses located in the *bodegas* (wineries), and this process marks the beginning of vinification. Furthermore, the Palomino grapes are now pressed as soon as possible after picking, to ensure their freshness, instead of being 'sunned' for a day on esparto grass mats to concentrate their sugar, as they were in the past; this process is still used for the PX and Moscatel grapes, which are sunned for two to three weeks – logically, these are the first grapes to be picked.

One difference in the making of Sherry from that of other wines is that the crushed grapes are dusted with gypsum before pressing; this has been done for many years 'to help the lees settle' – modern science confirms this as a process to convert bitartrates (which cause cloudiness in wine) to insoluble tartrates which settle out.

Classification

In the heat of the Andalusian summer the must ferments furiously for up to 72 hours, during which time practically all the sugar is converted into alcohol; the strength is now 11–12% vol. Fermenta-

1 From: Valencia, Utiel-Requena, Almansa, Alicante, Jumilla, and Yecla.
2 (a) Jumilla and Yecla. (b) 13% to 17% vol. alcohol. (c) Monastrel.

tion continues slowly and quietly for another six to seven weeks, as in other regions; but whereas in other regions the barrels are kept topped up and sealed, in the Sherry country they are allowed to ullage and are kept only loosely stoppered. The capacity of a butt of Sherry is generally taken as nearly 500 litres (108 gal) – yet the capacity of the cask is 600 litres, one-sixth being left as ullage so that the air can contact the new wine or *mosto*. The effect of air on the mosto varies in the three months before it is classified and fortified: some butts are affected by acetobacter, become sour, and are distilled; some oxidize, develop a 'nutty' nose, and darken; these are classified as *rayas* – the name for the strokes // marked on the cask – and fortified to 18% vol. to prevent them oxidizing further. Others have a delicate nose; of these some develop a film-yeast (flor) and are classified as potential *finos*; these are only fortified up to 15.5% vol., as a greater fortification would kill the flor. Fortification, incidentally, is never done directly with high-strength spirit, as this would be too much of a shock for the wine; the spirit is always first diluted with an equal quantity of mosto and allowed to settle down for a while before being added to the bulk.

The traditional method of vinification has its drawbacks – too much potentially good wine turns sour, and it is wasteful to have to turn this into spirit for fortification when spirit can be obtained cheaply elsewhere. Moreover, the traditional system of classification takes no note of the market for Sherry. If the flor does not appear (and it does not always) then sooner or later the price of fino Sherry will probably rise, usually when the market for it is expanding. Knowledge of the composition of wine and of the flor yeast (which is a variety of *saccharomyces*) enables the oenologist to produce wines in accordance with projected market requirements – by controlling nature to the extent that the original composition of the grape-juice allows. Temperature control of fermentation influences the production of flavour-giving aldehydes: choice of cultivated yeasts ensures purity: fermenting in sealed tanks controls oxidation: maturation of new mostos in sealed tanks aids classification in the early stages: and inoculation of suitable mostos with flor yeasts controls production of fino and amontillado sherries. All these things can be done, and many of them are already being done at great expense by leading

1 (a) How did Sherry get its name? (b) Approximately what size is the Sherry region?
2 From which two towns may Sherry be exported in bulk?
3 Of the Jerez soils, (a) one is sandy; (b) one has 40% chalk and (c) one is clayey. Name them and say which is best.

firms, yet different casks of *añada* (vintage) wines will still display minor but significant differences in style.

Soleras

Re-classification of the butts of añada wine continue every six months; these are kept, unassembled, as single vineyard wines until their exact classification is confirmed. Only then can they be ascribed to one of the shipper's solera scales, being near enough to it in character to refresh it without changing it; in turn the *criaderas* through which it will pass adjust the añada wine in turn with the character of a particular solera.

Flor, the characteristic of fino Sherries, feeds on the wine (particularly its glycerine content) in the presence of air, and is active in the spring, when the vine flowers, and in the autumn when the grapes are gathered. Lest this should sound too romantic – reflect that the temperatures at these times are about 16°–17°C (60°–70°F). At other times the flor sinks to the bottom of the cask, there to be called 'mother of wine'. As flor continues through all the criaderas of a solera, 'mother' gets a little sour after some years and has to be cleared out.

Refreshing a solera not only 'educates' the younger wines by bringing them into contact with progressive majorities of other wines, and replaces the wine taken from the final butt or shipping solera, but also, in the case of finos, provides additional food for the flor, without which it would die and the wine become amontillado; it stands to reason, therefore, that finos are young wines and that there are less criaderas in a fino solera than in an amontillado or old oloroso solera. And different soleras of each of these types may have differing numbers of criaderas.

Manzanilla: One group of Sherries hitherto unmentioned is Manzanilla, probably the most popular in Spain and increasingly so in Britain. These are wines from grapes grown near Sanlúcar de Barrameda by the sea at the mouth of the river Guadalquivir. All types of soil are found here, and both fino and oloroso wines occur, but finos matured in this area take on an unmistakably salty aroma

1 (a) From the town of Jerez de la Frontera, named Xera by the Greeks. (b) About 70 km in diameter.
2 Cadiz and Puerto de Santa Maria.
3 (a) Arena. (b) Albariza (the best). (c) Barro.

not found in wines matured elsewhere, due to the different yeast varieties in the flor, which is also active the year round.

Preparation for Sale

So far, only the development of Sherry constituent wines of Palomino origin have been discussed; the PX and Moscatel grapes were left lying on grass mats in the sun for two to three weeks, during which time they lost enough water to double their sugar content. After crushing, the grapes are pressed strongly; the pomace (*orujo*) is soaked in one-year-old oloroso to dissolve the remaining sugar, and then pressed again. As the juice is immediately quenched with brandy, PX and Moscatel wines are mistelas, and are used for sweetening. *Dulce*, a more delicate sweetening mistela, is made from unsunned Palomino grapes.

Colour wine, another ingredient of Sherry blends, really is wine. The juice of PX and Moscatel grapes is boiled down to a dark, caramelized syrup, one-fifth of its original volume, and is then mixed with ordinary unfermented must; this very sweet mixture ferments very slowly to about 9% vol.

That describes the constituents of Sherry – Solera wines, PX and Moscatel, and *color* (colour-wine) – but not commercial Sherry, which is created by blending these constituents to replicate existing Sherries on the market, or to create new blends to satisfy new markets.

After blending, Sherry blends must be left so that their constituents can marry – in one or two cases this is done by creating a solera for the standard blend. The final wine must be crystal-bright, and remain unaffected by changes in temperature. To ensure brightness the wine is fined using eggwhites, or Spanish earth from Lebrija, a village in the north of the region where this smectic clay is found. This removes the proteins, which are a common cause of cloudiness in wine. Soluble bitartrates can cause haziness when the wine is chilled to the temperature at which the lighter Sherries will ideally be served: the problem has already been partly dealt with by adding gypsum to the must before fermentation; complete clarity is ensured by attemperating the wine before bottling – the wine is chilled to −4°C (25°F) and kept at that temperature for 8 to 10 days.

1 When is the Jerez vintage and what is the average yield?
2 After fermentation, sherries (a) go sour; (b) go dark, or (c) develop flor. How is each classified?

Montilla y Moriles

This region of about 1500 ha lies 35 km south of Córdoba in Andalusia and produces strong white wines from the Pedro Ximénes grape grown on calcareous soil. The wines resemble Sherry but are generally unfortified although, as they reach a natural alcoholic strength of 16–21% vol., they fall into the fortified wine tax bracket. At one time they were more popular than the wines of Jerez – so much so, in fact, that the Jerezanos named one of their styles of Sherry Amontillado, 'like the wine of Montilla'. Now that Sherry is more popular than Montilla, the Montilleros have tried in true Gilbertian fashion to label one of their wines 'Amontillado'; but this was swiftly quashed.

Malaga

Almost adjoining Montilla/Moriles to the south lies the ancient region of Malaga, famous for its dessert wines as Montilla is for its aperitifs. The smaller northern part of the region round Mollina produces the lightest Malaga – *blanco seco* from the Pedro Ximénes grape; but although this would seem to replicate the wine of nearby Montilla, it is quite different because, like all Malaga wine, it must be made in bodegas located in the coastal city some 50 km to the south.

The main vineyards of Malaga lie in the coastal range to the east of the city, which gave rise to the name 'Mountain' for one style of Malaga in the past. Here the grape is the Moscatel, which in this hot climate accumulates much sugar; nevertheless fermentation is very rapid, resulting in a wine with so much alcohol it needs no fortification, yet which remains sweet.

1 The first three weeks of September. About 70 hl/ha.
2 (a) For distillation. (b) As rayas. (c) As finos.

7
Portugal

HISTORY

Only in the third quarter of the twelfth century did Portugal become a separate kingdom; before that time her story is part of that of Iberia as a whole. Consequently her wine industry has been developed by the Phoenicians, later refined by the Greeks and Romans; as with Spain, Portugal suffered less in her wine industry at the hands of the Moors who succeeded the Romans, than did the Romans at home when overrun by the barbarians. After the separation, Spain was an unfriendly neighbour, however, and this enmity was responsible for attracting the early and enduring friendship between England and Portugal, which started with the Treaty of Windsor in 1386.

1 On what element of Sherry does flor feed? Can it do so in a sealed tank?
2 Why may there be less criadera scales in a fino solera than there are in other Sherry soleras?
3 (a) Name the Sherry matured near Sanlúcar de Barrameda. (b) What special quality does it have?
4 Name two fining agents used in Sherry production.
5 How is complete clarity of Sherry ensured?

Climate and Soil

Portugal differs considerably from Spain in her geographical features, which have a very direct bearing on the wine industries of both countries. From a narrow coastal plain the wine-producing regions of Portugal rise steeply from her Atlantic coast up to the high sierras, the major part of Portugal being mountainous. The terrain requires extensive terracing which in turn means that vineyard work is highly labour intensive – these regions are sparsely populated. The clouds from the ocean give heavy rain as they are driven over the sierras by the prevailing strong south-westerly winds, producing a wet climate in all but the summer months from June to September. Therefore vineyards in Portugal are concentrated in small regions. All this is in direct contrast to the vast arid plains of Spain where the vineyards are extensive and more use can be made of machinery. The soils of northern Portugal are of granite on the hilltops and schist in the valleys. These soils are ideal for the production of good wine and much of the Portuguese wine industry is centred in the north. The central Estramadura region north and south of Lisbon produces large quantities of light wine from soils and clay and limestone, sometimes overlaid with sand (which discourages *phylloxera*).

WINE REGIONS

The Junta Naçional do Vinho in Lisbon is responsible for coordinating the work of regional authorities (Institutos, Gremios, Comissãos, or Federaçãos), controlling quality and complying with EEC Regulations and Directives. The following four regional authorities control their individual regions:

Instituto do Vinho do Porto;
Instituto do Vinho do Madeira;
Federaçao dos Vinicultores do Dão; and
Comissão do Viticulturo do Região dos Vinhos Verdes

The production of all other wine (including wines from the other Regiãos Demarcadas – Algarve, Bairrada, Bucelas, Colares, Carca-

1 The glycerine content. No: it needs air.
2 Because finos mature more quickly than Amontillados and Olorosos.
3 (a) Manzanilla. (b) An unmistakable salty aroma.
4 White of egg and 'Spanish earth' from Lebrija.
5 By attemperating the wine, before bottling at −4°c for 8 to 10 days.

velos, Douro, and Setúbal) is controlled directly by the Junta Naçional do Vinho. The following paragraphs, in alphabetical order, describe these regions in more detail.

ALGARVE

The demarcated region lies along the whole of the south coast of Portugal, from Cape St Vincent to the Spanish border, extending about 20 km inland. With its hot yet equable climate, it is better-known to golfers than to wine lovers; its wines are full-bodied in character, red or white, and production is centred on Lagoa.

BAIRRADA

This region, on the lower Mondego river, has grown immensely in importance since this book was first published. Lying to the west of the Dão region across the Sierra Caramulo, it occupies the clayey foothills of that limestone range.

The wines are mainly red from a local grape called the Baga, which had previously a poor reputation when grown on the schistous soil of the Douro valley. These wines are very fruity and mature well in wood. Some white wine is made in the plains round Agueda in the north of the region; most are made sparkling. Red and white wines with a hint of resin are made at Buçaco from Bical, Fernão Pires and Malvasia grapes.

BUCELAS, CARCAVELOS, AND COLARES

These three demarcated regions lie in the peninsula of Torres Vedras between the river Tajo (Tagus) and the sea. Bucelas produces light, rather acid, white wines from the Arinto grape grown on limestone clay soil some 15 km north of Lisbon. Carcavelos is on the south coast of the peninsula and has now been swallowed up by the important resort of Estoril. It is sweet, fortified golden wine that is

1 Why is the wine of Montilla y Morilés unfortified, although of 16% to 21% alcohol?
2 What does 'Amontillado' mean?
3 (a) Does Malaga need fortification? (b) What grape is used?
4 What caused Portugal to sign the Treaty of Windsor with England in 1386?
5 Name in chronological order the five nations who helped to develop Portugal's wine industry.

seldom seen, although some is still produced in the villages just inland from the sea.

Colares lies on the Atlantic coast just north of Sintra; the strong Atlantic winds have blown sand over the clay vineyards, often to a depth of thirty feet or more, so that the Ramisco and some Malvasia vines have to be planted in deep trenches. The shoots have to be lifted out of the sand at regular intervals. Over the years, the sand has reached a depth of some ten metres, where the roots of the vines lie safe from the ravages of *phylloxera*. Fine red and white wines are made here from the black grapes of the ungrafted Ramisco vine; they are becoming rare, as the labour of replanting is too great, and the land fetches a high price for seaside building.

DĀO

This region takes its name from the river Dão (pronounced 'dun') which is a tributary of the Mondego river. The region consists of a high granite plateau, intersected by steep ravines and surrounded by ranges of hills. Like many other wine regions, its early concern was for the protection of home industry; in the thirteenth century, the citizens of Viseu (the principal town of the region) petitioned for an embargo on wines other than local produce while stocks of Dão lasted. Without doubt wine production in the Dão was much of a 'cottage industry' until the formation of the Federação dos Viniculturores do Dão in 1942, which has resulted in improving standards of quality and labelling. The vineyards cover some 15,000 ha of granite sand at altitudes ranging between 250 and 2000 m above sea level, the lower producing better grapes and wine.

The demarcated region is divided into three districts, of which the poorest one surrounds and 'protects' the two central districts, which lie respectively north-west and south-east of the river Mondego. Apart from the fact that the outer district has the more mountainous and remote territory with colder winters, there is little to choose in potential by virtue of geological and geographical characteristics. Growers in the central districts have traditionally used the micro-

1 Because they reach this strength without fortification.
2 'Like the wine of Montilla'.
3 (a) No. (b) Moscatel.
4 Her unfriendly relationship with Spain after becoming a separate state.
5 Phoenicians, Greeks, Romans, Moors, and British.

climate to better advantage, and have concentrated, through viticultural and vinification discipline, on production of superior wines.

The vines in the Dão are cultivated low, in marked contrast to the method of pruning and training in Minho e Douro. Both red and white wines are produced, with a preponderance of red. The wines are smooth and reminiscent of the southern Rhône, and when aged for four years have a fine bouquet; they are often aged in wood for far longer, when they tend to fade quickly. They come from the Touriga, Preto Mortágua, and Tinta Pinheira vines, the last being a local descendant of the French Pinot Noir. The white wines need to be drunk young, for they become soft and lifeless with age.

MADEIRA

Portugal has another fortified wine of international repute, taking its name from the island of Madeira (meaning 'forested' in Portuguese) where it is produced. The island, roughly 50 by 20 km in area, rises sharply out of the Atlantic some 600 km off the Moroccan coast. Madeira is beautiful but rugged: the hinterland is generally inaccessible and activity is mainly coastal. Vineyards perch precariously on ledges in the cliffs, hanging sometimes hundreds of metres above the sea, and everything has to be manhandled though some mechanization is possible in Camara de Lobas. The native wine was not originally very acceptable to the tastes of wine connoisseurs, and exports of unfortified Madeiras ceased in the middle of the eighteenth century.

The story of Madeira is romantic. When Captain João Gonçalves, known as Zarco (the cross-eyed), was sent by Prince Henry to claim the island for Portugal early in the fifteenth century, he started a fire to make a clearing, but the fire took hold; raging for seven years, it burned all the forests on the island, and infused the permeable volcanic soil with potash. The soil of Madeira is therefore among the most suitable for vines. Early settlers from Europe brought Malvoisie vines (probably from Cyprus) and sugar-cane to the island. From early days a small quantity of local brandy has been added to

1 Describe the soils of (a) northern Portugal; and (b) central Portugal (north and south of Lisbon).
2 What government body controls Portuguese wines other than Port?
3 What is the Portuguese equivalent of the Spanish Denominación de Origen: who awards it?
4 Which is the controlling authority for production of the wines of (a) Bairrada; (b) Dão?
5 What is the main grape variety used in the Bairrada district?

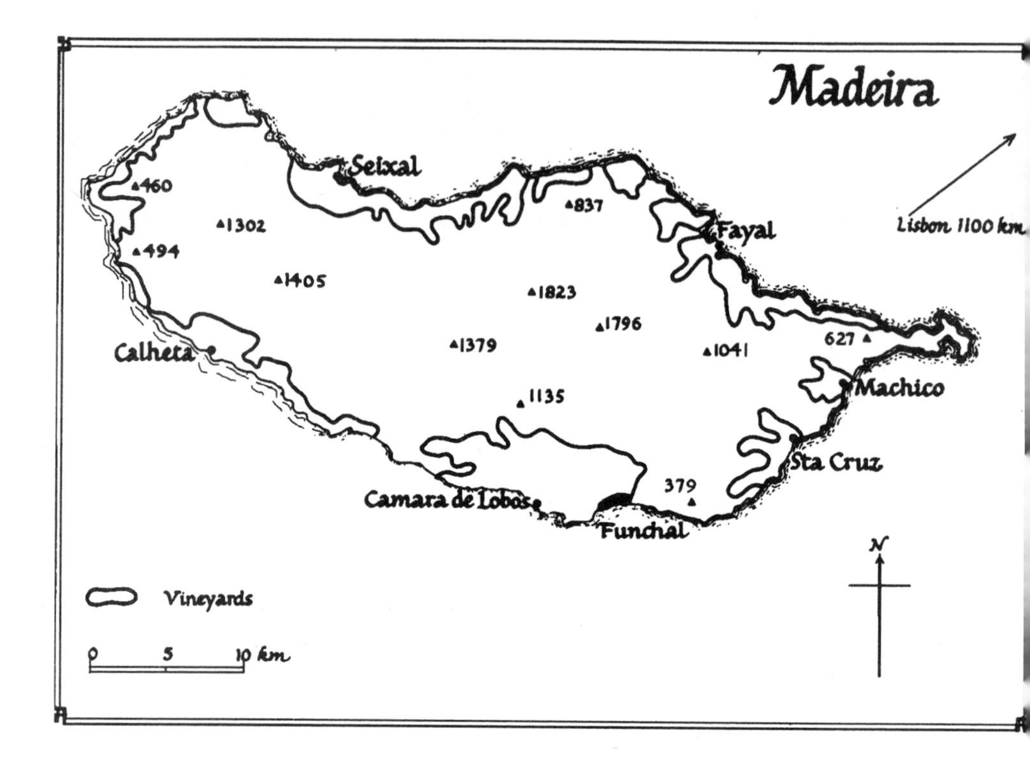

Madeira, rather as a preservative than for taste. But with a sugar-cane crop on the island, cheaper cane spirit became the accepted fortification until 1969. From then on, in view of Portugal's application to join the EEC, wine brandy supplied from the Instituto on the mainland has been and must be used.

As the island of Madeira was renowned among merchantmen as a source of fresh water and food, ships bound for the Far East often called at Funchal to top up their water barrels before venturing into the tropic doldrums where they might lie becalmed for weeks. As ballast to steady their empty ships in the South Atlantic gales, they shipped barrels of Madeira wine for sale in Australia or the East Indies – a much more profitable measure than the stones usually used for ballast. As the ships journeyed through the tropics, the wine was heated to a maximum temperature of about 45°C and cooled again later in the six-month voyage. It was discovered that this

1 (a) Granite on the hills and schist in the valleys. (b) Clay and limestone, sometimes overlaid with sand.
2 Junta Naçional do Vinho in Lisbon.
3 Denominação do Origem: awarded by regional authorities (Gremios, Comissãos, or Federaçãos).
4 (a) Junta Naçional do Vinho. (b) Federação dos Vinicultores do Dão.
5 [illegible]

natural cooking process (*estufagem*) gave a unique character to the wine.

Coincident with the Methuen Treaty of 1703, English wine-merchants had arrived on the island and built up an export trade with Europe and America which survived two disasters. The first was the American War of Independence during which the British attempted with only moderate success to restrict wine-shipments to British ships; the second was the onset of the plagues of *oïdium tuckerii* in 1852, followed by *phylloxera* twenty years later. The vines were all killed, and the British shippers who remained on the island went about the task of restoring the wine industry. After a false start with American vines, the grafting of European scions on American stocks was successfully adopted.

Viticulture

The vines in Madeira can only be intensively cultivated on the lower slopes of this precipitous island, and have to compete for the little space available with sugar cane, banana trees, and ordinary vegetables. For this reason the vine is grown on trellises up to a little over 2 m in height, with vegetables underneath, on terraces which are grouped in most irregular patterns to suit the irrigation system. Although heavy on the mountain tops, the rainfall tends to sink into the porous volcanic soil before reaching the lower slopes, and a network of aqueducts covers the island, the water being managed by a committee. Each vineyard owner will be allotted so many hours of water at a given time each week or fortnight, and must be ready to receive it when it is deflected to his vineyard – at whatever hour of the day or night.

Vines

The best-known vines of Madeira, which in England have given their names to styles of wine are: Sercial; Verdelho; Bual or Boal; and Malmsey; these, with Terrantez and Bastardo (now hardly planted) are noble varieties. Most Madeira is made from the Tinta Negra Mole (classed as a good variety), which has the ability to replicate the appearance and flavour of the other grapes, depending on where it is grown, and how vinified; some Moscatel is also made. The varieties

1 Describe the territory of the Dão region.
2 What vines are used to produce Dão white wine?

Isabella, Cunningham, and Jacquet are banned hybrid varieties; they are listed because they still comprise 50 per cent of the vines planted; the grapes or their juice cannot be bought by Madeira shippers, but may be made into table wine for local consumption.

Styles

Originally, the styles were Common Dry, Common Rich, London Market, London Particular; in America, the biggest market, wines were known by the name of the ship which brought them or the man who bought them. Gradually, the British adopted the names of the grapes – Sercial, Verdelho, Bual, and Malmsey in increasing order of sweetness; as most is now made from the Tinta Negra Mole, those grape-names are being replaced by brand names. 'Rainwater' is, however, a style rather than a brand name and is a light Verdelho, smooth and fairly dry, with a legendary origin. It is said that some casks for consignment to an American buyer were topped up before shipment from a butt of rainwater by mistake: the mistake was not discovered until after shipment, and the shippers waited for the reaction with some trepidation. Great was their surprise, therefore, when the buyer wrote back praising this new lighter Madeira with an even larger repeat order. The style continues!

Vinification

Some of the harvest is still crushed by bare foot in Jerez-type lagares, and the juice carried in goatskins to tankers at the nearest road; but communications have improved so that more grapes can be taken direct to central presshouses with modern equipment. After fermentation, the wines are stored in *estufas* heated to 45°C for not less than three months. Less-fine Madeiras are estufado in concrete vats heated by hot-water coils immersed in the vat; finer wines are stacked in scantling pipes outside the vats in the same building, where they may stay for up to six months. The lower temperature and longer period ensures that the wine does not acquire any 'roasted' flavour.

After *estufagem* the wines are cooled very slowly to 20°C, have their fortification adjusted to 17% vol. (finer wines are fortified before estufado), are filtered, and left to rest for one to two years.

1 A high granite plateau intersected by ravines.
2 Encrusado and Arsazio Branca Arinto.

Maturation and Styles

The wines then pass into stock *lotes*, which are vats, pipes, or collections of pipes each containing wine of different characteristics. From these lotes the shipper compiles his shipping lotes – the brand names recognized by customers. Some of the finest wines from individual noble varieties are reserved as potential vintage Madeiras, which take many years to develop, Sercial particularly; these spend not less than 20 years in cask, and it may be 30 years before some wines can be declared suitable as a vintage. Those that fail the test revert to stock or reserve lotes, used to replenish soleras.

After *oïdium* and *phylloxera* had hit Madeira, many shippers were forced to blend younger wines into their stocks of vintage Madeira to meet orders; they naturally chose wines as close to the original as possible for blending, and used the original vintage casks, thus creating soleras – which took their date from the original vintage date, possibly more than one hundred years earlier. Later, soleras fell into disrepute, so that they are now rare in Madeira, and strictly controlled by the Instituto do Vinho da Madeira, the Government body ruling the trade. This should not be confused with The Madeira Wine Company, which is a consortium of all the former English shippers.

PORT

Port is the most famous of all Portuguese wines. Both Portuguese and British law attribute an identical definition to it: 'Port is wine of the Cima Corgo and Baixo Corgo regions of the Douro valley which has been fortified by the addition of grape brandy, and matured in shippers' lodges in Vila Nova de Gaia or in the Alto Douro (Noval Champalimaud).

Not only do Portugal and Britain see Port in the same light: they have been partners in its development, and it is through their joint enterprise and skills that Port has become the finest of all the world's dessert wines. Following the Methuen Treaty of 1703 (under which English woollens were traded for Portuguese wines), a number of English wine-merchants settled in Oporto and many of their families have remained to this day. Originally the red wines of

1 (a) What is estufagem? (b) How did the process originate?
2 How are the vineyards of Madeira irrigated?
3 What is now the principal grape used in the production of Madeira?
4 What is the particular ability of the Tinta Negra Mole vine?

the Douro were coarse and heavy, unsuited to the English palate, but after the Treaty they were fortified to suit the British taste.

Soil and Climate

The wisdom of selecting the impossible conditions of the Upper Douro valley for planting vineyards has often been questioned; but the truth is that nothing else will grow in this rocky territory which, because of its barrenness, suits the vine admirably. The soil, if it can be called such, is a schistous shale in mountainous slopes of compressed clay protruding through a granite overlay. The soil, although rich in potassium, is poor in nitrogen and organic matter and lacks lime. Here the vineyards are planted and tended on man-made terraces in conditions which always seem to be extreme – summers with temperatures up to 41°C (105°F), excessive rains – 132 cm (52 in) in the first three months of the year, and bitterly cold winters. The ground can be so hard that explosives have sometimes to be used to make holes for the planting.

The Vines

Some forty varieties of vine are grown in the Douro region, which stretches downstream from the point where the river Douro enters Portugal from Spain, to the area around Régua near the confluence of the Corgo river with the Douro. The best vineyards lie near Pinhão on the Douro, about 32 km above Régua. The most widely used black grapes are the Roriz and Malvasia Preta, which give good colour and sugar and are resistant to heat; Sousão, used for the exceptional colour of its red juice; Tinta Francisca for its sweet high-quality grapes; Rufete, an early ripener; Touriga, and Mourisco Semente. Some white grapes are grown in the cooler atmosphere of the upper reaches, the most popular being Malvasia Corada, Malvasia Fina, and Rabigato. With few exceptions, these vines are grafted on *phylloxera*-resistant stocks.

1 (a) 'Cooking' the wine. (b) When sailing ships loaded with wine from Madeira were becalmed in the tropics and the wine reached a temperature of about 45°C.
2 By a network of aqueducts, controlled strictly by a special island committee.
3 Tinta Negra Mole – a cross between Pinot Noir and Grenache.
4 Its wine will take on the characteristics of wine made from other grapes, according to where it is grown.

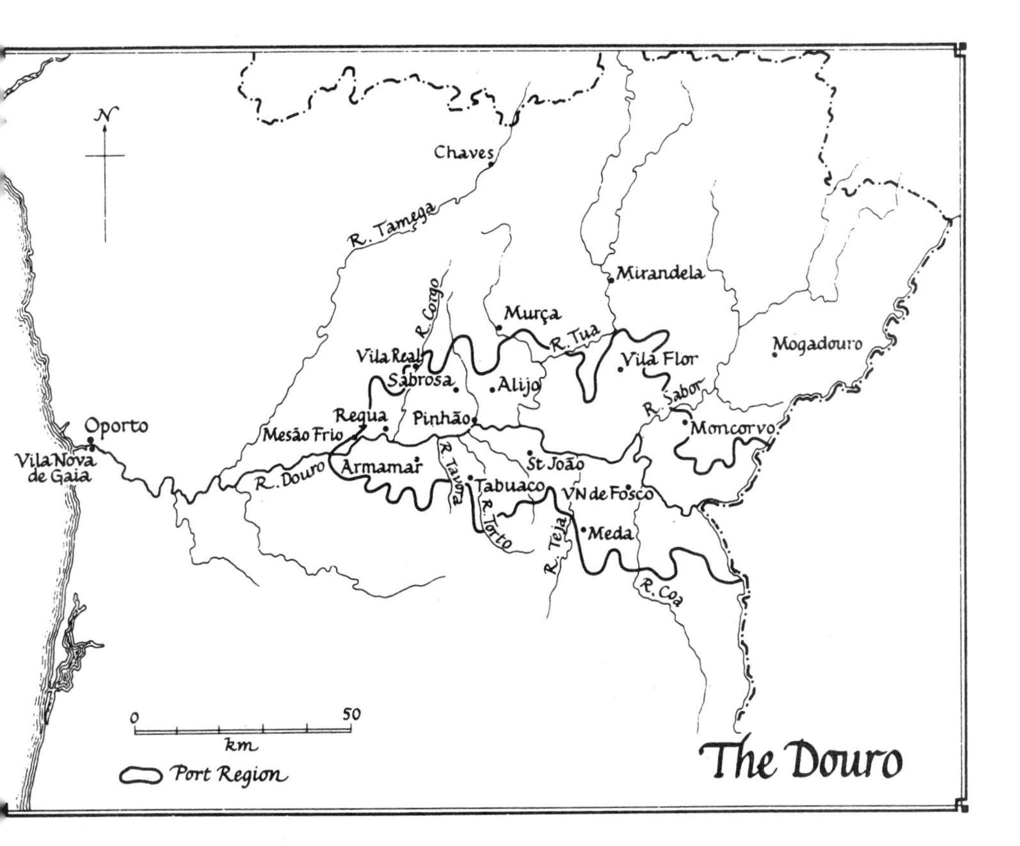

Vinification

Fermentation: At harvest the grapes are collected in tall baskets and brought in by man – no animal can negotiate the slopes – to the *quintas* (wineries). Treading of the grapes is now rare, but this process, known as 'cutting the lagar' (trough), is still practised in a few quintas, and receives a premium. This exhausting process is carried out to the strains of rustic music by 'chains' of treaders who work in four-hour shifts to start the fermentation. When fermentation has produced the required colour from the grapeskins the sweet must is run off into *toneles* (vats) which have previously been filled to one-fifth of their capacity with brandy. The alcoholic strength of this mixture, being over 17% vol., inhibits further fermentation, and so

1 Name the four styles of Madeira wine, mentioning the two dry wines first.
2 Which Madeiras are generally fortified before estufagem, and which after?
3 When are Vintage Madeiras declared?
4 Why are Madeiras of the early nineteenth century still in perfect condition?

the unfermented sugar remains in the Port, leaving it sweet. This feature of production contrasts with the Sherry method, where fortifying spirit is not added until *all* the sugar has been fermented, thus producing a completely dry wine. The modern alternative method of fermenting Port is called 'autovinification', a process which permits hygiene and temperature control. By this process, carried out either in a circulatory or siphon vat, the grapes are crushed and a proportion of the stalks removed. The circulatory vat (see p. 30), made of stainless steel, has a central column with an archimedean screw, which rotates and carries the *manta* (grape skins and stalks) down to the bottom of the vat continuously. The siphon vat, as the name implies, uses the pressure of CO_2 gas produced by fermentation to force the must into an upper open chamber which, when full discharges the ferment back with considerable force onto the manta, beating the colour out of it. This cycle continues noisily until the required colour is obtained.

Fortification: After autovinification, the must is drawn off into a separate vat containing sufficient grape spirit (aguardente) of 77% vol. to halt the fermentation. This brandy must (by law) come from the Casa do Douro (vinegrowers' co-operative). Early in the new year, the crude Port (by whatever process made) is run into 'pipes' of 522 litres capacity, and is taken down by road or rail to mature in the shippers' Lodges at Vila Nova de Gaia, on the opposite bank to Oporto, at the mouth of the Douro. The old *barco rabelo*, a single-sailed boat which was used for transporting the pipes before roads and railways were built, cannot be used now because of the numerous hydroelectric dams erected along the length of the river. It was always a hazardous journey, because of the rapids.

Maturation

The maturation of Port can be a long process; individual treatments and blending methods determine the varieties and ages of Ports on the market. New rules introduced in 1975 decreed that all dated Ports must be Portuguese bottled. Of course, large quantities of dated Ports of earlier years, bottled in UK, remain on the market; and a vast quantity of undated Ports – ruby, tawny, and white – are still shipped in bulk and bottled in London or Bristol. There are four

1 Sercial, Verdelho, Bual and Malmsey.
2 Fine wines are fortified before estufagem, less-fine wines after.
3 When they are shown to have developed vintage characteristics in cask, which may not be for 30 years.
4 Because estufagem stabilizes the wine.

categories of dated Port, of which two are named 'vintage' and three are Ports of a single year. There are four Ports blended from several years, of which one must be bottled in Portugal, while the other three may be shipped in bulk. The rules regulate the legal age at which each category shall be bottled, and the description which must appear on the label. Wines of different types in the Lodges are divided into *lotes*, and are blended using two measures – the *almude* of 25 litres and the *canada* of 2 litres: blends are made up in quantities of 537 litres with 21 almudes and 6 canadas.

Vintage Ports are wines of one exceptional year, matured in cask for two years and then left to mature undisturbed in bottle. These must be decanted off their heavy sediment before drinking and can be drunk enjoyably at any time after 10 years and up to 35 years or more. When laid down, vintage ports have a whitewash mark on the top-side of the bottle, so that they may be handled properly until decanted. Late-Bottled Vintage Ports are bottled 4–6 years after the harvest; Ports with 'Date of Harvest' are declared after the third, and bottled after the eighth year, following the harvest.

'Wood Ports' include: Ruby which usually matures for four years in wood before bottling for immediate drinking; old ruby, which matures for seven years; white, made from white grapes, but otherwise similar to ruby; Tawny, named from the hue it develops during eight to ten years' maturation in wood; young (and cheap) tawny, a blend of ruby and white ports; and Oporto-bottled Port of Indicated Age – fine old tawny – which may be sold with ages indicated at '10', '20', '30', or 'over 40' years only. Whatever the descriptions of Port marketed by a shipper, he may not ship more than one-third of his stock in any one year. Vintage, Late Bottled Vintage, and ruby Ports are usually served at the end of a meal and should be at room temperature. White Port (especially Dry White Port), taken as an aperitif, is much improved when served slightly chilled, as also are fine tawny ports, whether with indicated age or not.

Authorization

Port is a protected name for fortified wine from the Alto Douro. The officially demarcated area is divided into the Baixo Corgo, where

1 Name the organization which promotes and markets Madeira.
2 What is the soil of the Upper Douro Valley?

two to three pipes per 1000 vines are allowed, and the Cima Corgo where only one pipe per 1000 vines is permitted. Vineyards are divided into classes, and depending on their classification may have varying quantities of must allowed per vine. The elements governing authorization of vineyards, and their classification, are shown in Table 5.

Table 5
Elements governing authorization of Corgo vineyards

(a) *Elements*	*Points Awarded* Minima	Fixed	Maxima
Productivity	−900	–	+120 (for lowest production)
Altitude	−900	–	+150 (for lowest altitude)
Geographical position	− 50	–	+600
Soil Schist	–	+100	–
Granite	–	−350	–
Mixture	–	−150	–
Upkeep	−500	–	+100
Grape qualities	−300	–	+150
Gradient	−100	–	+100 (for steepest gradient)
Position in relation to climate conditions	0	–	+ 60 (for best shelter)
Age of vines	0	–	+ 60 (for oldest)
Distance between root and root	− 50	–	+ 50 (for greatest distance)

(b) *Classes of Vineyards*	*Total Points Awarded*	*Permissible Usage for Port in Litres per 1000 Vines*
Class A:	1200 or more	600
Class B:	1001–1199	600
Class C:	901–1000	590
	801–900	590
Class D:	601–800	580
Class E:	501–600	580
	401–500	580
Class F:	below 400	260

Note: Points per class are subject to annual review.

1 The Madeira Wine Company.
2 Schistous shale of compressed clay overlaying granite.

Although 170,000 pipes of wine can be made in the region each year, only some 50–60,000 pipes are authorized for Port. Less than 10 per cent of Port production is consumed in Portugal.

The controlling bodies are the governmental Ministerio de Economia and the Instituto do Vinho do Porto; the Casa do Douro (farmer's union) which was formed in 1932 to regulate the trade; and a shippers' association, the Gremio dos Exportadores do Vinho do Porto.

MOSCATEL DE SETÚBAL

A blend of black and white muscat vines is used to produce Moscatel de Setúbal in this region, 30 km to the south-east of Lisbon; fermentation is arrested by adding brandy to achieve a strength of about 16% to 20% vol. alcohol. Maturation is a long process in which the wine darkens from amber to brown as it intensifies in muscat flavour and bouquet; the grape-skins are steeped in the strengthened wine to develop the scent. Setúbal Superior sells at twenty years old.

VINHOS VERDES

This name is translated as 'green wines', but does not refer to the colour of the wines (80 per cent are red) but to the fact that they are made from underripe or 'green' grapes. The region lies between the river Minho, which forms the northern border with Spain, and the river Douro. Here the grapes are grown 2.5 m above the ground on trellises to provide cover for food crops beneath; the grapes are denied reflected heat from the soil, yet are protected from direct sun's rays by foliage above. They contain less sugar and more malic acid than normal at harvest time, and for this reason sustain a heavy malolactic fermentation after the alcoholic fermentation has taken place, and after bottling. The resulting production of carbon dioxide in bottle gives the vinhos verdes a light sparkle which enhances the

1 What is the manta?
2 (a) How many categories of 'dated' Port are there? (b) How many are Ports of a single year?
3 (a) How long are vintage Ports matured in cask? (b) How soon are they ready for drinking?

natural delicacy to be expected in a wine relatively low in alcohol (just under 10% vol.). The vines of the region are the Azal Tinto, Borracal, Espadeiro, and Vinhão, all black grapes, from which the red wines are fermented on the skins complete with pips and stalks: the Vinhão predominates. The white grapes are the Loureiro, Alvarinho, Azal Branco, and Douradinha. The harvesting starts at the end of September with the white grapes.

The region has six districts: Monção in the north; Lima, Braga; Penafiel; Basto, and Amarante. Granite prevails throughout and gives a flinty astringency to the wines. In the hillier districts, terracing is used; here mechanization is impracticable, but animal manure provides a compensating bonus. The red wines are as pétillant as the white, and have been described as 'looking like foaming Ribena, and tasting like scrumpy cider'. Both red and white are very good complements to the local fatty foods – such as grilled herring-sized *sardinhas* – while the pale greeny-gold white, which is not quite as acid as the red, makes an excellent aperitif. Vinhos verdes, red or white, are best served slightly chilled, but the red is really only acceptable in very hot weather, like that of a Portuguese summer.

DOURO

There are four districts: Mijo; Lamego; Sabrosa; and Vila Real. The Douro region is famed for Port, yet only a small proportion of the grapes grown there may be made into Port. Two-thirds of production in the Douro Valley is for light wine, and together with grapes grown in the Tras os Montes ('across the mountains') Vinho Região (VR) in the north-east of the country are made into wines for the table. These may be red, white, or undemarcated rosé, of which the rosé is the most popular in England, being made slightly sweet and semi-sparkling.

UNDEMARCATED VINEYARDS

Besides the Regiäos Demarcadas, whose wines have the right to carry a *Selo de Origem*, there are many others as yet awaiting this distinction.

1 Grape skins and stalks which have to be immersed in the fermenting must.
2 (a) Four. (b) Three.
3 (a) Two years. (b) From 10 years up to 35 years or more.

Beira Alta: Lafões and Pinhel lie north of Dão, and make similar wines to those of that region.

Estramadura VR: North of Lisbon, white wines are made in Torres Vedras, and red in Cartaxo.

Palmela: To the south of the Tejo, between Lisbon and Setúbal, the limestone Sierra Arràbida stands like a wall between the sea and the vineyards of Azeitão and Palmela, Ribatejo VR. The topsoil being of sand, vines can be grown ungrafted. The district, hitherto undemarcated except for fortified wines, produces much strong rosé wine, and red Periquita, which has strength and bouquet. Periquita is now eligible to receive a Selo de Garantia from the controlling body of the Moscatel de Setúbal region; the origin of its grape name is unusual and may be unique, for it took it from the brand name of the wine!

It is worthy of note that Portugal's *vinhos rosados*, or rosé wines, now have Denominação de Origem. They account for some 45 per cent of Portugal's wine export. The most popular varieties are pétillant, being made so by carbonation, and not by secondary fermentation, as in the case of the vinhos verdes. The grapes used for the rosés are mainly the Alvarelhão, Bastardo, Mourisco, and Touriga. The rosé wine industry is highly mechanized and conducted on a vast scale, the plants and their storage tanks being reminiscent more of oil-refineries than wineries when viewed from the air.

Language of the Label

Portuguese wines are described as *Tinto* (red), *Clarete* (light red), *Branco* (white), and *Rosado* (rosé); they are *Seco* (dry) or *Doce* or *Adamado* (sweet) and *Espumante* (sparkling). *Vinho de mesa* is 'table wine' and *vinho de consumo*, ordinary; *Reserva* is better quality. *Maduro* is old or matured, *Engarrafado na origem* means estate bottled, and *Garrafeira* means private cellar or top quality with additional ageing, and may be an inter-regional blend providing no RD claim is made. *Vinha* is a vineyard, *Quinta* a farm estate, and *Colheita* the vintage. *Região demarcado* is a demarcated area, and *Denominação do Origem* the coveted name of origin endorsed by state seal.

1 How many pipes per 1000 vines may be made into Port in (a) Baixa Corgo and (b) Cima Corgo? (c) What proportion of total output does this represent?
2 Name the two Portuguese government bodies controlling Port production.
3 Name the fortified wine made 30 km south-east of Lisbon. What is its strength?
4 What are vinhos verdes and why are they so-called? What colour is the wine?

8

Other Community Wines

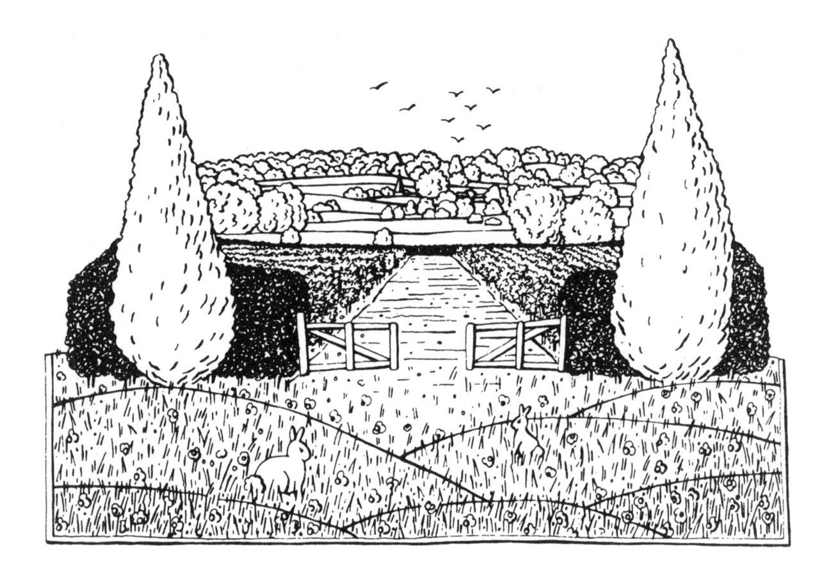

Since the first edition of this book appeared in 1979, the three great wine-making countries of the EEC – France, Italy, and Germany – have been joined by three others – Spain, Portugal, and Greece. Among such great company it is easy to forget that two of the remaining six countries – Luxembourg and the United Kingdom – also produce wine commercially, albeit only in small quantities.

1 (a) Two to three. (b) One. (c) About one-third.
2 Ministerio de Economia and the Institut do Vinho do Porto.
3 Moscatel de Setúbal: about 16% to 20% vol. alcohol.
4 'Green' wines made from under-ripe green grapes 80% red, 20% white and attach importance to soil and vine variety.

GREECE

History

In ancient times, Greece exported wine to all the known world; she also exported the knowledge of vines, their cultivation, and the production of wine from their fruit. Much of the language of wine is of Greek origin – oenology itself. Over the years, her influence waned, and her fame was taken up by others – her Monemvasia became Malmsey.

Like many of the Balkan states, Greece suffered Muslim Turkish occupation for five centuries which, although it curbed wine-making, kept the grapes growing for the table. With independence, wine-making restarted with the local grapes, and many native wines with individual characteristics survive; however, modern methods and varieties have been adopted, and the wine industry in modern Greece is flourishing.

Soils

Wine is produced throughout the country and in most of the islands, on soils which are mainly limestone on the mainland, and volcanic in the islands. One island, Thira in the Cyclades, previously known as Santorin, literally exploded in ancient times; the clouds of volcanic dust (*loess*) – richly fertile soil for vines – may have been the basis for the Tuniberg vineyards of Baden and enriched many other western vineyards. This volcano is still active, and last erupted in 1925.

Regions

Under Common Market regulations, twenty-six 'Appellations' have so far been delineated: but the details are still in a fluid state.

1 Which Portuguese wines have the right to bear the *Selo do Origem*?
2 (a) What proportion of the wines of Douro and Tras os Montes is unfortified? (b) Which of these is the most popular in England?
3 Name the Estramadura districts in which (a) red wines, and (b) white wines are made.
4 Where are the vineyards of Azeitão and Palmela situated, and what protects them?
5 What do the following Portuguese label details mean? (a) clarete; (b) adamado; (c) Vinho de mesa; (d) maduro; (e) Vinha; (f) colheita

Northern Vineyards

The best wines come from Macedonia in the north where it is cooler, particularly from the area around Naoussa in the highlands adjoining the Yugoslavian border and in the Halkidiki peninsulas. The wines are mostly red, from the Xynomavro and Limnio grapes, with some Cabernet Sauvignon making its appearance in the vineyards round Porto Carras in the Halkidiki. The most modern producers are to be found here; their wines have deep colour and good acid to balance their high alcohol content.

In the north-west of Greece on the Albanian border, the region of Epirus produces light white Zitsa from the local Debina grape.

Southern Vineyards

The Peleponnese region produces over one-third of Greece's annual output of 5 to 5.5 million hl, and specializes in heavy full-blooded reds from the Agiorgitiko grape grown in the Nemean valley and from the Mavrodaphne grape grown round Patras and in the south. This region is also responsible for much of the large production of branded wines such as 'Demestica' and 'St Helena', eagerly received by tourists who cannot appreciate retsina or the equally resinated rosé *Kokinelli*.

Island Vineyards

Of the islands, Crete produces mostly red wines in quantity, as do Corfu and Levkas in the Ionian Sea. Of the other large islands, Rhodos (Rhodes) close to mainland Turkey produces white wines, of which the best is Lindos. Travellers may note that Rhodos is a free port without Customs duties. Samos produces a sweet wine from the Muscat grape – often fortified – with world renown. Smaller islands of the Cyclades also produce sweet wines, although Santorini from volcanic Thira produces a notable dry white from the Assyrtico grape. Lesbos is unproductive.

1 Those from the Regiäos Demarcadas.
2 (a) Two-thirds. (b) Light rosé wines sold under brand names.
3 (a) Cartaxo. (b) Torres Vedras.
4 Between Lisbon and Setubal, protected from the sea by the limestone Sierra Arrábida.
5 (a) Light red. (b) Sweet. (c) Table wine. (d) Old or matured. (e) Vineyard. (f) Vintage.

Retsina

The forementioned wines are unresinated, possibly because they are red, and preserved by their tannin. But more than half of the white and rosé wines of Greece are treated with resin from Aleppo in Syria, the practice having originated to protect the wines from oxidation. The Greeks through long tolerance grew to like the taste – did it persuade the Muslims that it was *not* wine? – and continue the practice. Most of the retsina comes from Attica, the region to the north and west of Athens, and from the Peloponnese peninsula to the south, from a rather dull grape called Saviatano.

LUXEMBOURG

The wine industry of the Duchy of Luxembourg is concentrated on the river Moselle (the eastern frontier), where for a distance of 64 km the vineyards alternate with cherry and plum orchards along the slopes above the river. Luxembourg was part of France in the late eighteenth century and following the Revolution the vineyards were divided into small parcels, just as they were in Burgundy. Consequently there are to this day some 1600 vineyards in the 1200-ha region, most of less than half a hectare each. Indeed, there are only about six vineyards of over 3 ha.

The wines of Luxembourg closely resemble those of the upper Mosel-Saar-Ruwer region of Germany which lies across the river frontier. The vineyards are protected from the prevalent westerly winds by the Ardennes. The soils are generally suitable for vine growing; the best vineyards lie on the chalk and marl soil of Remich in the centre and at Grevenmacher at the north of the 64-km-long region. The recommended and authorized grape varieties of Luxembourg include Rivaner (Müller-Thurgau), Elbling, and Riesling principally, with lesser use of Auxerrois and Ruländer, Silvaner, and Pinot Blanc.

The wines of Luxembourg are all white, the still wines having the light and fruity character associated with Alsace. There is a signifi-

1 How many of the EEC countries produce wine commercially? Name them.
2 Describe the soils of Greece (a) on the mainland and (b) in the islands.

cant production of excellent sparkling wine in Remich, and a pétillant wine called 'Perlwein' is bottled with crown corks. The population of 300,000 – augmented by Common Market officials and tourists – manages to drink most of the country's production (averaging 150,000 hl per annum). There is, however, a well-established export to Belgium, with some going also to Holland and Germany.

Luxembourg has a 'Marque Nationale' for quality wines, established in 1935, and certified by a small neck label on each bottle. Luxembourg qwpsr may be classified as Vin Classé, Premier Cru, or Grand Premier Cru. Specific table wines may be permitted by decree to indicate their locality name; as with the vins-de-pays of France, these may then be labelled with the wine variety, vineyard name, vintage year, and an indication of bottling on the premises.

Cherries and plums from the Moselle valley are converted to fruit spirits; the cherries to Kirsch, and the plums to Quetsch and Mirabelle; all are distilled in the Duchy.

ENGLAND AND WALES

Climate and Soils

Although the British Isles lie just north of the fiftieth parallel and might hence be considered to be outside the wine-producing zone, the warm Gulf Stream moderates the climate so that the vines may be grown in suitable sheltered spaces. As the English vine-grower thus starts with the disadvantage of insufficient warmth (the average annual heat sum is only 850 degree days), and in addition suffers from an unsettled weather pattern with relatively high rainfall, a good harvest cannot be relied on in more than three years out of five. Even in good years, musts need enrichment with sugar. The English or Welsh grower produces white wine, seldom red: but, with careful viticulture and vinification, his wines can equal German or northern French wines in at least one year out of five. At their best, they have a

1 Eight: France, Germany, Greece, Italy, Luxembourg, Portugal, Spain, and the United Kingdom.
2 (a) Limestone. (b) Volcanic.

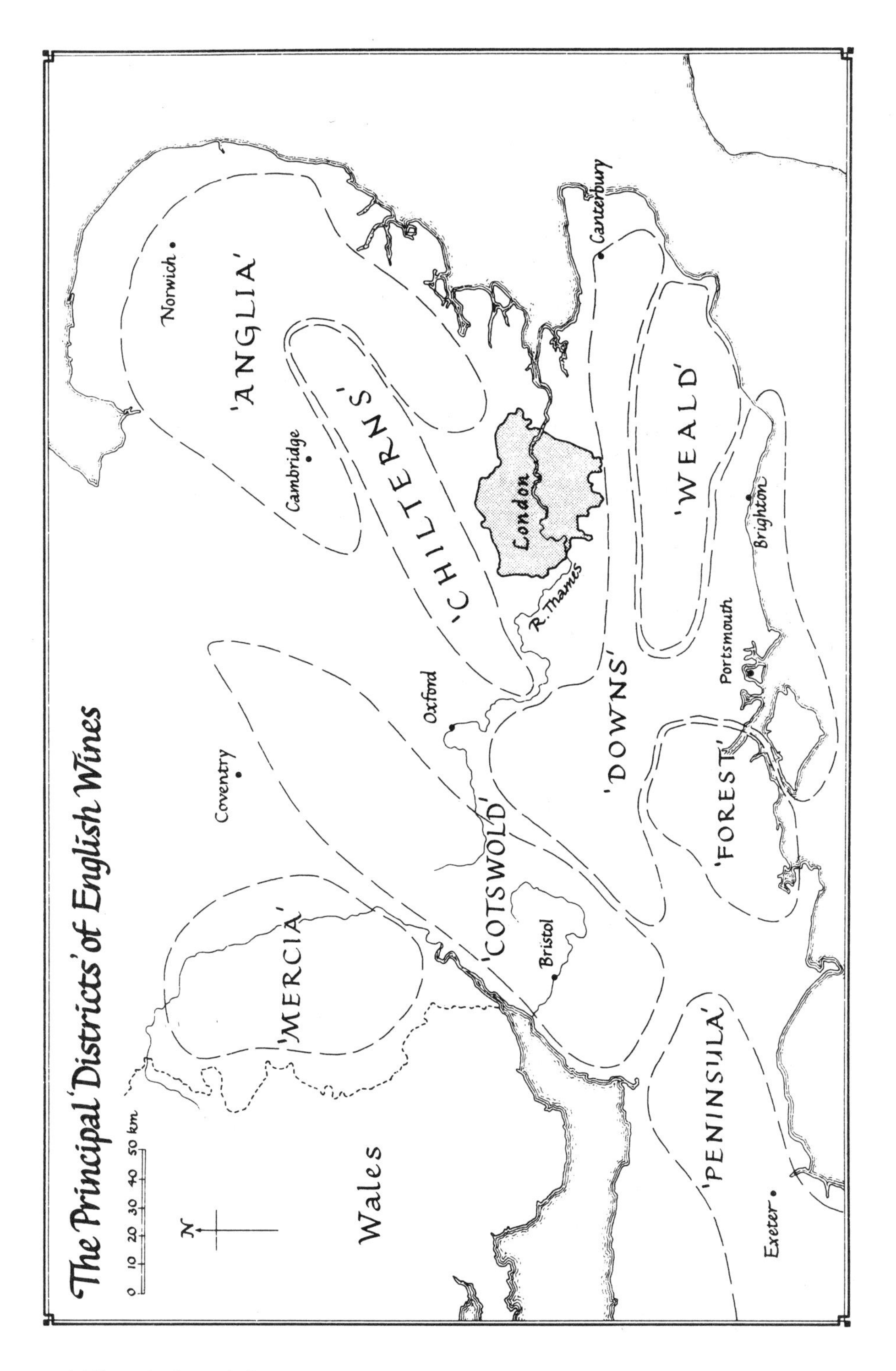

1 Where do Greece's finest wines come from?
2 In what type of wine does the Peleponnese specialize?
3 Describe Kokkineli.
4 Name the main Greek wine-producing islands.
5 (a) Where are the vineyards of Luxembourg centred? (b) How many are there? (c) Name the best grapes. (d) What colour are the wines?

powerful flowery bouquet. The very considerable increase in wine consumption and wine interest since 1947 has encouraged the revival of the English wine-producing industry, and the English Vineyards Association, which represents its joint interests, already has a membership of over 300 commercial vineyard proprietors. This compares with about 130 vineyards known to have existed in the medieval era, largely developed by the Church and dating from the Roman occupation. Henry VIII withered wine production in England (there was none in Wales) by dissolving the monasteries.

The United Kingdom contains all the most suitable soils for viticulture: granite, limestone, gravel, chalk, and clay. However the vagaries of the climate make it almost impossible for growers to make their living from the vine alone; many have other sources of income, and treat wine-making as a hobby; others, realizing that tourism is a prime British industry and that vineyards are tourist attractions, have tours, tastings, and shops to help balance their books.

Mercifully, the EEC does not yet recognize English and Welsh vineyards as other than 'experimental', whereby the grower is spared hours of paperwork and continual visits from Ministry Inspectors. Not surprisingly, the average (median) size of vineyard in the UK is larger than that of other Community countries, for the man who grows a vine or two on his allotment and adds his grapes to his neighbour's does not have to register his vines as he would have to do in France or Germany.

There are further benefits in being classed 'experimental', as grape varieties and viticultural or vinification practices as yet unapproved by the EEC can be tried out in the UK 'laboratory'. Nevertheless, the growers are highly professional in viticulture and vinification.

Grape Varieties

In 1974, an official EEC list of recommended and authorized vines for cultivation in the UK was published (Reg. 925/74). Vines omitted from the list were not banned, but are to be considered experimental. Recommended varieties are Müller-Thurgau, Wrotham Pinot, and Auxerrois. Authorized varieties that are being

1 Cool Macedonia in the north, adjoining the Yugoslavian border and in the Halkidiki peninsulas.
2 Heavy, full-bodied red wine from Agiorgitiko and Mavrodaphne grapes.
3 A resinated rosé wine of Greece.
4 Corfu, Crete, Levkas, Rhodos (Rhodes), Samos, Thira (Santorini).
5 (a) Along the River Moselle. (b) 1600. (c) Müller-Thurgau, Elbling and Riesling. (d) All

widely cultivated are Bacchus, Chardonnay, Huxelrebe, Kanzler, Madeleine Angevine, Ortega, and Seyve-Villard 5/276 (Seyval Blanc).

New additions which have proved successful are Reichensteiner [Müller-Thurgau x (Madeleine Angevine x Sangiovese)] and Schönburger [Pinot Noir x (Chasselas Rose)], and many other experiments are taking place in close co-operation with oenological centres at Geisenheim and elsewhere.

Viticulture

To avoid damage by grey rot in the cold damp climate, the English grower plants resistant varieties in well-spaced rows, and trains them on an open style such as Guyot, Lenz Moser, or Geneva Double Curtain. Low-trained Guyot was originally the most popular training method, but spring frosts made a higher form desirable; the preferred style is now the Geneva Double Curtain which has the advantages of height to make spraying and picking easier, and a good spread to make maximum use of the available sunshine (see diagram on p. 16).

Although so far north, English vineyards escape the severity of continental winters because of the moderating influence of the sea: nevertheless, the young emergent shoots are vulnerable to air-frost in April and May. The grapes are ready for picking between the last week of September and the end of October. Gathering them in has been an occasion for family and friendly help, but a few of the larger vineyards are now paying agricultural wages for the work.

Vinification

Apart from small 'vineyards' in private gardens, there are over 300 vineyards of average size 2.4 ha; most rely upon co-operatives for vinification. Inevitable delays in transporting uncrushed grapes to a distant winery encourage oxidation, with consequent loss of flavour in the wine, although the methods used by the co-operatives (based on German practice) reduce ill-effects to a minimum.

1 What is Perlwein?
2 What is the 'Marque Nationale'?
3 How can England grow grapes for wine although north of the 50° latitude?
4 How often can the English winemaker rely on a good harvest?

British Wines

This subject is introduced here because the terms 'English Wines', and 'British Wines' have created some confusion. The former are true wines, being made from freshly-gathered grapes in their country of origin, while the latter are 'made-wines' manufactured from grape-must (concentrated or not) from other countries, or from imported dried grapes which have been 'reconstituted' by soaking in water. It should be noted, however, that English Wines may be sweetened by the addition of unfermented must ('sweet-wine') imported from another EEC country without prejudicing their status.

Made-Wines: These were formerly known as 'British Sweets' and, if they did not exceed 10% vol. alcohol as most did not, could be sold in grocers' and chemists' shops without a licence. Many are sophisticated with aromatics (which have to be enumerated on the label) and called 'tonic wines'. One of the most popular is ginger wine – a made-wine base of soaked currants flavoured with ground ginger root – a good accompaniment to melon in the summer, and to whisky – 'Whisky-Mac' – in the winter.

Fruit Wines: These are also made commercially in Britain as they are in New England in USA. Apple Wine, distinguished from cider by its higher strength (over 8.5% vol.), Blackcurrant, Redcurrant, Blackberry, Cherry, Gooseberry, Parsnip, Wheat-and-Raisin, all containing enough sugar for fermentation, are made commercially, besides being made by thousands of housewives for their families. But the last word must surely lie with the Birch Wine made and marketed by Mrs Fraser at Moniacke Castle, Inverness: the sweet sap of the silver birch tree, fermented, yields a dry white wine with an elusive vegetal bouquet which well complements the local salmon – it is also reputed (though not advertised) to prevent baldness! She also makes, as 'Scottish Wines' a light red from imported grape-must, and a very acceptable white from must 'imported from England'. The lawyers may argue about whether the latter is 'Wine' or 'Made-Wine'.

1 A slightly sparkling white wine often bottled with crown corks.
2 Luxembourg's classification for quality wines, established in 1935, and recognized by EEC.
3 The Gulf Stream moderates the climate.
4 In not more than three out of five years.

British Wines are not nowadays all sweet; Moniacke is but an example. For the modern tendency is for these wines to be prepared for the table, light and dry, with their lower alcoholic content bringing them tax advantages which should place them within the reach of all.

1 Approximately how many commercial vineyards were there in England (a) in the medieval era, and (b) in 1986?
2 Why should English vineyards prefer to remain 'experimental'?
3 Name three EEC-recommended vine varieties for cultivation in England.
4 What is becoming the most popular training style in England, and why?
5 What is the average size of English commercial vineyards?

9
Aromatized Wines

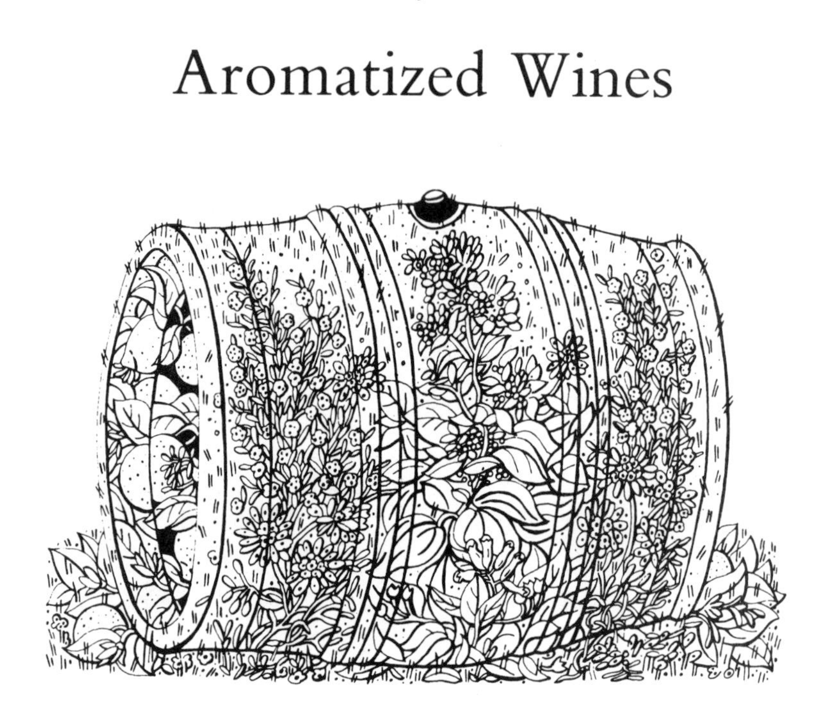

Wines have been aromatized or flavoured with spices, herbs, and plants for a very long time. In the fifth century BC, Hippocrates, 'the father of medicine', was prescribing wine flavoured with cinnamon and sweetened with honey for his patients; this was, appropriately, known as Hippocras, and the name persists to this day for the flavoured mead of Cornwall.

The ancient Greeks flavoured their wines with the resin of pine-trees, though this may have been unintentional, for they sealed the seams of their goatskin containers and the stoppers of their amphorae with this resin. Doubtless it counteracted the flavour of goat, but it remained popular and to this day much Greek wine is resinated – retsina. The medicinal properties of herbs and spices were

1 (a) About 130. (b) Over 300.
2 If they were not, they and *all* growers of vines in the country would become overwhelmed with EEC paperwork.
3 Müller-Thurgau, Wrotham Pinot, and Auxerrois.
4 Geneva Double Curtain, because it is high enough to avoid frost, makes work easier, and makes best use of sunshine.
5 2.4 ha.

exploited by medieval monks and alchemists in the preparation of liqueurs based on spirits; but these were also in far greater use as preservatives for food, and it was probably this property which ensured their continued use for flavouring wine; however, aromatized wines were consumed for their pleasant taste as well as for their medicinal functions.

Flavouring: The herbs used vary considerably: few wines are aromatized with only one, though one may predominate; Suze is such an example, being redolent of gentian, as St Raphäel is of quinine. Centaury, marjoram, hyssop, tea, coriander seed, orange and lemon peels, orris root, and camomile figure among the plants, while nutmeg, cloves, and vanilla join cinnamon in the spices. One other herb deserves special mention, for it gives its name to all the aromatized wines that include it – the flowers of wormwood, *artemisia absinthium*, which gives a delicate and unmistakable aroma. The Latin name conjures up the foul spirit Absinthe, now banned in most countries; but this was flavoured with a distillate of the leaves and stalks of the plant, whereas Vermouth (a corruption of the German *Wermut* tr. Wormwood) is flavoured with an infusion of the flowers only.

VERMOUTHS

To most people, aromatized wine means Vermouth, and they know two main types – French and Italian. Small matter that the dry, light-coloured Vermouth is also made in Italy: or that the sweet reddish-brown Vermouth is also made in France – they will always be referred to respectively as 'French' and 'Italian'! Recently a third type known as 'bianco' or 'blonde' has emerged, and is white, sweet, and flavoured predominantly with vanilla and cinnamon.

The classic methods of making vermouths differ between France and Italy; to be more specific, between the Languedoc region of France and the remainder of France and Italy.

1 Distinguish between English wine and British wine.
2 (a) What were 'British Sweets'? (b) What are they now called? (c) Are they now all sweet?
3 What are 'Fruit Wines'?
4 Give some examples of British Fruit Wines.

'French'

In the Languedoc region the wines — white wine from the Picpoul and Bourret grapes (which produce a very dry wine) and from the Clairette grape (giving a more full-bodied wine) – are fortified with brandy from their natural strength of 10–11% vol. up to 15%, and are put out in the open in casks to 'weather'. They are left out in the blazing sun of two summers and the cold winds of two winters, and during this time develop a slight bitterness, becoming *maderisé* – like the wine of Madeira. Following weathering, the wines from the various barrels are blended, and this blending is the process which determines the quality of the vermouth. Some of the wine is then fortified up to 50% vol., the better to extract the flavour of the herbs; the remainder of the bone-dry, bitter, wine is blended with mistelle, to give it body and some sweetness. Mistelle is juice from the Muscatel grape that has been prevented from fermenting by addition of brandy to bring its alcoholic strength up to 15% vol.

In Languedoc, the flavour of the herbs is generally extracted by maceration, that is to say, the herbs are left to soak in the strong (50% vol.) portion of the wine for two to three weeks in large vats. After this period, the wine is run off into casks, and the concentrated pressings from the sodden herbs are equally distributed among the casks. As a quicker alternative, the wine may be infused with the herbs, by continually passing it through a filter of them in rotating steel drums. The sweeter, weaker base wine is then blended with the stronger, highly aromatized wine to produce the final vermouth of about 18% vol. and left for a further period to marry completely.

Even after marrying, the vermouth is not ready for the market: it is cloudy, requiring fining with bentonite. It contains too much tartrate, which will make the wine cloudy in cold conditions, and this can only be prevented by chilling the wine to -8°C for a period to precipitate the tartrates. And, despite the weathering, it may still be unstable, requiring pasteurization before bottling.

Vermouths are made in other regions of France, those of Chambéry in Savoie on the upper Rhône valley being the most important. The wines for these are not, as for the vermouths of Languedoc, exposed to weathering, as this would spoil the delicate nature of the wines. One of the Chambéry vermouths is also flavoured with Alpine strawberries, making an unforgettable combination.

1 English wine is EEC table-wine made from fresh grapes grown in England. British wine is 'made-wine', produced from imported dried grapes or concentrated must.
2 (a) Made-wines with less than 10% alcohol, so that they could be sold without licence. (b) Made wines. (c) No.
3 Alcoholic beverages from natural vegetable products other than grapes.
4 Apple wine (over 8.5% vol.), cherry, gooseberry, parsnip, wheat-and-raisin, blackcurrant, redcurrant, and blackberry.

'Italian'

The manufacture of vermouth in Italy is centred on the city of Turin, where the industry was started in 1756 by Antonio Carpano. He was the first in the world to make vermouth commercially, and the firm continues to this day. Italian vermouths were originally made from the local Moscato d'Asti, flavoured with local mountain herbs, but the industry has grown so much that the local wine supplies are insufficient. Nowadays, the base wine – always white – comes from Puglia or Sicilia, and is of a natural strength of 15–17% vol. In Italy, the base wine is not weathered to maderize it, this process being peculiar to Languedoc. The herbs are macerated or infused in spirit, which may be hot or cold, and the aromatized spirit is then added to the wine. The colour of the classic 'Italian' vermouth comes in part from the herbs and also from the addition of caramel. The sweetness in the bianco as well as the dark 'Italian' vermouths comes from pure cane-sugar, in contrast to the French vermouths which use mistelle. In order to 'fix' the flavourings, and to ensure a stable and uniform product, the vermouth is pasteurized; it is, of course, refrigerated and filtered also to ensure its brightness at all temperatures.

Uses

Vermouths are wines, and are able to stand on their own as such: chilled, with a slice of lemon or orange, they make a perfect aperitif which will stimulate the palate and the digestive juices without dulling them with too much alcohol. To mix them with spirits, as in cocktails, diminishes their charms, although it may well enhance those of the spirits! They can also be made into refreshing long drinks by adding soda, and a particularly pleasant one is compounded by adding a very small quantity of *crème de cassis* (blackcurrant liqueur) to dry white vermouth and soda.

As with many other wines, aromatized wines and vermouth in particular are invaluable in the kitchen. Any recipe which calls for dry white wine and a *bouquet garni* may be amply satisfied by the substitution of dry vermouth, for vermouth is no more than wine that has been intimately combined with a bouquet garni. Chicken cooked in a mixture of dry vermouth and white wine is a dish for

1 For what did medieval people use aromatized wines?
2 What herb is essential to vermouth?

gourmets: and fish *fumet* with dry vermouth the focus for a perfect sauce for cod or sea-bass.

The aromatization of wines may seem complicated, but it can nevertheless be practised at home, as it is in many homes in Germany. '*Maiwein*' is made – in May, as the name suggests – by macerating the flowers of sweet woodruff (*Galium odoratum*) in wine. And very good it is, too.

1 Medicine, and for preserving foods.
2 Wormwood flowers (*Artemisia absinthium*).
(Questions continued on page 245.)

PART THREE

Wines of Third Countries

10

Central and South-Eastern Europe

The wine-lover gives less attention to the vast area of central and south-eastern Europe than is warranted. For here eight countries in latitude 40°–50°N have been making wine for more than a thousand years. Between them they produce 33 million hl of wine each year. Of the countries which make up this part of the world (known variously – at various times – as the Holy Roman Empire, the Austro-Hungarian Empire, and the Balkans) all but Switzerland and Greece have the river Danube in common. Their vineyards are widespread and produce a rewarding variety of wines at competitive

1 How are the flavours of herbs imparted to French Vermouth?
2 (a) What base wine was originally used for vermouth production in Italy? (b) Where does the base wine come from today, and why?
3 How are Italian 'Bianco' and 'Sweet' vermouths sweetened?
4 What is Maiwein? How and where is it made?

prices. The Danube rises in the Black Forest of Germany and flows east through Austria to Budapest before turning south across the Hungarian plain to Belgrade in Yugoslavia. Thence it runs east dividing Bulgaria from Romania to flow out in a broad delta into the Black Sea. These countries are watered by many important tributaries of this great river, which at the same time drain the vineyards and provide a means of transport for the wine.

SWITZERLAND

This mountainous country contains the headwaters of the Rhine, the Rhône, and some tributaries of the river Po. Its wines take after the characters of French, German, or Italian wines, depending upon the

1 By cold maceration.
2 (a) Moscato d'Asti. (b) White wine from Puglia or Sicilia, as there is not enough local wine.
3 With pure cane sugar.
4 A flavoured beverage made in the month of May, by macerating flowers of sweet woodruff (*Galium odoratum*) in wine, in German homes.

region and language spoken. A compact and interesting wine industry flourishes, centred mainly in the French-speaking cantons in the west of the country. The vineyards range up the Alpine foothills, and reach a height of 1100 m above sea level.

Of approximately 10,000 ha under vines in nineteen cantons, the major vineyards lie along the shores of Lake Geneva in the canton of Vaud, around Lake Neuchâtel, and along the upper reaches of the Rhône in the canton of Valais. Vines are also grown in the German-speaking cantons of the north and in the Italian-speaking canton Ticino in the south.

Vaud

This canton lies on both sides of Lac Léman and has wide slopes extending down to the lake – ideal for vines. The region is divided into three districts: on the north side of the lake, Lavaux to the east of Lausanne and La Côte to the west; and Chablais to the south-east, controlled by France. The principal vineyard of Lavaux is Dézaley, where golden still dry wine is made from the Chasselas, here called Dorin, and light red wine from the Pinot Noir, here called Salvagnin. The dry white wine of Aigle, in Chablais, has a gunflint bouquet.

The canton of Geneva also produces wine from the Chasselas (Perlan).

Valais

The canton of Valais stretches south and east from Lac Léman along the Rhône valley into the mountains; the principal wine is Fendant, again from the Chasselas grape, and red Dôle from the Pinot Noir and Gamay. Visitors to the region may also find astringent wines from the local Humagne grape.

1. Give the approximate annual wine production of the eight countries of central and south-eastern Europe.
2. In what latitude do these countries lie?
3. Name the natural feature of central and south-east Europe most important to wine production.

Neuchâtel

In the most northerly canton of French-speaking Switzerland, the light and bright Neuchâtel white wines from the Chasselas grow on a jurassic limestone. Experiments have led also to the production of very acceptable Neuchâtel pétillant wines *sur lie* (matured on the yeasty sediment). Vinification of the Pinot Noir grape *en blanc* produces a faintly-tinged wine with the appropriate name of Oeuil-de-Perdrix.

Other Cantons

In the German-speaking cantons of Zürich and Thurgau, through which flows the Rhine tributary river Thur, red wines similar to those of neighbouring Baden are made from the Pinot Noir (Blauburgunder or Klevner) and drunk locally under the name of Klevner.

In the Italian-speaking canton of Ticino, Merlot is grown and yields sound 12% vol. red wine of which some is exported under the name Viti.

A broad and interesting selection of the wines of Switzerland find their way to a big auction near Lausanne every November, and attract considerable world interest.

AUSTRIA

Only the territory towards the eastern end of Austria is suitable for vine-growing, for here the mountains drop first into foothills and then down to the flat Danubian plain. Near the Czechoslovak and Hungarian borders, Austria produces about 3 million hl of wine every year. This includes a range of red, white, and rosé wines divided between *Qualitätswein* (quality wine) and *Tafelwein* (table wine), although these are more commonly called Qualitätswein and Tafelwein. A major revision of the Austrian Wine Law has recently taken place (not encompassed in this edition) to which reference should be made for regional and district boundaries and names.

1 33 million hl.
2 Between 40° and 50°N.
3 The river Danube.

The Austrian grower Lenz Moser was responsible for developing the individual system of vine growth, contrasting with the traditional method of closely-planted rows; and now the Lenz Moser system which permits mechanical viticulture is being adopted widely abroad. The area of vine cultivation is approximately 50,000 ha, of which 61 per cent is in Lower Austria, 32 per cent in Burgenland, 5 per cent in Styria, and 2 per cent in Vienna; there are about 40,000 vineyards. The principal vines planted for white wines are the Grüner Veltliner, a green grape accounting for nearly a third of all Austrian wine; the Wälschrizling (or Italian Rizling) and Müller-Thurgau account for some 9 per cent each, the balance coming from a multitude of local varieties. The principal vines for the relatively

1 In which two cantons are most of the best Swiss wines produced? What is the main white wine called?
2 Name the Swiss wine districts surrounding Lac Léman (Lake Geneva).
3 What is the principal vineyard of Lavaux, and what wines does it produce?
4 Describe Dôle.

small proportion of red wine (about 12 per cent) are the Blauer Portugieser and Gamay.

Lower Austria

Three of the four wine regions are sub-divided into districts (Vienna being too small for such treatment). Niederösterreich in the north-east has eight such districts: Wachau; Krems, Langenlois; Klosterneuburg (both east and west of Vienna); Retz; Falkenstein; Gumpoldskirchen; and Vöslau. The soil varies considerably through these districts, with schist, gravel, sand, loess, and chalk all contributing to a wide variety of wines including Austria's best white wines. 30 per cent of the vineyards are planted with the Grüner Veltliner, for this is first and foremost a white wine region. The white wine of Gumpoldskirchen, one of the better known outside Austria, is medium-dry, strong, and well-balanced; it is made from the Rotgipfler and Zierfandler grapes.

Burgenland

The second largest region, Burgenland, produces about one-third of all Austrian wine. Burgenland has four districts: the first is Neusiedlersee Hugelland on the western side of the Neusiedlersee, which produces full wines reaching their apotheosis in Mittelburgenland Ausbruch (a category between Beerenauslese and Trockenbeerenauslese wines) from the Furmint grape, not widely planted elsewhere; and Neusiedlersee to the north and east of this great lake, where white wines are made from Wälschrizling, Traminer, Neuburger, and Muskat Ottonel vines. Sudburgenland stretches south from the Neusiedlersee towards the Styrian border, where two notable red wines are made from the Blaufrankisch (Limberger) and Blauer Portugieser grapes.

1 Vaud and Valais: Fendant.
2 La Côte and Lavaux to the north, and Chablais to the south-east.
3 Dézaley; dry white Dorin from Chasselas grapes, and light red Salvagnin from Pinot Noir.
4 Red wine made from Gamay and Pinot grapes.

Styria

Styria is very small by comparison – a rustic hilly area bordering Yugoslavia – and is divided into three districts. Südsteiermark (Southern Styria) – with 900 ha under vines the most productive of the three – is planted mainly with Wälschrizling and Müller-Thurgau grapes for white wine production. In Sud-Osteiermark (East Styria), excellent white wines are made from Wälschrizling, Traminer, and Ruländer grapes on volcanic soil, while Westeiermark (West Styria) produces light red and white wines from local grape varieties.

Vienna

There is virtually no export of the wine made in Vienna; the region is, however, renowned for the custom of serving *heurige* (young) wines in the *Heurigen* or wine bars of Grinzing, a northern suburb of Vienna. These young wines are not produced in sufficient volume to satisfy the ever-growing tourist demand so, by special permit, wines from outside may now be brought in. The heurige wines are made mainly from Gruñer Veltliner, Nussberger, and Rheinriesling grapes.

CZECHOSLOVAKIA

Almost the whole of Czechoslovakia is mountainous; consequently, her wine production is diminutive and is restricted to the valleys of three rivers: the Elbe, Danube, and Oder. The population of over 15 million people drink half as much again as their 1.5 million hl production, so that Czechoslovakia is an importer rather than an exporter of wine. The wine is of good quality and bears resemblance to the wines of Germany and Austria, while a little enclave down in the south-east corner of the country produces wines similar to Hungarian Tokay – made just across the border. The soil in the vine-growing regions is mainly loess, characteristic of mountain valleys.

1 What is Austrian Tafelwein?
2 Name the predominant white grape of Austria.
3 Which are the best wines of Styria?

YUGOSLAVIA

Yugoslavia is the biggest of the Balkan countries; it stretches 900 km from Austria to Greece and 450 km from the Adriatic to Hungary and Romania; it has two alphabets and four languages, and is divided into six Republics plus two autonomous provinces in Serbia; it also grows more original vine varieties than any of its neighbours, including one with the engaging name of Grk.

Annual production is about 7.5 million hl; annual consumption in 1983 was nearly 7 million hl. The wine-producing vineyards cover 280,000 ha, divided mainly between four of the republics: Serbia 42 per cent; Croatia 30 per cent; Macedonia 15 per cent; and Slovenia 9 per cent. Although nearly all the vineyards are in private hands, most of their grapes are vinified by the ten state co-operatives. The vineyards are divided into six main wine regions.

Serbia – Kosmet – Macedonia

Here, in river valleys stretching all the way from Vojvodina and Banat on the Romanian frontier down to Albania and Greece, over half the wines of Yugoslavia are produced: the wines of this region are mostly drunk at home. The soil is mainly loess with some limestone farther south. Serbia produces a good white wine, Smederevka, from the vine of that name (which in turn takes its name from the town of Smederevo), sometimes blended with Wälschrizling. There is a full red wine called Prokupac (Prokoopat), from the grape of that name, and Ružica is a good rosé from the same grape. The vineyards of Kosmet were only replanted ten years ago after being ravaged by *phylloxera* at the turn of the century, and the vine types are nearly all western European. A wine made from Pinot Noir is exported to West Germany as Amselfelder Spätburgunder. In mountainous Macedonia, red wines are mainly produced from Prokupac and Vranac grapes.

Vojvodina

This region has some 25,000 ha of vineyards; the best lie in the Fruška Gora hills on the right bank of the Danube north-west of

1 Table wine.
2 The Grüner Veltliner.
3 White wines made from Grüner Veltliner, Wälschrizling and Müller-Thurgau.

Belgrade, where Plemenka is made from the Bouvier and other local vines; Bermet, a type of red Vermouth, has been made for 200 years. Blending wines, made largely for export, come from the Kadarka and Pinot Noir vines and some Gewürztraminer.

Slovenia

The finest white wines of Yugoslavia come from the vineyards between the Mura and Drava rivers, where the soil is lime and marl. Among the recognizable vine-names are Silvaner, Laskirizling, Müller-Thurgau, Traminer, Pinot Blanc, and Furmint (Sipon); a sweet wine named 'Tiger's Milk' is made from Ranina (Bouvier) grapes. The red wines made on the Sava banks use Gamay, St Laurent, Merlot, and Barbera grapes. The 23,000 ha of vineyards are divided into four districts; a Wine Institute improves technology and marketing methods.

Croatia

Mainly white wines are produced in the northern part of this Republic known as Slavonia, between the rivers Sava and Drava, from a district vineyard area of over 40,000 ha. The wines are strong and big, but are not generally regarded as equal to those of Slovenia. Pinot Blanc, Pinot Gris, Traminer, and Sauvignon are among the vines cultivated.

Istrian Peninsula and Dalmatia

The western and southern parts of the Republic bordering the Adriatic, known as the Istrian Peninsula and Dalmatia, are romantic and beautiful with a perfect climate for wine-making. Their vineyards yield strong red and white wines, a number of sweet dessert wines, and some sparkling wine. It comprises three districts. In Istria and the Kvarner in the north, big, deep, well-balanced wines come from the Gamay, Pinot Noir, Cabernet, and Merlot grapes; and sweet white wines from Malvasia and Muscat. Around the

1 What are 'Heurigen'?
2 Name the three rivers in whose valleys Czechoslovak wines are produced. How much is exported?

towns of Dubrovnik and Zadar (claimed to be the home of Maraschino), red wines called Plava Mati and Postup, whites called Pošip, and rosés called Opol are made, all but the last (which is made from Plavać) being named after their vines. Imotski, in the Dalmation hinterland, produces red and white wines of little distinction from the raisin grape Dingaĭ.

Hercegovina

This small region, on limestone subsoils, produces Yugoslavia's most prized wine – Zilavka Mostar, made from the white Zilavka grape. Mostar is the provincial capital. Some red wine is made from the Blatina grape.

HUNGARY

Hungary is the exception among the central European countries in that, surrounded by mountains, it has few itself; it has a vast plain, and Balaton, the largest lake on the continent. The vine came to Hungary from the Caucasus, with nomadic tribes moving through the Caucasus and Ukraine at the same time as the Phoenicians and Greeks were moving the vine west by the Mediterranean. Here, between latitudes 46° and 48°N, the climate is kind to the vigneron, and the soil varies through basalt and loess in the west and sand in the centre to clay and volcanic rock in the higher north-eastern territory. In all, Hungary has nearly 200,000 ha of vineyards, producing about 7.5 million hl of wine every year, much of it excellent in quality. Exports are controlled by the state (Monimpex), but many vineyards are in private hands.

The country divides into four regions. The Danube runs from north to south through the centre of Hungary, and to its west, are two regions, the Small Plain (Kisalföld) in the north-west, and Transdanubia lying between the rivers Danube and Drava, running down to the borders of Yugoslavia. The Great Plain (Alföld), to the east of the Danube, is by far the largest region; and beyond this plain,

1 Young (one-year-old) wines; or the wine bars in which they are served.
2 Elbe, Danube, and Oder. None.

in the foothills of the Carpathians, the North Massif region contains the Tokaj and Eger vineyards. These four regions are sub-divided into fourteen registered districts.

The Small Plain

This region in the north-west is not divided into districts, and is registered as Sopron, where Beaujolais-like red wine is produced from the Kékfrankos (Limberger) grape. This is best drunk young and fresh, like Beaujolais. The vineyards of Sopron have been tended through the centuries by the great families who were originally responsible for spreading the fame of Hungarian wine.

Transdanubia

The three districts in this region which surrounds Lake Balaton – Balatonfüred-Csopak, Badacsony, and Somló – have a reputation for white wines of outstanding quality, and a further two districts, Mór and Mecsek, also produce good dry white wine from the Ezerjó grape. These wines tend to be sweet after the German fashion, so it is not surprising to find Wälschrizling and Silvaner vines with indigenous vines such as the Ezerjó and Kéknelu. Better wines have been made among the vineyard slopes of Somló – a volcanic mountain – since the days of Hungary's first king, St Stephen, in the eleventh century, and still are.

The Great Plain

Over half the vineyards of Hungary are sited on the sand dunes of the Great Plain, where the wines are generally (with some rare exceptions) common rather than of quality, and are divided about equally between reds and whites. Kadarka (named for the vine) makes up the bulk of the red wine, and Olaszrizling (Wälschrizling) and Mezesfeher much of the white: all these wines are soft and light, and feed a substantial export market, notably to Germany.

1 What is Grk?
2 (a) Name the rivers between which Yugoslavia's finest white wines are produced. (b) What is the soil there? (c) Which is this region?
3 What is 'Tiger's Milk'?

The Northern Massif

The eastern region of Hungary is the home of the fabled Tokay, made on volcanic slopes north of the Bodrog river, close to the border with Czechoslovakia and about 60 km from Russia. Production is concentrated in the town of Sátoraljaújhely (pronounced Shatoc-ralyee-owyee-haylye) – meaning something like 'the place where our tents have rested', on the lines of the 31-lettered station in Anglesey! Wisely, the merchants chose the name of a smaller and more pronounceable town, Tokaj (Tokkoy) to describe their wines. This wine starts with grapes affected by noble rot from the Furmint and Hárslevelü vines. Carefully selected grapes which have been affected by noble rot, called *aszú* berries, are crushed by their own weight only in tubs called *puttonyi*; their juice, Tokay Essence, is collected in casks called *gönci*. The aszú berries are then pounded into a paste, either by treading or using a machine simulating the treading action. The resulting aszú paste is added to the normal wine must in proportions which determine the final quality (and sweetness) of the resulting Tokay Aszú. As a *puttonyo* contains 25 kg of aszú paste and the capacity of a *gönc* is 140 litres, it follows that a 'Tokay of 5 puttonyi', where 125 kg of paste are mixed with 140 litres of must, will be very rich. The wine is filtered into gönci after several days' fermentation in vats, to mature for five or six years in small tunnels hewn into the volcanic mountainside. The maturation period in cask is the number of puttonyi plus two, so that a Tokay of five puttonyi will mature in cask for seven years. During this period the little casks are left unbunged, and no wine is added to replace ullage.

never reaches more than 13.5% to 14% vol. alcohol; this is even more true of Tokay Eszencia, to which remarkable curative properties are ascribed: assisting longevity in man and woman, it is near to immortality itself, for occasionally bottles from the seventeenth century are sold at auctions!

A wine labelled Tokay Szamorodni, 'as it comes', is made from grapes from which the aszú grapes have not been separated. It is to Aszú as Graves is to Sauternes – drier but eminently drinkable.

Another very different wine is found in the Northern Massif: a red wine made from the Kékfrankos and Kadarka grapes called Egri

1 A vine variety of Yugoslavia.
2 (a) The Mura and Drava. (b) Lime and marl. (c) Slovenia.
3 A sweet white wine from Slovenia made from the Ranina (Bouvier) grape.

Bikaver – 'Bull's Blood of Eger'. It is not the only red wine to be called 'Bull's Blood', but it is a fine full-bodied wine.

ROMANIA

The importance of Romania's wine industry may be measured by its annual production – about 8 million hl – which ranks seventh of the seventeen producing nations in Europe. The Romanians are enthusiastic wine drinkers, each head of population accounting for some 29 litres per annum; nevertheless they export about 1 million hl, mainly to East Germany, Poland, Czechoslovakia, and Austria, but with some also to France, the UK, Sweden, Belgium, and Holland. The product varies, and the best is distinguished.

The vineyard regions of the country are divided by the Carpathian Mountains, which sweep down from the northern border with Russia in a south-easterly direction until they reach the southern half of the country, when they turn west. Inside this 'J', is the province of Transylvania, containing the Tirnave region, generally devoted to white wines and producing fine Fetească.

Across the mountains to the east is the province of Moldavia, now mostly appropriated by Russia, but including the region of Cotnari, where fine well-balanced dessert wines of 13 to 15% vol. are made from the Grasa grape; fine wines are also made from other native varieties. The third fine wine region lies in the Murfatlar Hills, just inland from the port of Constanta on the Black Sea, where the Muscat grape has been cultivated since the grape came to Europe; other vines, particularly Chardonnay, are also grown here with success.

Between Cotnari and Murfatlar, on the eastern slopes of the Carpathians, the vineyards of Focsani and Nicoresti yield the greatest production of Romania. Red Babească and white Fetească grown on a sandy soil give good wines.

Further south, outside the bend of the Carpathian Mountains, lie the vineyards of Dealul-Mare, where Cabernet Sauvignon is widely grown and improvingly vinified in recent times, which has been reflected in the quality of exports. The regions of Argas and

1 Which Dalmatian town is claimed to be the home of Maraschino?
2 What is Yugoslavia's most prized wine, and where is it produced?
3 Where do the districts of Balatonfüred-Csopak, Badacsony, and Somló lie? What grapes are used to produce white wines there?

Dragasani lie further to the west beyond the Ploesti oilfields and produce red and white wines from a variety of grapes. Finally, in the far west beyond the mountains lies the sandy Banat plain shared with Yugoslavia and Hungary, where red wines are made from the Kadarka (here Cardarca) grape.

BULGARIA

Most of Bulgaria is mountainous, but its two fertile plains, the Danube and Meritza valleys, lying between latitudes 42° and 44°N, are ideally situated for viticulture; the vineyards are located on the tributaries of these two rivers rising either side of the east-west range of the Stara Planina which divides them. Bulgaria's wine history goes back to very early days, deriving from ancient Thrace. During five centuries of Muslim Turkish rule, wine-making withered, but soon revived to produce 2 million hl by 1940. By 1977, Vinprom, the state wine organization, had concentrated an increased area into collectives, and by 1985 production had steadied at a figure exceeding 4 million hl.

Wine laws have also been instituted to control production and quality, and have set up about thirty 'regulated wines' for specified locations and vine varieties; five of these regulated wines are for Cabernet or Cabernet/Merlot blends, and ten more are for other western varieties.

The native red grapes are Pamid, Melnik, and Mavrud, of which the last two, being of greater quality, are included in the list of regulated wines. Also included is the Kadarka of Hungary (here called Gamza) which is sometimes blended with the Russian Saperavi. Generally speaking, these red wines are made in the western half of the country, south of the Stara Planina, but also in the north.

Native white grapes are Dimiat, which is heavily planted but does not yet rate a regulated wine, and Misket which produces a muscat-flavoured light wine, although unrelated to the Muscat varieties. Dimiat wine is soft and perfumed, and is often made into brandy. Of

1 Zadar.
2 Zilavka Mostar, a white wine from the Zilavka grape; Mostar is the provincial capital of Hercegovina.
3 In Transdanubia, surrounding Lake Balaton. Olasrizling (Wälschrizling), Silvaner, and Ezerjó grapes.

imported varieties, Chardonnay rates two regulated wines. The most heavily-planted white grape is the Russian Rkatsiteli, from Georgia, which yields dry, alcoholic white wines of some character.

1 (a) In what region of Hungary is Tokay made? (b) What are aszú berries? (c) What are puttonyi? (d) What are gönci?
2 Where and for how long is Tokay Aszú matured?
3 What is 'Bull's Blood of Eger'?
4 What is Cotnari?
5 Which vineyards produce the most wine in Romania? Name the grape varieties.

11

The Levant and North-West Africa

THE LEVANT

CYPRUS

The wines of Cyprus have an age-old history emerging from Greek mythology, for Greek and Phoenician colonies thrived there, and the early wines developed an international reputation which increased with every voyage of discovery. Her grape varieties were introduced to Madeira, and reputedly to Marsala and to Tokay in Hungary. An early Cypriot wine called Nama, made from dried grapes and

1 (a) The Northern Massif. (b) Grapes affected by noble rot. (c) Tubs weighing 25 kg in which the aszú berries are crushed. (d) Casks of 140 litres capacity in which the must is fermented.
2 In small tunnels hewn into the volcanic mountainside. Maturation is for five or six years.
3 Egri Bikaver, a full-bodied red wine made from the Kékfrankos and Kadarka grapes in the Northern Massif.
4 A Romanian wine region in the north-east producing a well-balanced dessert wine of 13% to 15% vol. made from the Grasa de Cotnari.
5 Focsani and Nicoresti, on the eastern Carpathian slopes. Red Babeascà and white Feteascà.

intensely sweet, was renamed Commandaria by the Knights Templar, and in 1363 Commandaria was taken at the famous 'Feast of the five Kings' given at Vintners' Hall by Henry Picard, Vintner and Lord Mayor of London, to honour King Peter of Cyprus. (The Wine and Spirit Education Trust was the first occupant of adjacent 'Five Kings House', named to commemorate this banquet.)

The vineyards, mostly on the southern slopes of the rainy Troödos mountains, but also near Nicosia and the Makheras mountains, are planted with traditional Mavron grapes for red wine, and Xynisteri and Alexandria Muscat for white, on a largely volcanic soil. Cyprus has always been free of *phylloxera* and new varieties introduced to the island have to spend long periods in quarantine. Cyprus produces slightly under 1 million hl of wine each year, of which about one-quarter is light white wine with the balance divided between red wine and Cyprus Sherry.

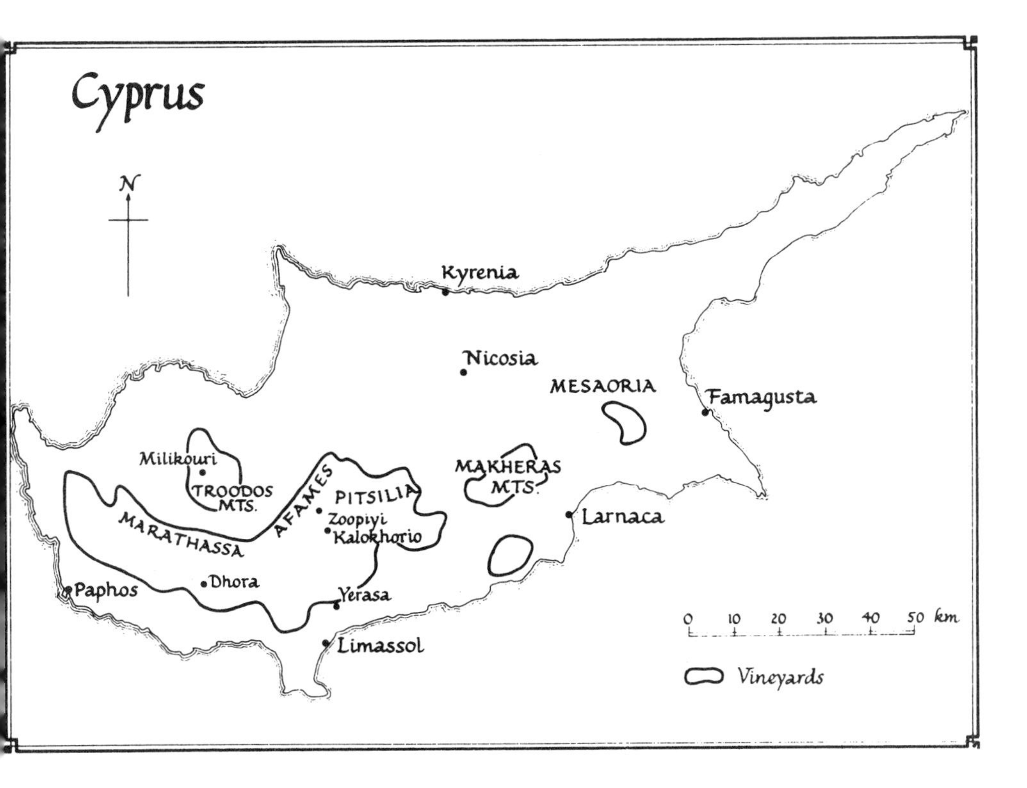

1 What caused Bulgaria's wine industry to wither from AD 1400 to 1900?
2 (a) What are Bulgarian 'regulated' wines? (b) Are they restricted to wines from indigenous vines?
3 What is Rkatsiteli?

The full-bodied red wines from the Mavron grape are typified by Afamés, and this grape is also used for the rosé Kokkineli. The local yeasts ferment to a high degree and produce full-bodied dry white wines; when these are weathered in cask in the sun they assume a nutty flavour akin to Sherry. Cyprus Sherry has been popular in England for a long time. Nowadays, the taste is for lighter Cyprus Sherries, and the wines are inoculated with flor to produce fino styles.

Cyprus Sherries have been imported into England (often in bulk tanker-ships), both fortified to high strength and at low strength; blending these wines together after payment of duty achieved savings not now possible under the new excise regulations. Much of the crop is fermented at co-operative wineries, and is exported from the port of Limassol on the south coast.

Turkey

In the story of wine, Turkey must rank as an anomaly. For here the vine – and wine – flourished in the third millennium BC, long before the French vineyards were planted. Yet the Turks, through their conversion to Islam in the eighth century, lost the art of making wine, and recent attempts to restore a wine industry have been slow to take hold. At the dawn of the twentieth century Turkish wines were to be found in western Europe, but the Second World War destroyed the trade. Despite the decree of 1928 removing the Moslem faith as the established religion of Turkey, the conversion to free manufacture and drinking of wine has been slow (annual consumption is still less than half a litre per head).

Most grape growers send their produce into the cities for wine-making, either to the government-controlled co-operatives and the Ataturk Farm Administration, or to private enterprises. Only 3 per cent of the grapes of Turkey are converted to wine, so that from a total area of 300,000 ha of vineyards, something less than half a million hl of wine is produced annually, of which around 5 per cent is exported. The remainder of the vineyards produce grapes for the table or for raisins, indicating perhaps that the Turks prefer to eat grapes than to drink wine.

1 Muslim Turkish rule.
2 (a) Wines of specified locations and vine varieties. (b) No, there are a number of western vines in the government-approved list.
3 The most heavily planted white grape in Bulgaria, yielding dry, alcoholic white wines of character.

A dry white wine, Trakya, is produced in Thrace and around the Straits of Marmara by Istanbul; and a red, Buzbag (pronounced Boozhwar!), in Anatolia in the south, opposite Cyprus. Although the vine is widely cultivated in many varieties, these two wines are the ones most likely to be seen in England.

The Soviet Union

The wine-producing areas of the Soviet Union are restricted by nature to two identifiable regions. The first, in the latitude 45° to 50°N, embraces the Republic of Moldavia (which borders Romania) and the Southern Ukraine immediately north of the Black Sea. The second lies farther east in Georgia, on the northern slopes of the Caucasus mountains, between the Black and Caspian Seas, in latitude 40° to 45°N. The important thing to remember about the Soviet wine industry is that its state organization was not conceived until 1950, yet already by 1965 over 1 million ha of vineyards had been developed. Possibly one-third of that area has been devoted to wine-making (in Moldavia and Georgia) from ancient times. Certainly the industry has been developed to suit the Russian sweet tooth, with a predominance of sweet red, white, and rosé wines, some good sparkling wines, and a range of fortified wines similar to – and named after! – Port, Madeira, and Sherry. A number of French vines, like Cabernet, Aligoté, Merlot, Malbec, and Pinot, have been added to Saperavi and Romanian Fetească. Riesling and Traminer are also grown in Moldavia where, because of the latitude, white wines should present less problems than red. However, the most widely planted variety remains white Rkatsitely, which yields wines of high alcohol and residual sugar but with a good balancing acidity.

Syria and Lebanon

This is the land where grape juice was first fermented to make wine. Yet progress has been gradual. Syria has some 75,000 ha under vines in the regions of Aleppo, Damascus, and Homs; Lebanon has

1 Which country's grape varieties were introduced to Madeira, Marsala, and Tokay?
2 Why are new vines for Cyprus quarantined for long periods?
3 What are the traditional grapes of Cyprus for (a) red, and (b) white wines?

25,000 ha, mostly in the valley of Beka'a; in both countries, as in Turkey, grapes are marketed primarily for the table, or as dried raisins or grape juice. The very small resulting wine production, from French vine stocks, is made by the Christian population; *phylloxera* is a recurring problem which grafting has failed to solve. With a strength of 9% to 10% vol. alcohol, the wines, with the notable exception of Château Musar in the war-torn Beka'a Valley, rank as ordinary; there is also a little sparkling wine production.

Israel

Israel is another country whose history shows that vines and wines existed as early as 3000 BC. For practical purposes, however, the modern wine industry of Israel began with establishment of an agricultural school at Mikveh in the late nineteenth century. Here a vineyard was planted with Alicante, Carignan, Bordeaux, and other varieties; these were followed a few years later by German grape varieties, including Riesling, planted in the Carmel region by the German Templars. Baron Edmond de Rothschild then became the patron of Israel's wines, introducing new vines, combating and overcoming *phylloxera*, and developing two wineries at Rishon le Zion and Zikhron-Ya'aqov which he eventually handed over to the growers. These wineries today produce three-quarters of all Israel's wines. So that it can be used for religious celebrations, most of the wine is prepared under rabbinical supervision to make it kosher.

Current production is about 400,000 hl per annum, of which over one-third is consumed locally and the balance is exported. The Israeli Wine Institute was established at Rehovot in 1957, and has done much to improve the quality of Israel's wide variety of wine types. These range from dry and sweet red and white wines for the table to sparkling wines (made by the méthode Champenoise), fortified wines, and aperitifs. The oranges of Jaffa are famous, and an excellent orange liqueur, Sabra, is widely exported.

1 Cyprus.
2 To exclude *phylloxera*, from which Cyprus has never suffered.
3 (a) Mavron. (b) Xynisteri and Alexandria Muscat.

EGYPT

Today there would be no wine industry in Egypt but for the vision and purpose of one man – the Egyptian Nestor Gianaclis who, at the beginning of the twentieth century, searched for and discovered the chalk soil on which the vines of ancient Egypt (whose wine was so much praised by Horace, Pliny, and Virgil) had flourished. In thirty years Gianaclis achieved the near-impossible, and produced from the Mariout vineyard in the Nile Delta wines which were acclaimed in Paris as worthy of comparison with Rhône and Burgundy wines. Today the Gianaclis vineyards of about 15,000 ha are still cultivated and produce some 50,000 hl of wine each year. The main export is to the Soviet Union.

MALTA

Red wines of average character are made here and on the neighbouring island of Gozo. A small quantity of sweet white wine is also produced, some being exported to Britain. The vines are cultivated along the south coast of Malta, and yield some 30,000 hl of wine from an area of about 2000 ha.

NORTH-WEST AFRICA

Tunisia, Algeria, and Morocco share a common history of Phoenician and Greek settlement, Roman occupation, Moorish conquest and, in modern times, French dominion and recent independence. Through all these changes cultivation of the vine and production of wine has fluctuated; it reached its peak under French dominion. The colonists produced vast quantities of wine (at one time Algeria was the third largest producer in the world) of high alcoholic strength but very ordinary quality, which was shipped back to France for reinforcing the low-strength wines of the Midi. With independence and the departure of the colonists, production flagged; and with the

1 What Turkish wines might be found in the UK?
2 Name two principal wine-producing regions of Russia.
3 Name the most widely-planted Russian grape variety, and describe the wine made from it.

emergence of the EEC wine regime under which blending of these wines with EEC wines is forbidden, this market has virtually closed down.

Tunisia

The vineyards are situated in the north-east of the country, forming a semicircle west, south, and east of the city of Tunis. They stand on the typical sandy soil of North Africa in the valleys of the Oued Medjerda and Oued Miliane. *Phylloxera*, which did not strike until 1936, reduced the area of Tunisia's vineyards by half. Today the volume of production averages about 400,000 hl from an area of 35,000 ha. Tunisian wines mature rapidly in the hot climate and the red wines accordingly tend to maderize to an onion-skin colour, with a characteristic bitterness. To correct this, the vines originally brought over by the French (Alicante-Bouschet, Carignan, and Cinsaut for red wines: Clairette de Provence and Ugni Blanc for white) have been supplemented by the Cabernet, Nocera, and Pinot Noir for red wines and Pedro Ximénez, Sauvignon, and Sémillon vines for white. The classification Vin Supérieur de Tunisie was established in 1942 for wine at least one year old having an indication of vintage, but was applied generally to the wines of Tunisia from any grape variety. The Appellation Contrôlée Vin Muscat de Tunisie, which followed five years later, was general to Tunisian fortified wines of 17% vol. alcohol, but they had to be made from the Alexandria Muscat, Muscat de Frontignan, or Muscat de Terracina grapes, to which rectified wine spirits were added. Tunisia produces red, white and rosé wines, the last being among the best of the North African rosés.

Algeria

Algeria's wine production under French occupation dating from 1830 grew very considerably: at the peak 15 million hl of wine were

1 Trakya (white) and Buzbag (red).
2 Moldavia and Georgia.
3 Rkatsiteli; fairly balanced sweet white wines of high alcohol.

produced from 350,000 ha of vineyards. In 1977, output was one-third of this and was further reduced to 1 million hl by 1980. Here again the local Moslem religion forbids drinking, and the enormous labour force in the vineyards, representing no less than 35 per cent of the total employed in the country, work solely for the export industry. When Algeria gained independence in 1962, the French market virtually dried up, and it was necessary to find a new market: the Soviet Union provided one. There has subsequently been a tendency to rationalize production, so that the area under vines today is only about 200,000 ha.

The vineyards stretch along the Mediterranean coastal plain through three regions, Oran in the west, Alger in the centre, and Constantine; Oran and Alger also have plantations in the hills amounting to about 60,000 ha and here better quality wine is produced, which under French rule ranked as VDQS.

Algerian wines tend to be soft and fat, but of sound quality with a high alcoholic content (11% to 15% vol.) and low acidity. Most of the wine is a deep red and comes from the Alicante-Bouschet, Aramon, Carignan, Cinsaut, Morrastel, and Mourvèdre vines. Clairette, Faranan, Listan, Macabeo, and Ugni Blanc are cultivated for white wines. The even climate produces no variation in vintage from year to year.

Morocco

The wine history of Morocco is even shorter than that of Algeria, for the French who entered the scene in 1912 did not develop the vineyards until after the First World War. Between 1919 and 1956 (when Morocco regained independence), about 80,000 ha were cultivated as vineyards and remain today. Under Moroccan law, wines must now have a strength of at least 11% vol. alcohol for export. Exceptionally for a Moslem country, 50 per cent of annual production is consumed at home, and the remainder is consumed in France and other EEC countries, and in America. Wines of good quality are produced, mostly red, and come from the same grape

1 Which has the greater area planted with vines, Lebanon or Syria, and by what proportion?
2 Name Lebanon's best vineyard.
3 Who is called 'the patron of Israel's wines'?
4 What is Sabra?

varieties as those cultivated in Algeria. The main vineyards are in the districts of Meknès-Fez, Oujda-Taza, and Rabat-Casablanca. The wines of Casablanca are in such demand that they seldom survive to mature properly.

1 At 75,000 ha, Syria has three times as much land given to vineyards as Lebanon.
2 Château Musar in the Beka'a Valley.
3 Baron Edmond de Rothschild.
4 An orange liqueur made in Israel from Jaffa oranges.

12
North America

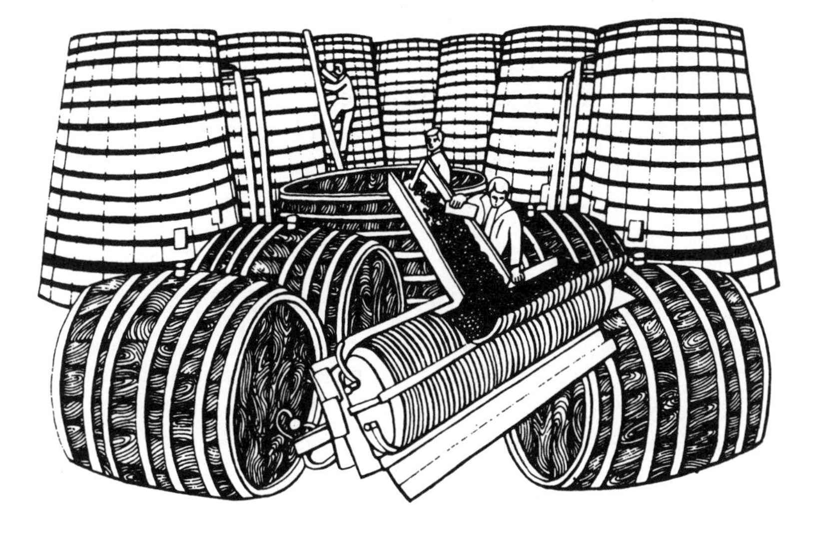

The genus *Euvitis*, from which all wine-producing vines spring, has existed in North America as long as man; not the true vine, *Vitis Vinifera*, but different species such as *V. Riparia, V. Labrusca, V. Rupestris*, and *V. Berlandieri*. These are hardy species, more resistant to heat, cold, and pests; they grow naturally from Ontario in the north to Texas in the south. The only trouble is that their juice, and wine made from it, has a dreadful foxy pungency, some say like a maybush in bloom. Yet it is popular, and 'wild' vine varieties of Noah, Concord, Catawba and others are widely cultivated around the Finger Lakes district of Upper New York State, in neighbouring Ontario, and on the banks of the Ohio river; also, in recent years, in the Okanagan Valley of British Columbia, where the winters are so cold that they have to bury the whole vine in the fall.

1 What effect does climate have on Tunisian wines?
2 Where are Tunisia's vineyards situated?
3 How has Algeria's wine production varied since Independence?
4 Name the present market for Algerian wine exports.
5 (a)How long has Morocco been making wine? (b) What minimum strength must it have for export? (c) How much of its wine is consumed at home?

HISTORY

Vitis Vinifera crossed the ocean, from the Old World to the New, coming to California via Mexico with the Spaniards, and here found an almost ideal homeland. The Jesuit priests carried on the tradition of wine-making by planting vineyards wherever their missions were established throughout southern California. The variety of *Vitis Vinifera* which they brought from Spain was appropriately called the Mission, and produced the sweet altar wine needed. The missions and their wines flourished until, in 1830, the Mexican government put down the missions and with them, their vineyards. There is an interesting parallel here with the earlier events of Roman and English history in which religion was responsible for the survival and development of wine production.

UNITED STATES

Joseph Chapman and Jean-Louis Vignes (a Frenchman from Bordeaux) came to California at this time, bringing varieties of vine from France to supplement the Mission grape which, for the table, produced only mediocre wine. Nevertheless it survived as an important Californian wine up to the end of the nineteenth century.

In 1851 a Hungarian soldier of fortune, Agoston Haraszthy, arrived on the scene and introduced a selection of European vines. Haraszthy's faith in California's potential for wine production convinced Governor John G. Downey, who commissioned him to bring further vines from Europe. He returned in 1861, with 100,000 cuttings from 300 varieties of French, German, Italian, and Spanish vines, which were planted in the Sonoma Valley, and later distributed to growers throughout the state. Two years later Haraszthy founded the Buena Vista Vinicultural Society at Sonoma and gave his farms and vineyards there to the Society. He is rightly acclaimed as the 'father of Californian viticulture'.

In the last 130 years the Californian wine industry has experienced successes and set-backs. The 1849 Gold Rush brought prospectors who paid suitably inflated prices in gold dust for the 'liquid gold'.

1 They mature rapidly, maderize easily, and have a characteristic bitterness.
2 In the Oued Medjerda and Oued Miliane valleys, encircling the southern approaches to Tunis.
3 Production, once third highest in the world, has been drastically reduced.
4 The Soviet Union.
5 (a) Less than seventy years. (b) 11% vol. (c) 50%.

But when the lodes ran out, the market failed, and had to be revived with tax advantages. The next blow came in 1870, when *Phylloxera*, indigenous to the wild vines of the East Coast, struck the noble *V. Vinifera* vines from Europe and ravaged the vineyards which had been established in Napa, Sonoma, Sacramento, and Yolo. The industry survived, however, by adopting the grafting system which had been found to work in France. In 1880, the Californian State Board of Viticultural Commissioners was formed to advise and control the industry: much valuable work has resulted in appropriating vines to their best sites and soils, and in controlling pests and other maladies. With the natural scourges overcome, the industry prospered; quality and production improved; but the hardest blow of all fell in 1918, when the Volstead Act amended the Constitution and prohibited the manufacture and sale of alcoholic drinks. Exceptions were made for production of wine for medicinal and sacramental purposes (an excuse, if one were ever needed, for praising God for the gift of wine!). But Prohibition, far from weaning people from 'the demon drink', merely led them to prefer and consume easily-carried, concentrated spirits, and made worse the ill that it set out to cure. Following repeal of the Act in 1933, great demand for wine distorted the market, and poor wines at rich prices were common.

California

California stretches from the latitude of Rome in the north to that of Benghazi in the south, and might therefore be expected to produce wines heavy in alcohol and tending to flabbiness unless fortified. It does produce such wines, but also produces, from the grape varieties of northern Europe, light, delicate wines – both red and white – which rival their originals. The secret lies in the topography and climate.

Topography and Climate

Reduced to its simplest terms, California consists of a series of ranges and valleys running NNE to SSW between the Pacific Ocean and the Rocky mountains; there is ample rainfall in winter, which fills the

1 Name two Euvitis vine species.
2 Where in America is the 'wild' vine cultivated for wine-making?

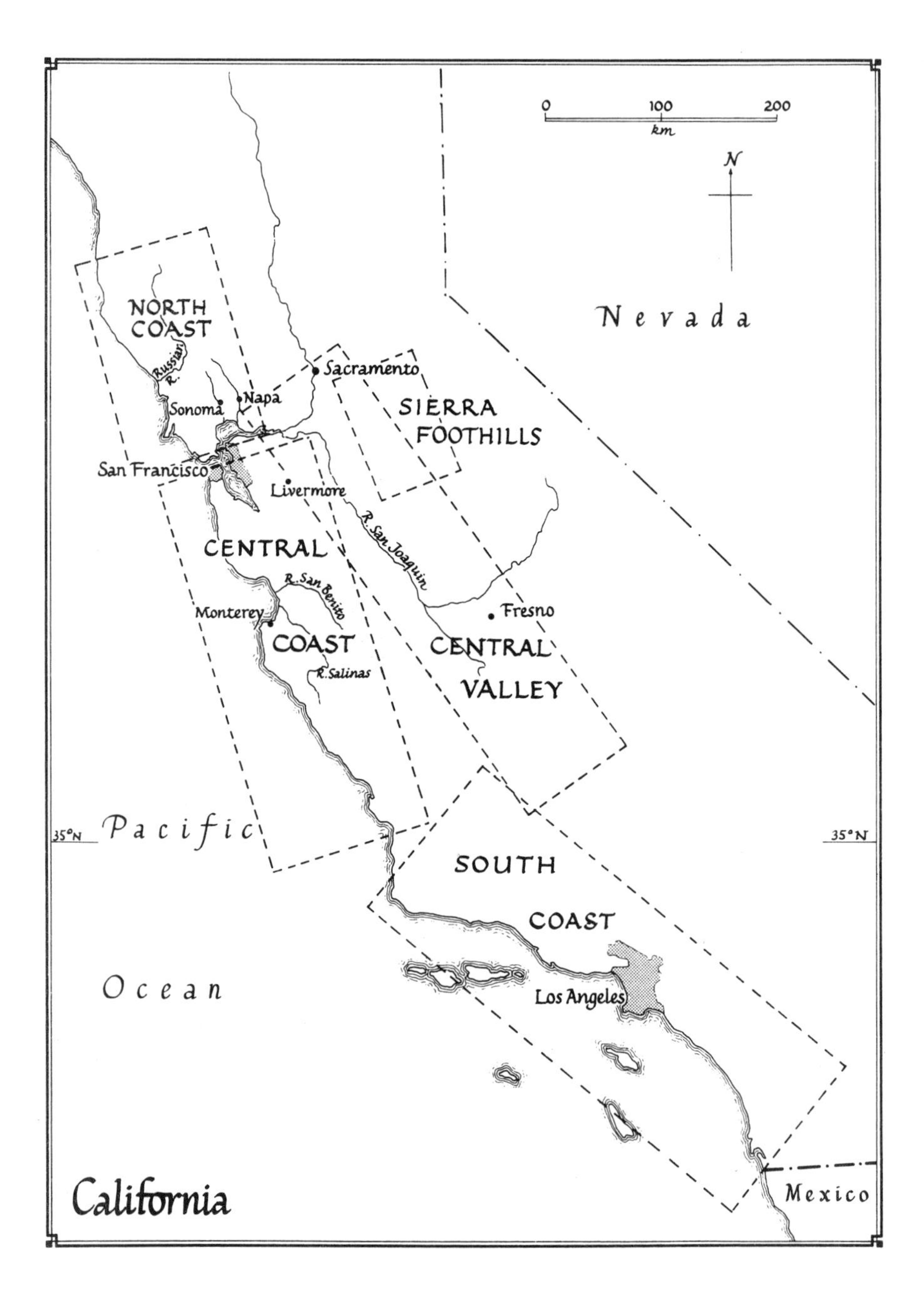

1 From: *V Riparia, V Labrusca, V Rupestris, V Vinifera*, and *V Berlandieri*.
2 In New York State, Ontario, and on the Ohio river.

reservoirs, augmented by snow-melt from the Rockies in spring. The summers are as hot as in the Mediterranean.

Frost Precautions: The cold in winter is severe and spring frosts are common, even in California, although to a greater extent in Oregon to the north. Because of this, little wine is made in Oregon, even though the latitude is that of France, and expensive precautions have to be taken in California. These include the 'smudge-pots' so often seen in northern France, but since Californian anti-pollution laws forbid the emission of smoke, these have had to be modified to consume their own smoke, with consequent loss of efficiency. Another method used is the installation of giant wind-machines in the vineyards, driven by electricity; but these methods are now giving way to the cheapest method of all, devised by the Germans – to spray the vines with water when frost is expected to damage the young shoots, so they are given a protective coating of ice; moreover the latent heat given off during freezing of the water moderates the cold. As Californian vineyards have to be irrigated by spray in the summer anyway, this involves no extra capital expense.

Vineyards and other fruit crops take priority for water during the summer, when it is common to see signs saying 'Save water – drink wine', and 'Share a shower with a friend!' The intense heat of the summer sun is moderated in the coastal regions because the sea is icy, flowing southwards along the coast from Alaska and the Aleutians; when the prevailing cold north-westerly wind carries the cold moist air from the sea over the coastal mountains, fog forms, particularly on summer mornings. This cools the valleys near the sea and produces the ideal warm humid climate for vines.

The valleys nearest the coast are coolest, like those of the Sonoma, Napa, and Russian Rivers (collectively called the North Coast); it is also cool further south in the Livermore region and in Santa Clara/ Central Coast region which centres on Monterey. Light, delicate red and white wines are made in these regions, mostly going under familiar varietal names. Further inland, in the Sacramento and San Joaquin valleys, the climate is much hotter, as the coastal fogs have been burnt up, and in these regions the accent is on bulk production of 'jug' wines and on fortified wines.

1 (a) Whence did *Vitis Vinifera* first come to America? (b) Who were responsible for carrying it? (c) Name the original grape variety.
2 (a) Who was the Hungarian who did most to establish California's wine industry? (b) When did he arrive in California?
3 Name four early districts of California ravaged by *phylloxera*.
4 (a) What Act prohibited the manufacture and sale of alcoholic drinks in the USA? (b) What exceptions were allowed? (c) When was it repealed?

Soils

The soils of California do not include chalk or limestone; clay and schist are found, particularly in the Napa Valley, with gravel at Livermore and granite sand in the Monterey district. But what California lacks in lime is more than compensated by the expertise of the growers and wine-makers; the departments of Enology at Davis and Fresno are counted among the foremost research stations in the world, and their teaching is reflected in the care with which even the smallest establishments prepare their wines.

Naming of Wines

With a wider circulation of their wines throughout the world, which is greatly to be encouraged, the Californians are moving away from 'generic' naming (e.g. Burgundy) towards varietal or Californian district labelling. As these wines can stand comparison with any in the world, such honesty can only be rewarded with success.

Approved Viticultural Areas

Until 1980, Californian wines took their geographical identities from the counties in which the vineyards lay. However, the Federal Government has now taken a hand in defining a number of Approved Viticultural Areas ('AVAs') in the state, and this process is continuing. AVAs are indicated here in **bold type** and are equivalent to the European AOC, QbA, DOC, DO, and similar delimited areas. Major viticultural counties are grouped into larger regions, as shown in the map on page 272.

NORTH COAST REGION

Lake County: Here, **Clear Lake** and **Guenox Valley** produce mostly red wine from Cabernet Sauvignon.

***Mendocino County*: Anderson Valley**, **Cole Ranch**, **Potter Valley**, and **McDowell Valley** produce mainly red wines from

1 (a) Europe. (b) Spanish colonizers. (c) Mission.
2 (a) Agoston Haraszthy. (b) 1851.
3 .Napa, Sonoma, Sacramento, and Yolo.
4 (a) The Volstead Act, 1918. (b) Wines for medicinal and sacramental purposes. (c) 1933.

Carignan, Cabernet Sauvignon and Zinfandel, with some white from Colombard.

Napa County: **Howell Mountain** (eastern hills), **Napa Valley**, and **Carneros** (south). All types of wine are produced in these areas.

Sonoma County: **Northern Sonoma** includes **Alexander Valley**, **Chalk Hill**, **Dry Creek Valley**, **Green Valley-Sonoma**, **Knights Valley**, and **Russian River Valley**, which all drain to the Pacific.

Sonoma-Carneros County: **Sonoma Mountain** and **Sonoma Valley** drain to San Pablo Bay by Napa. Production is fairly evenly divided between red wines from Zinfandel, Cabernet Sauvignon, and Pinot Noir and white wines from Sauvignon, Colombard, and White (Rhine) Riesling.

Solano County: **Green Valley-Solano** and **Suisun Valley** lie south of Napa on San Pablo Bay.

CENTRAL COAST REGION

Alameda County: **Livermore**.

Monterey County: **Carmel Valley** which lies near the Pacific, and **Arroyo Seco** and **Chalone**, which both lie inland, in the Salinas Valley, produce red wines from Cabernet Sauvignon, Pinot Noir, Zinfandel, and Petit Syrah; and white wines from Chenin Blanc, Chardonnay, White (Rhine) Riesling, and Sauvignon Blanc.

San Benito County: **La Cienega**, **Paicines**, and **Limekiln**. The County lies inland from Monterey.

San Luis Obispo County: **Paso Robles**, **Templeton**, **York Mountain**, and **Edna Valley**. This is a pioneer grape-growing county, producing 40 per cent red from Zinfandel and Cabernet Sauvignon; and 60 per cent white from Sauvignon Blanc, Chenin Blanc, and Chardonnay.

1 What is the simple up-to-date method used in California to combat frost?
2 What types of wine are made in the Sacramento and San Joaquin valleys?

Santa Barbara County: **Santa Maria Valley** and **Santa Ynez**.

Santa Clara County: **Santa Cruz Mountain** (extends also into adjoining Santa Cruz and San Mateo counties).

SOUTH COAST REGION

San Bernadino County: This County has no AVAs; it is one of the original vineyard areas now being swallowed up by the city of Los Angeles.

Riverside County: **Temecula**.

San Diego County: **San Pasqual** – very small.

CENTRAL VALLEY REGION

This, the largest region, comprising the San Joaquin valley to the south and the Sacramento valley to the north, produces 80 per cent of Californian wines. However, it has but three AVAs:

Madera County.

Yolo County: **Clarksburg** and **Merrit Island** are on the banks of the Sacramento River.

Other Counties: No AVAs have as yet been granted for areas within the following: Fresno, Kern, Merced, San Joaquin, Stanislaus, and Tulare.

The Palomino grape is widely grown, and wine labelled 'sherry' is made from it; but this wine is nearer in character to Madeira, perhaps because the 'sherry-making' process in California is often one of heat-treatment.

'Port' is also produced, and it is worth noting that when this is made from grape types found in the Douro, for example the Touriga and Sousão, the flavour is characteristic of Port, whereas if it is made from the Grenache grape, it has a flavour not unlike that of AC Banyuls, which employs the same grape.

1 Spraying the vines with water.
2 Mainly fortified wines.

SIERRA FOOTHILLS REGION

This region lying to the north-east of the Central Valley was established in the Gold Rush days for thirsty prospectors. It now produces red wine from Zinfandel, and white from Sauvignon.

Amador County: **Shenandoah Valley** and **Fiddletown**.

El Dorado County.

Calaveras County: As yet, no AVAs have been granted.

North-Western States

It has been said that the country in Oregon is like England, and its climate is similar though warmer. A wine industry is developing in the hills east and south of Portland in the valley of the Willamette River. The wines are light and well balanced (if a little acid). Pinot Noir and Chardonnay do well.

There are also vineyards in Washington State, in the dry upland valleys of the Columbia basin, between the high ranges of the Cascades and the Rockies. Also further south, in Idaho, there are vineyards on tributaries of the Snake River.

All vines in these states are *Viniferas*.

North-Eastern States

Indigenous vines have influenced the taste of Americans in the eastern states for over 200 years. The principal growing regions in New York State are centred on the Finger Lakes, immediately south of Lake Erie, and the Hudson River valley. The Ohio River Valley has a noteworthy wine industry, now grouped around Columbus and Springfield. A range of red and white light wines is produced.

In these regions, the native vines Niagara, Catauba and Concord varieties prevail, together with the hybrids Dutchess, Aurore, Vidal,

1 Do California wines generally take their names from district, producer, or vine variety?
2 What does AVA stand for, and what is its European equivalent?
3 Name the counties in the North Coast Region of California.
4 Name and locate the AVAs in Monterey County.
5 Is more red than white wine made in San Luis Obispo County?

Seyval, and Maréchal Foch. The introduction of European *viniferas* is slowly finding a market. Some wine is produced on a non-commercial basis in the New England States of Massachusetts, Rhode Island, and Connecticut.

CANADA

Canada's first vineyard was planted by John Schiller, a German, at Cooksville, Ontario, in 1811; by 1890 some 2000 ha of vineyards were thriving in the Province. But public feeling against alcoholic drinks became steadily stronger, and Prohibition in all provinces except Quebec was introduced early in the First World War.

After repeal in 1927, government-monopoly liquor stores were established in all provinces and purchasers required an annual licence in order to buy any alcoholic beverage. This retarded development for 25 years, and only after the Second World War did appreciable progress begin: immigration brought European expertise and with it a taste for European wines.

Ontario

Ontario's advantages included Canada's natural wine centre – the fertile Niagara Peninsula – lying between Lake Ontario and Lake Erie and running west to Hamilton. Thus practically all the vineyards licensed under the post-Prohibition legislation lay in Ontario. The lake waters guarantee the Peninsula mild winter temperatures, and the growing season averages 175 days.

Like neighbouring New York State and Ohio in the United States, wines were originally made from native vines. However production is increasingly concentrated on wines from *vinifera* varieties which has greatly improved quality. A full range of wines is made, including sparkling, crackling (semi-sparkling), dry and medium table wine, and fortified wines.

1 Vine variety.
2 Authorised Vineyard Area: AOC, QbA, DOC, DO, etc.
3 Lake, Mendocino, Napa, Solano, Sonoma, and Sonoma-Carneros.
4 Carmel Valley near the ocean; Arroyo Seco and Chalone inland in the Salinas valley.
5 No; 60% of production is white wine.

British Columbia

This Province became a wine producer in 1934 when William Bennett, a Kelowna trader, started producing wine commercially from imported grapes. When later Bennett became Premier of British Columbia, the incorporation of a majority of grapes grown in the province became mandatory and, of course, the federal licensing system controlled vineyards and wine consumers.

By the mid-1970s the fertile Okanagan Valley contained some 1400 ha of vineyards. The valley, although beyond the northern limit for commercial growing, has a growing season averaging 180 days with irrigation from the Okanaga River. Production includes good varietal red and white wines and some fortified.

MEXICO

Spanish missionaries brought vine and wine to Mexico 450 years ago, but it was only after Mexico became independent of Spain in the nineteenth century that wine production progressed beyond fulfilling the needs of the church. Even then there was trouble ahead.

First help came from James Concannon, a Californian wine-maker, who persuaded President Porfirio Diaz to adopt and plant *vinifera* vines throughout Mexico. Concannon's plan was completed in 1904, when another Californian, Antonio Perelli-Minetti, took over from him and created Mexico's largest vineyard, Rancho El Fresno. But disaster struck in 1910 when the 10-year Mexican Revolution began. Perelli-Minetti retired wisely to the United States during the Revolution, though he later returned with helpful advice.

Mexican Wine Regions

The vineyards of Mexico after the Revolution fell into disrepair, and the wines were adulterated. Finally the Association Nacional de Vitivinicultores decided that the only way to regain credibility was

1 Name the largest region in California, and its three AVAs.
2 Where is the centre of the wine industry in Oregon?
3 Where are the vineyards situated in Washington State?
4 Name the main wine-producing areas of Eastern USA.
5 Mainly, what sort of wines are made in Eastern USA?

to forget the past and regard the Mexican wine industry as the youngest in the Americas, and this has succeeded.

Although Mexico lies in the hot sub-equatorial region south of latitude 32°N her vineyards nearly all lie some 1500 m above sea level on her Central Plateau. For every 300 m above sea level, the temperature drops by 1.7°C, so that the altitude of the Mexican vineyards compensates for their southern latitude. However, night temperatures can be low, and rainfall is variable, causing drought and flooding, and calling for extensive irrigation.

The eight vineyard regions of Mexico stretch south from Ensenada, at the boundary with the United States, via Hermosillo, Delicias, Torreon, Saltillo/Parras, Zacatecas, and Aguascalientes to San Juan del Rio, north of Mexico City. Nine tenths of the country's grapes (mostly Ugni Blanc) are converted to brandy.

Table wines are varietal from Cabernet, Grenache, Merlot, Sauvignon Blanc, and Muscat.

1 Central Valley Region: Madera County, Clarksburg and Merrit Island in Yolo County.
2 East and south of Portland in the Willamette Valley.
3 In the Columbia basin between the Rockies and the Cascades.
4 Finger Lakes and Hudson River Valley in NY State; Columbus and Springfield in Ohio.
5 Red and white light wines from non-vinifera species or from hybrids.

13
The Southern Hemisphere

The arts of viticulture and vinification were taken to the Southern Hemisphere by explorers in the fifteenth and sixteenth centuries. In all of these continents ideal natural conditions exist for vine growing and wine production.

It must be remembered that in the Southern Hemisphere the summer ends in February, and autumn in May, so the grapes ripen between February and April, and the harvest starts in March.

1 Which provinces in Canada produce wine commercially?
2 Where is Canada's natural wine centre?

AUSTRALIA

History

Australia's wine story had a shaky start. In 1787, Captain Arthur Phillips set sail from England with a fleet of eleven ships in which vines were included among 'plants for the settlement'. Unfortunately these, planted near the sea at Sydney, proved to be infected and fell victim to anthracnose; but a second attempt, three years later, succeeded in establishing a 1 ha vineyard at Parramatta, 19 km inland. The first name in Australia's wine history is that of Captain John McArthur, for he was the first to develop wine production on a commercial scale. The story of his early treatment by Governor Bligh has all the savage elements of early colonialism. Despite having his brandy stills confiscated and returned to England, and being tried in the criminal courts for illegally attempting to produce spirits, McArthur survived to develop his winery with his two sons.

In 1824 James Busby, an enthusiastic young Scot, came to Australia cherishing an indomitable faith in the future of Australian wines. He was granted some 800 ha of land 160 km north of Sydney, and there he produced a manual to spread the practical knowledge he had gained by experience. His greatest contribution was made after a tour of research through the European vineyards: he brought back 678 varieties of which 362 were struck successfully at the Botanical Gardens in Sydney. These vines were to propagate the vineyards of Australia.

Until well into the twentieth century Australia's wine exports were of the Burgundy type. A crisis arose after the First World War, when steps were taken by the Australian Bruce-Page Government to create employment for returning soldiers. A programme of grape-growing was embarked upon, despite the warnings of the industry, and a massive overproduction of grapes resulted. As so often happens in this world, the disaster led on to a success. On the advice of the Federal Viticultural Council, subsidies were granted on exports of fortified wine, and producers also enjoyed a drawback of excise on the fortifying spirit, making a total government contribution of four shillings per gallon. They also benefited from British Imperial Preference. As a result, exports of Australian wines rose

1 Ontario and British Columbia.
2 On the Niagara Peninsula of Ontario.

steadily to a ceiling of nearly four million gallons per annum. Australian wine exports have receded slowly since then to about half that figure in 1977; now 'Australian Burgundy' is a thing of the past, and in its place have appeared a range of lively red and white wines, strong and healthy, and strangely but commendably different from European wines from the same vine varieties.

Climate and Soil

The Australian climate and soils are kind to the vine-grower. The former is even and the latter endless in variety, providing all the space and opportunity for selective development. Here one variety of vine planted in differing but carefully chosen soils, can produce wines of widely differing characteristics. As the centre and northern regions of the continent are tropical or too dry, wine producing is restricted to the states of New South Wales, Victoria, and South Australia, with small contributions from Perth in Western Australia, and Roma in Queensland.

The early vineyards were established near the cities, but recently new cooler climate vineyards have been established in southerly areas such as Tasmania, Southern and Central Victoria, and Coonawarra in South Australia. Other cooler vineyards have been established at higher altitudes in the Barossa and Adelaide Hills where 500 m height can mean a month later in ripening. While these cooler areas have made a great improvement in quality, the hotter, high-yielding vineyards will continue to prosper by supplying the bulk wine that represents 60 per cent of the market at present.

Grape Varieties

In the past, most production was of heavy fortified wines, and quality was no more than ordinary; but during the last thirty years there has been a revolution in Australian viticulture and vinification which has yielded wines of superb quality.

Rhine Riesling, Sémillon – confusingly called Hunter River Riesling in New South Wales – and Shiraz (French Syrah) have always

1 Explain how wine can be made in Mexico from vines growing at latitude 23 North.
2 What proportion of Mexico's wine production is consumed as light wine? What happens to the rest?
3 In what months (a) does the vine flower, and (b) are grapes harvested in the Southern Hemisphere?

been popular; but while Shiraz by itself does better in hot climates, it blends magnificently with Cabernet Sauvignon if grown in a cooler climate. Logically, because it *is* the grape of Hermitage, Shiraz is known by that name in Australia: confusingly, the Cape winegrowers use the name for their clone of the French Cinsaut, which they have also crossed with Pinot Noir, giving the name Pinotage to the offspring. Pinotage is unknown in Australia, but Pinot Noir is a relative newcomer, as are Chardonnay and Sauvignon Blanc; there is

1 Vineyard temperatures are reduced by their height above sea level.
2 10%. The remaining 90% is distilled to brandy.
3 (a) November/December. (b) March/April.

some interest in Cabernet Franc and Merlot, but the main research is to find disease- and frost-resistant clones for the risky 'cooler climate' vineyards.

Vineyard Regions

The comparative lack of AC-type controls has allowed the Australian wine industry to develop rapidly and to use new production techniques. Moreover, as blending is commonplace, and wines are more likely to be sold under the name of their producer, individual districts matter little – at present.

South Australia

This state produces over 60 per cent of all Australian wine. The main districts are Adelaide Metropolitan and Southern Vales in the hills just south of the capital and the Barossa Valley to the north. The best vineyards are those new ones planted higher in the Barossa Range, in the Clare/Watervale district further north, and in cooler Padthaway/Keppoch and Coonawarra well south of the capital.

Barossa is famed for its Shiraz, and Clare with its limestone soil for its Rhine Rieslings which improve considerably with maturation. Coonawarra has a very limited area of fine red soil over limestone, and produces fine wines from Cabernet Sauvignon and Chardonnay grapes. Langhorne Creek is noted for fine dessert wines.

New South Wales

This, the original state of Busby fame, is still the second-largest producer with 3500 ha under vines, and has a promising future with wines made in the best modern conditions; the best of these are the Rhine Riesling and the Sémillon (Hunter Valley Riesling), which both improve with several years in bottle. Hunter River Valley was developed by Busby. Other districts include Forbes, Mudgee, and Rooty Hill in the north, and Riverina and Corowa (renowned for dessert wines) in the south.

1 Who was James Busby, and what did he do for Australian vineyards?
2 Describe the current trends in Australian winemaking.
3 How much of the present Australian Market is for fine wines?

Victoria

Although Victoria at one time had the largest wine industry, which had been established by Swiss immigrants in the classic European style, it was the only state to suffer from the dreaded *phylloxera* which, starting at Geelong near Melbourne, spread north to Bendigo; although few of the original vineyards have been reclaimed, the state is noted today for high-quality red and white wines for the table.

The principal centres are Rutherglen, on the Murray River opposite Corowa, and Milawa, the Goulburn Valley, and Yarra (just east of Melbourne). The last three, particularly, at the southern end of the Great Divide, are gaining repute for their red and white table wines, while Great Western maintains its own reputation for fine sparkling wines.

NEW ZEALAND

Although the first vines were planted in 1819 and James Busby, pioneer of the Australian Wine industry, tried to run a vineyard, a combination of wars, natural disasters, governmental interference, and prohibition campaigns strangled the Trade. In the Depression of the 1930s the viticultural research station at Te Kauwhata was put up for auction – there were no bidders (it is now going strong again)! In 1939, total production was under 1 million litres: in 1986 more than that quantity was exported, and total production in the previous sixteen years had jumped to 50 million litres.

Fortified wines, which used to dominate the market, have now dropped to a 15 per cent share; table wines, which only twenty years ago were called 'Hock' if white and dry, 'Sauterne' (sic) if white and sweet, or 'Claret' or 'Burgundy' if red, are now being produced in modern wineries from sound varieties. Much is blended and sold as cask wine – 'bag-in-the-box' to the English, 'Château Cardboard' to the New Zealander. However, excellent bottles of New Zealand Cabernet Sauvignon and Riesling can be found in Europe, and it is noteworthy that the UK imported more wine from New Zealand in 1986 than did its neighbour Australia.

1 A Scot who successfully raised 362 varieties of European vine in Australia.
2 New plantings in cooler climates away from the cities, where finer wines can be made.
3 About 40%, and increasing.

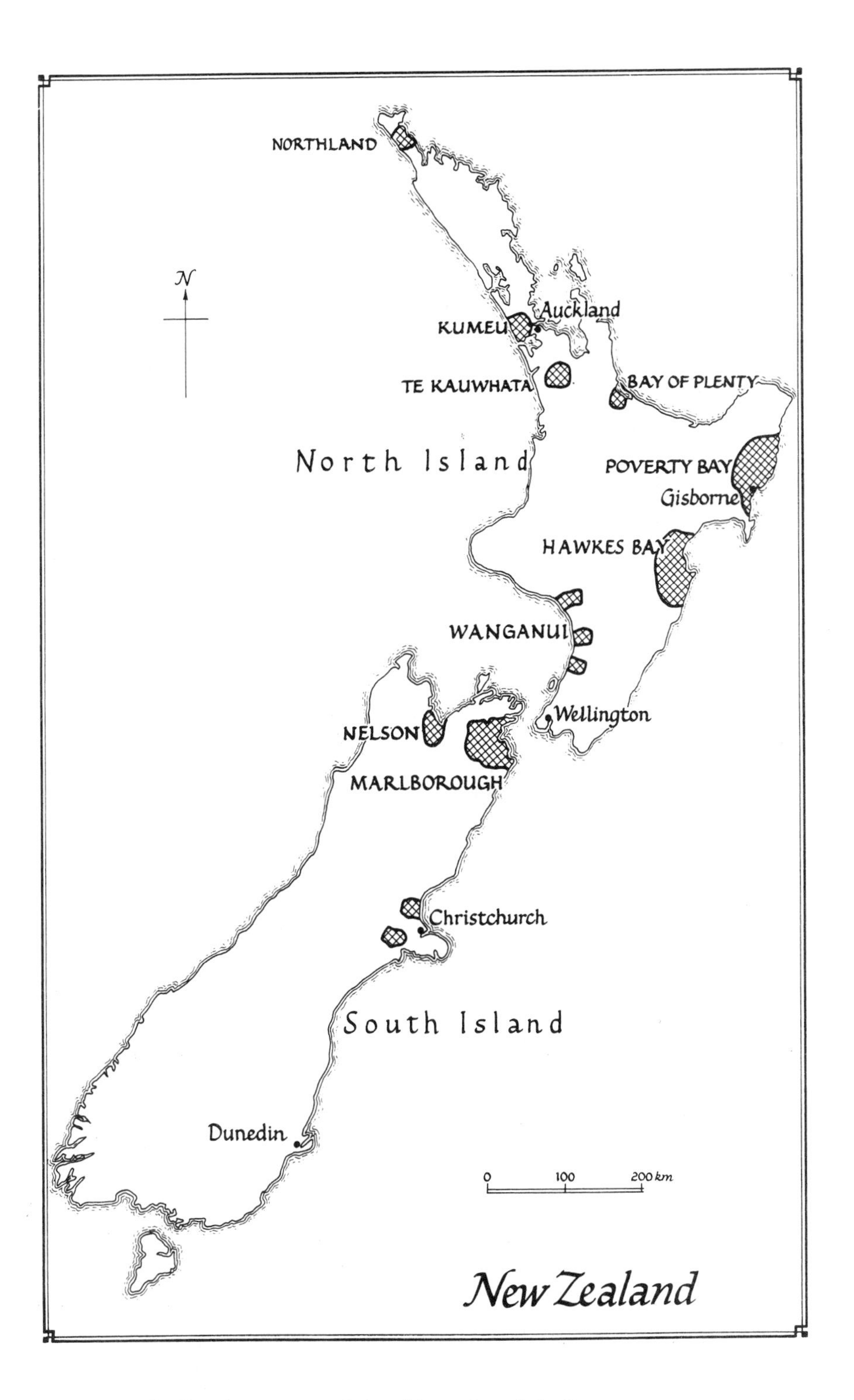

1 Contrast the Hermitage grape varieties of Australia and the Cape.
2 What is the current line of Australian research?
3 Which Australian state produces the most wine?
4 (a) Name the best districts of South Australia for wine production. (b) Which makes outstanding Riesling?
5 (a) Name the district founded by Busby. (b) In which state does it lie? (c) What is the area of its vineyards?

Vineyard Areas

In the North Island the small vineyards in the Auckland region remain at about 500 ha; the vineyards in the traditional Poverty Bay and Hawkes Bay regions on the east coast have stabilized at about 1500 ha each. However, in the South Island new vineyards in the Marlborough region have grown to 1000 ha in a short time and are developing well; other vineyards have been planted even further south on the east coast at Christchurch and Otago. As in Australia, the finer wines come from the cooler regions, but the need to take precautions against frost and disease puts their price up.

THE CAPE

History

The first vineyard of the Cape was planted by Jan van Riebeeck as soon as the Dutch settlers arrived in 1652. In 1679, Simon van der Stel, Governor of the Cape of Good Hope, planted vines at Constantia, 20 km south of Cape Town; thirty years later Constantia wines were exported. In the meantime, French Huguenots arrived and set about the task of improving the Cape wines, and, true to their traditions, succeeded. The wines of the Cape, and of Constantia in particular, became reputed abroad. When, in 1805, the British took possession of the Cape, a considerable trade with England began and developed for the next forty years, offsetting the loss of European wines through the Napoleonic Wars. But when Gladstone removed preferential tariffs, this seriously affected the South African trade.

To make matters worse, *phylloxera* followed in 1885 and devastated the vineyards. The long process of replanting with grafted vines led eventually to a state of serious overproduction and uneconomic prices; this brought about the formation (in 1918) of the Co-operative Wine Growers' Association (KWV). The Wine and Spirit Control Act – passed six years later – empowered the Association to fix prices for all distillation of Cape wine.

1 Cape Hermitage derives from Cinsaut; Australian Hermitage from Syrah.
2 To find frost and disease resistant clones of classic European varieties.
3 South Australia, with more than 60% of the national production.
4 (a) The Barossa Valley, Clare/Watervale, Padthaway/Keppoch, and Coonawarra. (b) Clare.
5 (a) Hunter River Valley. (b) New South Wales. (c) About 3500 ha.

Soil and Climate

In the Coastal Plain the soil ranges between Table Mountain sandstone in the west and granitic compounds on the mountain slopes farther east. In the Karoo, shales predominate, deriving from the red soil, but alluvial sandy soils are found in the river valleys where most vineyards are located. The regional climate is mild, of mediterranean characteristics with, however, a considerable variation in rainfall between the coastal areas (which have 850 mm annually) and the more mountainous country inland, where rainfall is less than half as much. The yield in grape harvest is under 12.5 tonnes per hectare near the coast and 20 tonnes per hectare in the Little Karoo, where the vines are grown under irrigation.

Vine Varieties

The most common cultivars in the Coastal Region are the Cabernet Sauvignon, Shiraz, Cinsaut (called Hermitage), and Pinotage (a cross between Hermitage and Pinot Noir) for unfortified red wines, and Steen (a clone of Chenin Blanc), Sémillon, and Cape Riesling for whites. In the Breede River Valley region and the Klein Karoo district the Cinsaut, Trebbiano, Colombard, Palomino, Muscat d'Alexandrie, and Steen are most commonly used. For fortified wines, the Palomino is being succeeded by the Steen for production of the Sherry types, while Souzão and Tinta das Baroccas are used to produce port-type wines. Most fortified wines are produced in the Boberg and Breede River regions.

Labelling

A new system of demarcation was introduced in 1972 and a straightforward method of labelling devised, so that purchasers might see on one label the origin (area or region of production), vintage, vine variety, individual estate, and quality of the bottle's contents. The Wine and Spirit Board ensures compliance with these regulations.

1 Why did Victoria decline from being Australia's greatest wine-producing state?
2 Which district of Victoria produces the best sparkling wines?
3 When were the first vines planted in New Zealand, and when did production start expanding?

The Cape winelands are divided into three regions and seven lesser detached districts under an Act of 1972 and subsequent regulations. The Coastal region comprises seven districts, the Breede River Valley region three; Boberg is a fortified wine region for the districts of Paarl and Tulbagh in Coastal region.

Wines of Origin, corresponding to EEC qwpsr, must bear the name of the region or district on the label as well as the name of the producer – and 100 per cent of the wine must come from that region or district. The vintage may also be shown if 75 per cent of the wine is of one year, and the name of the cultivar (grape variety) if 75 per cent of the wine is made from that grape *and* a tasting test shows the grape characteristics.

Wines of Origin bear a numbered seal on the neck of the bottle; the background is white, or gold if the wine has been certified by the Wine and Spirit Board as 'Superior'. One to four horizontal bands appear, the top three coloured, the bottom background; from the top, their markings and meaning are:

Origin (blue):	Wine comes from the Region or district named on the label.
Vintage (red):	Wine is of the vintage stated on the label.
Cultivar (green):	Wine comes from, and is characteristic of, the variety shown on the label.
Estate (black):	Wine comes from grapes grown and vinified on the estate named on the label.

Coastal Region

This region lies between the sea and the north-south mountains of the Winterhoek, Drakenstein, and Simonsberg ranges. With the prevailing westerly wind from the sea, the climate is mild and mediterranean in character; it has from 600 to 850 mm rainfall, mostly during the winter months from June to September.

1 It was the only state affected by *phylloxera*.
2 Great Western.
3 In 1819; production started expanding in the 1970s, and is increasing.

Swartland

The vineyards are situated on the eastern side of this open and undulating plain, in the foothills of the mountains where the soils are generally gravelly, with sand or clay. The wines are mostly red and full-bodied, with some sweet and medium-sweet white, and some fortified.

Estates: Allesverloren, and three co-operatives.

Tulbagh

This district lies to the east of Swartland in a deep valley with very stony soil. Production is almost entirely of white wine, principally from Steen (Chenin Blanc) and Cape Riesling (Cruchen Blanc) vines. Some is fortified for dessert wines.

Estates: Montpellier, Twee Jonge Gezellen, Theuniskraal, Lemberg, and two co-operatives.

Paarl

This district lies south of Tulbagh at the foot of the Drakenstein Mountains; the soil is mainly granitic, with some sandstone in the south; lime has to be applied to the soil, as it does not occur naturally. Although most of the production is of white wine, including excellent fortified aperitif and dessert wines, substantial quantities of fine red wine are made, and good brandy.

The great wineries of KWV and Nederburg are situated in the distrct; although the latter has its own vineyards, the bulk of its production comes from farms in this and neighbouring districts.

Estates: Villiera, Landskroon, Fairview, Draaihoogte, De Zoete Inval, Douglas Green, Morris, Union Wine, and seven other co-operatives.

1 Where are the new vineyards being planted in New Zealand?
2 Who planted the first vineyard in the Cape, when, and where?
3 Which has produced wines longer, Australia or the Cape.
4 What is the KWV, when was it formed, and what does it do?

Franschhoek

Originally part of Paarl, this has now been declared as a separate district, lying at the upper end of the Drakenstein valley on granitic soil. Production is mostly of white wine.

Estates: Backsberg, Boschendaal, l'Ormarins, La Motte, Welgemeend, Bellingham, and three co-operatives.

Stellenbosch

The most important, and the second oldest district of the Cape, Stellenbosch lies to the south of Paarl on the western slopes of the Simonsberg and the banks of the Eerste river, extending south to the sea at False Bay. The granite soils of the north and east favour red wine production, while the sandstone soils of the south are better for white wines; lime has to be applied to the soil in the river valley.

Production is dominated by Stellenbosch Farmers' Winery (owners of Nederburg in Paarl), the Bergkelder (owned by the Oude Meester Group), and Gilbeys, all of whom have their headquarters in Stellenbosch; these take wine from the co-operatives and blend, process, age, and bottle it. They also bottle wines from estates. The town is also the home of the Research Institute and the School of Agriculture.

Estates: Vergenoegd, Meerlust, Spier, Vredenheim, Jacobsdal, Audacia, Alto, Rust-in-Vrede, Klein Zalze, Blaauwklippen, Klein Vriesenhof, Overgaauw, Goedgeloof, Uiterwyk, Bonfoi, Neethlingshof, Zevenwacht, Hazendal, Middelvlei, Mooiplaas, Goede Hoop, Bellevue, Koopmanskloof, Hartenberg, Devonvale, Verdun, Oude Nektar, Simonsig, Muratie, Delheim, Kanonkop, Uitkyk, Le Bonheur, Rustenberg and Schoongezicht, Lievland, and six co-operatives.

Durbanville

This small district lies between the sea and Paarl/Stellenbosch, just to the north of Cape Town, where its vineyards are in danger of being

1 In the South Island; Marlborough, Christchurch, and Otago, with new areas being explored in cooler regions.
2 Jan van Riebeeck, in 1652, at Constantia.
3 The Cape, by about 100 years.
4 Co-operative Wine Growers Association (of South Africa): 1918: controls wine and spirit production and price.

swallowed up by the extending city. They lie on the north-eastern side of the Tygerberg, and produce red wines almost exclusively in a cool climate.

Estate: Meerendaal.

Constantia

This is the oldest district in the Cape, and lies on the southern slopes of Table Mountain. It is famed for the rich red dessert wine from Muscat grapes which is still made today, although the emphasis is now more on light red and white wines.

Estates: Groot Constantia, Klein Constantia, Alphen, Buitenverwachting.

Breede River Valley Region

This is watered by the Breede River, which rises in the mountains east of Tulbagh and flows south-east between the Riviersonderend and Langeberg ranges to reach the Indian Ocean some 200 km east of Cape Town. Lying in the hinterland, its climate is hotter and drier than that of the Coastal region, and irrigation is necessary at all times. The soils are rich in lime and grapes grow well, with high sugar; most of the Cape's fortified and distilling wines come from this region, from Muscat (Hanepoot) and Palomino grapes. However, cold fermenting techniques and the planting of new varieties are leading to the production of finer light wines.

Worcester

This district adjoins Tulbagh at the upper end of the Breede valley, where the soils are lime-rich sands and clays with alluvial loams.

Estates: Bergsig, Lebensraum, Opstal, and twenty co-operatives.

1 What are the regions of the Cape winelands? Are they all separate?
2 Which districts of the Coastal Region compose Boberg Region?
3 What are the Cape equivalents to EEC qwpsr, and how may their bottles be recognized?
4 What does the word 'cultivar' indicate?
5 What do the Wine of Origin neck-label bands indicate, and what are their colours?

Robertson

Lower down the Breede river, the vineyards are mostly situated close to its banks. Production is mostly sweet white wine from Muscat (Hanepoot) grapes.

Estates: Wonderfontein, Le Grand Chasseur, Mont Blois, Rietvallei, Mon Don, Zandvliet, Bon Courage, De Wetshof, Van Loveren, Excelsior, Ardein, Weltevrede, Goedverwacht, and seven co-operatives.

Swellendam

Still further down the Breede valley, the vineyards lie between the river and the southern foothills of the Langeberg for 60 km from Bonnievale to Suurdam. On the other side of the Langeberg lies the district of Klein Karoo, often bracketed with Swellendam, but not part of the region. There are no estates, but four co-operatives.

Marketing

Many Cape growers send their harvest to co-operative wine cellars (including the KWV) for vinification, although an increasing number produce their own Estate wines. Estate and co-operative wines are marketed through wine merchants.

Until recent times little was known in England of Cape wines, except for Cape Sherries although, until South Africa left the Commonwealth, these enjoyed preferential rates of duty in the UK; however, due to the activities of the KWV, wines for the table are becoming better known, particularly the whites. Of the white wines, the best is the Steen, and of the reds, Hermitage and Shiraz. Fortified wines of port- and sherry-type are produced in considerable quantity and to a highly acceptable standard.

1 Coastal Region, Breede River Region, and Boberg Region which is for fortified wines within Coastal Region.
2 Paarl and Tulbagh.
3 Wines of Origin; a neck-label with a white background, and up to four descriptive bands.
4 The Cape name for a vine variety.
5 Origin (blue), Vintage (red), Cultivar (green), and Estate (black print).

SOUTH AMERICA

Vineyards in South America are found over a wide range of latitude from 40° to 10°s, in Argentine, Brazil, Chile, Peru, and Uruguay. In Peru, the most northerly, the nearness to the equator is balanced by the height above sea-level of between 500 and 1000 m. To a lesser extent this also applies to Brazil's vineyards in the Sierra do Mar, nearly at the Tropic of Capricorn.

CHILE

The vineyards of Chile are exceptional and remarkable. They cover some 110,000 ha and stretch along the Pacific coast of South America for over 2000 km through climates ranging from arid in the north to very wet in the south. The Andes to the east, a prevailing on-shore wind from the southern Pacific, and the deserts in the north have succesfully prevented the entry of *phylloxera*, which has attacked vineyards in other parts of the continent.

Like California, Chile's shores are washed by a cold polar stream (the Humboldt current) producing coastal fogs which make fine wine production possible.

The original vineyards were developed by the Spaniards; later, French vignerons brought in by the Government of Chile planted the better vineyards from which South America's finest wines are now drawn.

Chile comprises three distinctive wine regions. The arid north, from the Atacama desert south to the Choapa river (about 150 km north of Valparaiso), produces fine fortified wines resembling Sherry, Port, or Madeira, but made principally from Muscat grapes. A quantity of this wine is distilled to produce Pisco brandy – the fruit spirit of South America which originated in Peru. In the central region, between the Aconcagua river in the north and the Maule river, the best wines are made, for here the climate is ideal for vine growing, although the vineyards need irrigation as they do in

1 Name the districts in the Coastal Region.

California. Bordeaux vines predominate, including the Cabernet Franc, Cabernet Sauvignon, Malbec, Merlot, and Petit Verdot for red wines, and the Sauvignon and Sémillon for white. The Bordeaux tradition extends to the methods of viticulture and vinification. The finest wines, long-lasting, balanced and strong, are made in the Aconcagua and Maipo valleys and have some similarity to good Spanish Riojas. Though the soil has more limestone here than elsewhere, the red wines are better than the white wines.

The southern region, from the Maule down to the Bio-Bio river, is planted mainly with the indigenous País vine, although some Cabernet, Malbec, and Sémillon are grown, with a lesser quantity of Riesling. Here the rain is heavy and only in the north of the region is irrigation necessary. The wines are ordinary and cater for domestic consumption.

The government controls of wine in Chile are strict. The volume of production is controlled, and surplus wine must be converted to fruit spirit or industrial alcohol. Exported wines must have 11.5% vol. alcohol for red or 12% vol. for white wines, and must be described, according to their age, as *Courant* for one year old, *Special* for two, *Reserve* for four (possibly the best) and *Gran Vino* for six years old or more.

ARGENTINA

The wines of Argentina are generally undistinguished and about 70 per cent of total production is *vin ordinaire*. In fact there are so far virtually no laws to control vintage dates, place names, or brand names, with the result that 'Margaux', 'Chablis', 'Rioja', 'Beaujolais', and 'Champagne' abound, and may have little or no relationship to the European originals, either in the variety of grape used or in the process of manufacture. Three-quarters of Argentina's wine is produced in the province of Mendoza, and the neighbouring province of St Juan to the north. Both stretch out into the vast plain east of the Andes foothills, where enormous wineries ferment wine in

1 Constantia, Durbanville, Franschhoek, Paarl, Stellenbosch, Swartland, and Tulbagh.

staggering quantities, which may account for the lack of character of the resulting product. The Uco valley is central in the Mendoza region, and here Malbec, Sémillon, and Tempranilla grapes are commonly used to produce the better quality wines, including the Criollas. The vineyards are mostly irrigated, and *phylloxera* is dealt with by flooding and drowning the pest, which does not do the vine much good: although ineffectual, this treatment tends to wash the pest from one region to the next.

The province of Rio Negro, south of Mendoza, accounts for some 5 per cent of the country's wine. Other important areas of wine production include Occidente and Norte (both in the plains under the Andes in the north-west of Argentina), Cordoba in the central region, and Litoral and Entre Rios in the east. Two-thirds of Argentinian wine is red, the balance being mainly white, with some rosé, sparkling, and fortified wines.

BRAZIL

The majority of the 836 million ha of Brazil is tropical – and unsuitable for wine production. But at the southern tip of the country a considerable wine industry has grown since the beginning of the twentieth century, largely initiated by Italian immigrants. Various *vinifera* vines have been imported and do well, but the most common vines in the country are Dutchess, Niagara, and Isabella. Some 70,000 ha is devoted to vine growing, mostly in the province of Rio Grande do Sul, which borders the Atlantic, and runs south into Uruguay and inland to Argentina. To the north are three smaller regions, São Paulo, Rio de Janeiro, and Minas Gerais, and a new fourth region, Santa Catarina, which ships its grape harvest to São Paulo for fermentation.

Good red wines are produced principally from Barbera, Bonarda, Cabernet, and Merlot grapes, and white wines from the Malvasia, Riesling, and Trebbiano.

Practically all of Brazil's output is consumed within the country, which previously relied on supplies from Chile and Argentina. The

1 Which districts in the Cape produce mainly white wines?
2 Name the districts in the Breede River Region.
3 What natural features have prevented *phylloxera* from reaching Chile?
4 What rivers mark the boundaries of the best wine region of Chile?

range of Brazilian wines is considerable and includes sweet and dry red and white wines, both still and sparkling, and a selection of sweet fortified wines and vermouths. As the wine industry has grown, local demand for wine has grown also; consequently exports remain relatively small.

1 Franschhoek, Tulbagh, Paarl (also much red), and all districts of Breede River.
2 Robertson, Swellendam, and Worcester.
3 Deserts in the north, the Andes to the east, and a prevailing on-shore wind from the southern Pacific.
4 Aconcagua and Maule.
(Questions continued on page 301.)

PART FOUR

Spirits

Introduction to Part Four

The history of distilled spirits is as long as that of wine or indeed of any alcoholic beverage; for it has always been in man's nature to seek out the secret of his existence and to create the means of prolonging it. Distillation was held to produce 'the water of life', commonly called spirit.

Spirits can be distilled from any alcoholic beverage – a weak mixture of ethyl alcohol and water. There are always two steps: stripping the bulk of the water from the mixture, and purifying the mixture which remains. The two steps may take place consequentially in one apparatus, but they are still distinct.

Production of spirits from fruit is simplest, because fruits contain sugars which can be fermented directly; these will be considered first. Grains used for spirit production must have their sugars (starches) converted by malting or other processes before they are capable of alcoholic fermentation.

Other sources of potentially fermentable sugar, such as agave, beet, cane, potatoes, or yams have to be treated before fermentation. Finally, spirits may be flavoured for medicinal or other purposes.

1 (a) Where is most of Argentina's wine produced? (b) Which grapes produce the best wines here?
2 (a) Who were responsible for developing Brazil's wine industry? (b) In what province is it largely located?

14
Fruit Spirits

Wherever fruit is fermented to produce an alcoholic wash, a fruit spirit may be distilled from that wash. The essential definition of a fruit spirit, therefore, is an unsweetened spirit made by distillation of fermented natural fruit juice. The most common fruit spirit, and certainly the oldest, is brandy made by distilling wine made from grapes. This liquid has had many names of which the earliest was elixir (El Ixr – 'the spirit'); the Romans called it *aqua vitae* – 'water of life' – and the French still call it eau-de-vie-de-vin – 'water of life of wine'. In the Middle Ages the Dutch merchants sailing from La Rochelle with cargoes of sea-salt from the region also brought back the distilled wine of the Charentes which they called *brandewijn* ('burnt wine'): the English corrupted this to 'brandy'. The Germans

1 (a) In Mendoza and St Juan provinces. (b) Malbec, Sémillon, and Tempranilla.
2 (a) Italian immigrants. (b) Rio Grande do Sul.

call it *Weinbrand* ('burnt wine'); the Spanish call it *aguardiente* ('ardent water'). There can be no doubt that the world's most popular brandies are French, and the finest come from Cognac and Armagnac.

COGNAC

Up to the seventeenth century, the Départements of the Charentais had developed a trade in inferior wine with England and the Low Countries; mostly the wine slaked the thirst of the seamen who came to buy salt. But at this time the practice of boiling down the coarse wine began, mainly to avoid the tax then levied on bulk. The

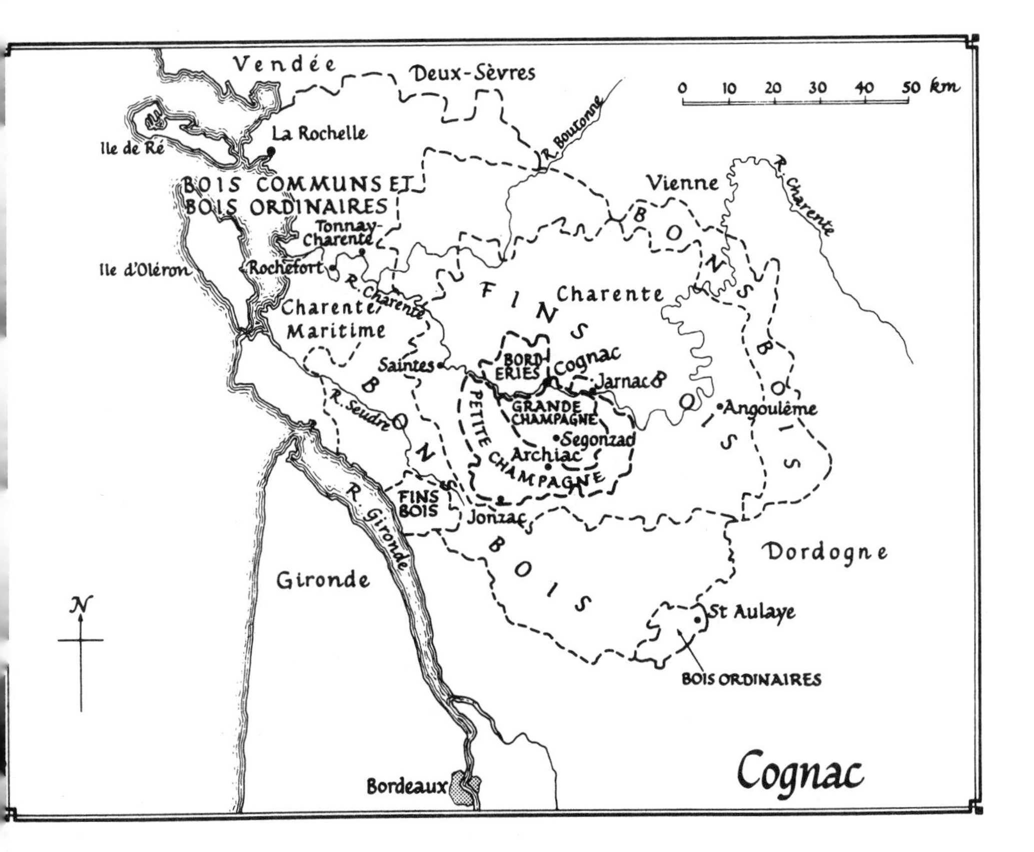

1 Why is little of Brazil's wine production exported?
2 What was distillation originally held to produce?
3 What are the two essential steps in distillation?
4 Why is the production of spirits from a fruit base easiest?
5 What main process enables grain to be fermentable?

amazing thing was that the spirit was so much finer than the wine from which it was produced. This soon attracted the attention of four shrewd traders: Mr Hennessy from Ireland; Mr Hardy and Mr Hine from Dorset in England; and Mr Martell from the Channel Islands. With their interest, the foundation of a great industry was laid. Today some 22,000 growers throughout the Cognac region provide wine for processing into Cognac. Less than 10 per cent of them distil their own wine, and the number is diminishing.

With a small part of the Département of Deux Sèvres, the Cognac region comprises the entirety of the two Départements of Charente and Charente Maritime that take their names from the river Charente which rises in the hills north of Angoulême and, passing through that city, flows through the gently rolling countryside to enter the sea at Rochefort. Halfway between Angoulême and the sea, the little town of Cognac straddles the river, giving its name to the region and the spirit produced there.

Climate and Soil

Lying just to the north of the Gironde and the Bordeaux region, Cognac enjoys much the same climate of warm summers and cool winters, with south-west winds bringing moist air from the Bay of Biscay; however, there are no coastal pine forests to protect the vineyards, and those near the coast suffer from the salty air; the atmosphere over the whole region is humid.

The region is divided into six districts of differing quality, with the best in the centre and the others encircling them. The quality and character of Cognac depends greatly on the proportion of chalk in the soil. The town of Cognac is at the centre of the region, where vineyards are most concentrated and the soil is best. Here is the rolling open chalk downland described by the French as *campagne* – champagne in the old spelling, and very similar to the great region of that name. Of the six districts, the central one with a soil of soft chalk lying just south of the river Charente, is given the name Grande Champagne (11 per cent of the vineyards of the region). Surrounding this district to the south, with a small portion to the north of the river, is the district of Petite Champagne (17 per cent of the

1 Brazilians drink it.
2 'The water of life'.
3 First, stripping the bulk of water from the wash; second, purifying the distillate.
4 Because the fruits contain directly-fermentable sugar to make a wash for distillation.
5 Malting.

vineyards), in which the soil, while still calcareous, is more like limestone. Just to the north of the town of Cognac itselif is the smallest district, Borderies (4 per cent), where the soil is of marly clay. These three districts, although in area only 35 per cent of the vineyards, produce most of the fine Cognac.

In ancient times, this downland area was surrounded by forests or 'bois' which were gradually cleared for cultivation. In general their soils are clayey with varying amounts of lime, but become sandier towards the coast. The three remaining districts of Cognac form concentric circles around the central group and are named in decreasing order of quality, moving outwards – Fins Bois, Bons Bois, and Bois Ordinaires et Communs. In these districts dairy farming predominates and vineyards are increasingly rare – on the Ile de Ré (which with the Ile d'Oléron is included in the Bois Communs) there is only one small vineyard.

The main vine species was originally the Folle Blanche – a Picpoul variant – but today the Ugni Blanc, for which the local name is the St Emilion des Charentes, accounts for over 98 per cent of production. Colombard, Folle Blanche, and a few other permitted varieties make up the balance, but individual plantings are tightly controlled. The St Emilion is a heavy cropper in the mild humid climate of the Charentes and, in contrast to the fine wine Appellations, the yield per hectare is not limited.

The white wine of Cognac is sour and acid, for the sun is not hot enough to ripen these particular grapes sufficiently; the Ugni Blanc flourishes further south in Algeria and California; in Italy, as the Trebbiano, it produces fine white Orvieto DOC. The wine for Cognac is made quite normally, except that the continuous screw press is not permitted nor any 'improvement' by addition of sugar. As soon as fermentation is finished the wine goes to the stills, for it must all be distilled before 31 March of the following year if it is to be provided with a certificate of age (otherwise it loses one year).

Distillation

By law, Cognac must be produced by a double distillation in the approved Charentais still, which, for the second distillation, must

1 Define a 'fruit spirit'.
2 What was 'brandewijn' and to what did it give its name?

not have a capacity exceeding 30 hl, nor contain a charge of more than 25 hl. The still must be fired by a 'naked flame', which excludes oil, steam or electric heating.

The most modern stills use gas, but coal briquettes are often used. The apparatus consists basically of a boiler (*chaudière*), capped with a head which may be one of two different shapes, either 'Turk's Cap' or 'Olive'. A swan-neck pipe leads from it to a condenser cooled with running water. Some stills may be fitted with a pre-heater, but this is not done in every case as it is an expensive piece of equipment and opinions are divided on its effectiveness.

The wine entering the still has an alcoholic strength of about 8–9% vol. After the first distillation, the condensate or *brouillis* reaches a strength of 30–35% vol. and its volume will be reduced by about 30%. The still is then emptied and cleaned, and a fresh lot of wine is run in. After this operation has been carried out three times, there will be enough brouillis for a full charge for the second distillation (called the *bonne chauffe*), during which the 'heads' and 'tails' are separated from the heart, and added to the next charge of brouillis. The eventual strength of the heart must not exceed 72% vol.

Maturation

After the two distillations, the water-white heart, raw and fiery in character, is put in barrels of oak to mature. This oak must come from the forests of Limousin around Limoges, some 160 km east of Cognac, or from those of the Tronçais east of Châteauroux further to the north; these oaks have a low tannin content and are porous, two qualities essential to the production of fine brandy. The oak should come from trees approaching a hundred years old which will give a mellowness to the spirit; young wood will give an undesirable sharpness. The wood must be air-seasoned for at least four years before being split into staves (never sawn).

It has been said that brandy owes as much to the oak in which it matures as to the grape from which it is made. Brandy starts as an almost colourless spirit and during maturation takes on colour, finesse, and much of its flavour from the tannin and other chemicals in the oak. At the same time brandy loses alcohol by evaporation

1 An unsweetened spirit made by distilling fermented fruit juice.
2 The name given by the Dutch to the distilled wines of the Charentes. The English shortened this to 'brandy'.

through the pores of the cask. Considering that the best brandy comes from wine of harsh quality it will be realized that the resultant spirit is very much a creation of man as well as nature.

The new brandy goes into a new barrel for one year; during this time the brandy will absorb nearly 2 kg in weight of oak products and will lose bulk by evaporation through the pores of the cask, to be replaced by air which itself will combine with the constituents of the spirit. After a year, the brandy will be transferred to an older barrel. Throughout maturation in cask, the spirit continues to evaporate – it has been reckoned that the evaporation loss in Cognac each year is equivalent to the total consumption of brandy in France – which the French phlegmatically call the 'angel's share'. As the loss consists largely of the undesirable volatile head element, the angels do not get a very good deal. This evaporation manifests itself in a fungus as black as night (*torula*) which appears in the chais and on their roofs.

During maturation, besides decreasing in volume, the brandy will change by oxidation, hydrolysis, acetylization, and absorption of tannin. The brandy is matured in *fûts* of 340 litres, topped up from a partially-full *chanteau* as necessary; old brandies will be transferred to 540 litre *tierçons,* where they will mature more slowly.

There are one or two popular misconceptions about brandy which it may be well to dispel forthwith. Brandy matures and improves only while it is stored in wooden barrels, and even there usually reaches its prime after about forty years, after which it becomes tainted with a woody smell: from the day that brandy is bottled (unlike wine) its condition is static – provided of course that storage conditions are right. This goes to show that 'Napoléon' brandy (even if it originated in the time of Napoléon I – or for that matter of any succeeding Napoléons, which is equally doubtful) can have no claim to maturity of more than forty years.

When fully matured (in warehouses spread throughout the region) the brandy is brought into the shippers' establishments for blending – a slow process in which the different brandies are given time to marry. At this point the brandy will have a strength of nearly 70% vol. which would take half a century to reduce by evaporation to the required shipping strength of 40% vol. To wait so long is an economic impossibility, and so distilled water is added before bottling; to reduce the shock of this addition, *faible* or weak brandy

1 Name the four 'shrewd traders' who discovered and encouraged the production of Cognac, and say whence they came.
2 Name the three districts of Cognac where the best brandies are produced.
3 Name the three outer districts of Cognac.
4 (a) Is the yield limited in Cognac? (b) Is chaptalisation allowed?

of 27% vol. is often used instead, and even then the reduction is tempered in stages of about 8% vol. at a time. At the same time caramel may be added to correct colour.

Cognac is an AC spirit, and there are separate Appellations for Grande Champagne, Petite Champagne, Borderies, Fins Bois, and Bons Bois. One condition applying to each is that it must be matured in a separate warehouse containing no other district's Cognac and, for AC Cognac, containing no spirit that would not be entitled to the AC. If the inspector were to find a bottle of Scotch, or even of AC Armagnac, all would become nameless eau-de-vie-de-vin. There is one further Cognac Appellation which must be mentioned – Fine Champagne; this is a blend of Grande Champagne and Petite Champagne Cognacs only, of which not less than 50 per cent must be Grande Champagne. Grande Fine Champagne, on the other hand, must be 100 per cent Grande Champagne, and is therefore a synonym for that Appellation.

The characteristics of the brandies from the six districts vary considerably. Those of Grande Champagne have superb finesse and bouquet and need to mature for at least fifteen years. Petite Champagne produces a slightly less distinguished brandy; it has less body than Grande Champagne and matures more quickly. Borderies produces brandy with the greatest body of all the Cognacs. In the outer districts the brandies, because of the richer or sandier soil, or proximity to the sea, lack the elegance and finesse of those from the calcareous centre.

Up to the maximum of about forty years (for the finest brandies) or ten (for the Bois), quality will vary with the length of maturation. AC Cognac may be sold as such for consumption in France after one year: if it is to bear a *marque* it must mature for three years, or four if this is VSOP. These initials stand for 'Very Special Old Pale', meaning a brandy that had not been treated heavily with caramel; the name is a relic from the days when OB – Old Brown – Cognac was popular.

The marque is usually ★★★ although some brands may have one, four, or five stars, and others a brand name; it does not denote the age, and indeed most brandies of this quality have an average age of about five to seven years, remembering that they are all blends. VSOP Cognacs are usually of ten to fourteen years age. Most

1 Hennessy from Ireland, Hardy and Hine from Dorset, and Martell from the Channel Islands.
2 Grande Champagne, Petite Champagne, and Borderies.
3 Fins Bois, Bon Bois, and Bois Ordinaires et Communs.
4 (A) No. (b) No.

shippers market other, finer blends to a restricted market at great price: they used to add V's and O's to VSOP until M. Martell thought that VVVSOOOP was a bit much for his fine Cognac, and renamed it 'Cordon Bleu'. These fine old brandies are generally known as 'liqueur Cognacs' because they are taken at the end of a meal as digestifs, although there is no trace of sweetening in them as there is in liqueurs.

Cognac imported into the UK cannot be offered for consumption until it has matured in cask for three years, and this led some British merchants to import it in cask and to mature it in their own cellars in, say, Bristol or London. Curiously, this produced great changes in style for, in the damp cold climate of England, the spirit lost strength but little bulk, whereas in warm, dry Cognac, it lost bulk but little strength. This system can also produce vintage Cognacs; the French will certify the date of harvest of a Cognac for only five years, even if it remains unblended, but the British will accept the certified age as long as it remains unblended. Hence, one can buy dated 'early-landed, late-bottled' Cognacs in the UK and these will be naturally broken down to low strength by the English climate; even to below 37% vol., at which other spirits have to be described as 'diluted with ....'. Not so 'late-bottled' Cognac: it only needs to be labelled with the actual strength at the time of bottling.

ARMAGNAC

On the other side of Bordeaux in Gascony, another AC brandy – Armagnac – has been perfected. The region is contained in the Département of Gers, with a small overspill into Landes, and is divided into the districts of Bas-Armagnac, Ténarèze, and Haut-Armagnac. The region remains medieval and recalls Dumas's d'Artagnan and his Three Musketeers of an earlier way of life which still seems to endure. Out of it comes the second-greatest brandy of the world.

Bas-Armagnac in the west has a sandy soil and gives the finest brandy of the region; this may be considered strange, especially as Haut-Armagnac has chalky soil yet produces a brandy only used for

1 In what still is Cognac produced?
2 Whence does the oak used in the maturation of Cognac come?
3 What is 'the angel's share'?

liqueurs, which reverses the logic of the pattern in Cognac. Ténarèze, placed geographically between Haut- and Bas-Armagnac, has clay in the soil and the brandies are lighter and develop more quickly than those of Bas-Armagnac. The region of AC Armagnac is deeper inland than Cognac and does not have such a maritime and moist climate. It is hotter, and as consequently the grapes ripen to give ample natural sugar, the brandies tend to be heavier in bouquet and richer.

The Vines

The authorized vines for Armagnac are mostly the same as for Cognac. Folle Blanche (Picpoul) used to be the chief variety, but Ugni Blanc has been steadily substituted, together with Colombard and Jurançon; but a strange newcomer, the BACO 22A or Baco Noir, which is a hybrid cross of Picpoul × Noah (*V. Labrusca*), has been tried with some success. The EEC '*Wine Régime*' does not allow the wine: will the 'Spirit Régime' allow the spirit?

But, apart from the vines, practically everything in Armagnac differs from Cognac – not only the soil and the climate, but the method of distilling, the strength, and the colour; even the oak in which it matures is different.

Distillation

Armagnac is distilled in a special type of continuous still. In this unusual alembic the vapours are refined by the wine, and the aim is to pass over as much of the original flavour and aroma as possible with the distillate.

The stills, of which even mobile varieties drawn by oxen tour the farms, are generally fired by wood. Three boilers are superposed above the fire; above them is a fractionating column of plates where

1 The Charentais pot still.
2 From the forests of Limousin or the Tronçais.
3 Spirit lost through evaporation during maturation, reckoned to be equal to all Cognac consumed in France.

vapour and wine meet. Two pipes connect this column to the cooling column; one is coiled within the cooling column in a *serpentine* to condense the vapours. The other allows wine (which is the cooling liquid) to flow into the boiler column, when the cooling column has been filled. To operate the still, the cooling column is first filled with wine; further wine passes into the boiler column, trickles though the fractionating column and fills the top boiler until it flows over into the middle boiler, and having filled that, fills the bottom boiler. The fire is now lit to heat the wine in the bottom boiler, vaporizing the ethers which bubble into the wine in the centre boiler and recondense. As the still heats up, these revaporize, and pass through the top boiler (when this is also at operating temperature) and through the fractionating column, into the condenser where they condense. Spent wine is allowed to drain from the bottom boiler, being displaced by pouring more wine into the condenser. The spirit is collected at the bottom of the condenser at a strength not exceeding 63% vol. The amount of wine that may be distilled in one day is limited to two fillings of the condenser.

This process absorbs all the tails in the spent wine, but allows some of the volatile heads to be included in the spirit issuing from the still – which will leach out during maturation. Thus Armagnac is produced in a single distillation, and instantly assumes a truly remarkable flavour and perfume, thought it is yet too young and raw to drink.

Maturation

Armagnac must be matured in barrels made of Armagnac oak; this black oak is becoming progressively more scarce and thus more expensive; but experiments have proved that no other oak will do. The resulting brandy is acclaimed throughout the world, yet the shippers are not known as household names, as are Martell, Hennessy, and Hine, for instance, of Cognac. To find the best Armagnacs it is necessary to visit the three centres of this ancient industry, Éauze in Bas-Armagnac, Condom in Ténarèze, or Auch in

1 (a) Name five ACs for Cognac, other than AC Cognac and AC Fine Champagne. (b) What is Fine Champagne?
2 How old must AC Cognac be (a) for sale in France; (b) to bear a 'marque'; and (c) to be sold as 'VSOP'?
3 What is the other AC French spirit from grapes, and in which Département is it made?

Haut-Armagnac. Whatever else the visitor discovers, he will be shown where d'Artagnan was born, and the work of the Company of Musketeers, who now promote the name and spirit of Armagnac.

Cognac and Armagnac are the only AC eaux-de-vie-de-vin of France, but there are many others which bear the lesser Appellation Réglementée (AR); this could loosely be described as a 'spirituous VDQS'. Among the eaux-de-vie-de-vin to look for are those of Champagne (Fine Marne), Bourgogne, Alsace, Faugères from Languedoc, and Aquitaine from Bordeaux.

CALVADOS

Normandy is too far north and too cold to grow grapes successfully, but supports many apple orchards. Here, in the ancient province of Normandy between the Somme and St Malo, in an area approximately 259 km square, a massive crop of apples and pears is fermented each year into rather acid green cider or perry; much of this is then 'burnt' into Calvados, a fruit spirit of very dry character, disconcertingly strong for the unwary. AC Calvados du Pays d'Auge, in the centre of the region, must be made from fruit fermented for at least one month and distilled twice in a charentais pot-still. Other named Calvadoses, such as Calvados-du-Calvados, or Calvados-de-la-Vallée-de-l'Orne, have Appellation Réglementée and may be made by distillation in continuous stills. Eau-de-vie-de-cidre may be made anywhere by any process. When young the spirit tends to be fiery and raw, and the best Calvados is matured in oak casks for several years, though one year is the legal minimum. Good Calvados has the character of fine brandy, yet retains the aroma and flavour of apples.

If entirely from pears, the spirit is marketed as eau-de-vie-de-poiré. Poiré is French for perry or fermented pear juice, and when distilled the spirit may also be known as Poire Williams in France, and Birngeist in Germany.

1 (a) Grande Champagne, Petite Champagne, Borderies, Fins Bois, and Bons Bois. (b) A blend of Grande Champagne and Petite Champagne, having not less than 50% Grande Champagne.
2 (a) At least one year. (b) Three years. (c) Four years.
3 Armagnac, in the Département of Gers.

OTHER FRUIT SPIRITS

Although fruit spirits are produced throughout the world, it is significant that a number of those enjoying international fame are produced in Alsace or Lorraine, respectively east and west of the Vosges mountains. One of them, Mirabelle, is distilled from the yellow plum (the mirabelle) in both provinces; Mirabelle de Lorraine has Appellation Réglementée and must come from plums grown in the region of Nancy and Metz. The juice of the plums is fermented with special yeasts and, after two months, the wash is distilled twice by the charentais process before maturation. Other fruit spirits, for which Alsace is the primary region in France, are Quetsch from blue plums which flourish between the rivers Ill and Rhine; Fraise, from wild and cultivated strawberries; Mûre, from mulberries or blackberries; and Framboise from raspberries – sold at a price which reflects the fact that 35 kg of raspberries are required to produce one litre of spirit. This contrasts with Kirsch, distilled in considerable quantity from cherries and flavoured with their stones, where the same weight of fruit produces 15 litres of spirit; the wild cherry harvested for Kirsch grows halfway up the slopes of the Vosges in the area to the north-west of Colmar. Kirsch is also distilled in Germany and Switzerland and is called Kirschwasser. It is used by the French, Germans, and Swiss in cooking; for culinary purposes, fine Kirsch is not used but rather Kirsch Fantaisie, in which the flavour-giving Kirsch is diluted with neutral spirit (usually from grapes). In Germany, fruit spirits similarly diluted are given the suffix -geist rather than -wasser, e.g. Himbeergeist from raspberries.

Pomace Spirits

And so returning to the grapes with which the chapter started, the universal cheap spirits of the wine-producing world must be mentioned. They can be superb, rivalling even Cognac in delicacy: they are seldom so, and are probably responsible for more cirrhosis of the liver and alcoholism than any other spirit: they taste of raisins and are made from pomace, the residue of skins and stalks left after pressing the wine out. They go by various names: *eau-de-vie-de-marc*

1 How is the logic of Cognac reversed by Armagnac?
2 What was the original vine principally used for Armagnac, and what vines are now being substituted?
3 (a) At what strength is spirit collected from the Armagnac still? (b) How is the distillation controlled?

(France), *grappa* (Italy), *tresterbranntwein* (Germany), *aguardiente de orujos* (Spain), *bagaçeira* (Portugal) – and are usually made by soaking the pomace in water in an open pit, allowing the remaining sugar to ferment, and then extracting the alcohol with steam.

This sounds crude, and often is, but there are fine marcs in France. Although they tend to have a greater tannin content and hence more bitterness, these pomace spirits can mature into extremely palatable brandies, for this is what they are. Each develops its own characteristics reflecting the qualities of the grape, and of the soil and climate in which the grape grows. Excellent marcs come from Burgundy, especially those from the Hospices de Beaune, Nuits-Saint-Georges, and Montrachet. Others of repute are eau-de-vie-de-marc-de-Champagne and eau-de-vie-de-marc-originaire-d'Aquitaine, the latter distilled in Bordeaux since the days of Henry II.

1 Chalky soils produce the best Cognacs but only the poorest Armagnacs.
2 The Folle Blanche (Picpoul); Ugni Blanc, Colombard, and BACO 22A.
3 (a) Not exceeding 63% vol. (b) It is restricted to two fillings of the condenser per day.

15
Grain Spirits

HISTORY

Grain spirits have been distilled in Scotland for over a thousand years. The art is thought to have been introduced by Christian missionary monks in the Dark Ages, though it made no appearance in the records until five hundred years ago as *aqua vitae*. The name slowly evolved through the Gaelic *uisge beatha* also meaning 'water of life'. By the year 1618 it was called *uiskie*, and in the twentieth century, by trade custom, became Whisky (without an 'e'), or simply Scotch.

Until the early 1830s all Whisky was made from malted barley, then fermented and distilled in pot stills; this was too strong in

1 (a) Where does the district of Calvados lie? (b) What is its main crop?
2 (a) Name the AC of Calvados. (b) How long must the fruit be fermented? (c) How must it be distilled?
3 (a) In what area of France are other fruit spirits made? (b) Give examples.

flavour for all but the hardy Scots. Blending of malt whisky with spirit from other grains made in a continuous still was introduced in the 1860s, producing a milder spirit which rapidly gained a market in England and then throughout the world as Scotch.

MALT WHISKIES

The pot still process for the production of malt whisky has four main stages: malting, mashing, fermentation, and distillation.

1 (a) In Normandy, between the river Somme and St Malo. (b) Apples and pears.
2 (a) AC Calvados du Pays d'Auge. (b) For at least one month. (c) Twice in a Charentais still.
3 (a) In Alsace and Lorraine (east and west of the Vosges mountains). (b) Mirabelle, Quetsch, Fraise, Mûre, Framboise, and Kirsch.

Malting

In this first stage, the barley is cleaned by a screening process before being steeped for sixty hours in water tanks. The barley is then spread out on a concrete malting floor, where it germinates in the course of eight to twelve days – depending largely on the season of the year. As the barley germinates it secretes diastase, the enzyme which will convert the starch of the barley into sugar. During germination the temperature of the barley rises, and it is necessary to rake or turn the barley to ensure an even humidity and thereby achieve a controlled rate of germination. When germination is sufficiently advanced, the green malt is dried in peat-fired kilns, whose 'reek' impregnates the malt and gives it flavour. The industry today is so vast that malting of the barley is ideally done at centralized maltings where it is automated in drums or Saladin boxes, and is continually monitored by biochemists.

Mashing and Fermentation

The dried malt is then ground in a mill; the resulting grist is mixed with hot water in the mash tun – a circular vat equipped with stirrers. Here the soluble sugar, maltose, is dissolved and any remaining starch is converted into maltose. This sugary wort is drawn off leaving a by-product called draff which is processed into cattle food. After two spargings with hotter water, and cooling to blood heat, the wort is fermented by yeast in tanks called washbacks, which may hold up to 100,000 litres; in about 48 hours the wort has been converted into wash of about 11% vol. alcohol. This low-strength alcohol also contains unfermentable solids in suspension and fermentation by-products.

Distillation

The purification and concentration of the alcohol is carried out by the process of distillation in large copper pot stills called wash stills. The wash is gently heated to a temperature of 79°c by which all the alcohol vaporizes, with some of the water and the other volatile fractions. The vapours rise up the still and pass through the Lyne arm into the condenser (a copper coil immersed in running cold water)

1 Give the (a) Italian; (b) Spanish; and (c) Portuguese names for eau-de-vie-de-marc.
2 For how long has whisky been distilled in Scotland?
3 What was whisky called (a) 500 years ago; and (b) 350 years ago?

where they condense. This distillate is known as low wines; the residue of yeast and water, two-thirds of the original wash, is known as pot ale. This residue is malodorous and deoxygenated, highly unsuitable for discharge into the rivers; but as it is full of valuable protein, it is mixed with the draff and dried to form the cattle-food known as dark grains. So valuable is this by-product, and so voracious the tax on spirits, that whisky has been called a by-product of the cattle-food industry! Were the Chancellor to transfer the taxes on whisky to dark grains, Europe might exchange its butter mountain for a whisky loch and be freeer and healthier.

The low wines are now passed into a different still, called either a low wines still or a spirit still, for further concentration. The first and last runnings in this second distillation (called foreshots and feints) are drawn off and collected; when the spirit passing over as vapour reaches an acceptable standard, it (the heart) is collected in the spirit receiver. This process of selection, carried out by the stillman with the aid of hydrometers and thermometers, visible to him but bound in a Crown-locked spirit safe, is necessarily arbitrary. Any usable potable alcohol included in the foreshots and feints is recovered by redistillation; and deleterious fractions included in the heart are leached out by maturation in oak – for a minimum legal period of three years. 'British Plain Spirits' issues from the stills; not until the three years are past may it be called Whisky. The residue left in the still is known as spent wash and has no food value; it is sprayed over the hillside.

The pot still process is essentially a batch process, as opposed to a continuous one.

Single Malt Whiskies

Malt whiskies are divided into four groups by geographical location of the distilleries, but all are made from malted barley only. The biggest, most important, and most northerly group, the Highland malt whiskies, are produced north of the Highland Line, an imaginary line drawn from Greenock to Dundee south of which, traditionally, the kilt is not worn. Lowland malt whiskies are produced south of that line. Islay malts come exclusively from the

1 (a) Grappa. (b) Aguardiente de Orujos. (c) Bagaçeira.
2 Over 1000 years.
3 (a) *Aqua vitae*, or uisge beatha. (b) Uiskie.

Isle of Islay, and the fourth and smallest group, the Campbeltowns, from the town of that name on the Mull of Kintyre.

The considerable differences in flavour, not only between the groups, but also between individual distilleries within each group, are put down to a number of factors. In the pot still process, secondary constituents from the barley are retained in the spirit, and phenolic substances from the peat used to smoke the malt are also retained and can be detected in the smoky flavour of malt whisky. The amount of secondary constituents, such as fusel oils and other congeners, will depend upon the shape of the still and the way it is operated. Another factor which contributes to variation in flavour is water, flowing off granite through peat into the Scottish rivers and streams, each of which has its individual mineral content.

GRAIN WHISKY

Barley is as essential to grain whisky as to malt because the enzyme diastase, produced in the malting of barley, is required for the conversion of starch to fermentable sugar. This enzyme works upon all starches; the mash for grain whisky consists of a lesser proportion of malted barley together with a greater proportion of unmalted grains, mostly maize, although unmalted barley, wheat, and millet are also used. However, maize requires pretreatment by milling and by cooking under steam pressure for three-and-a-half hours in a converter to burst the starch cells. The mash thus prepared is transferred to the mash tun for conversion into sugar by the diastase of green malt; the resultant wort is less concentrated, with a lower specific gravity than the wort for malt whisky.

Continuous or Patent Stills

For grain whisky production, the continuous still invented in 1830 by the Irish Customs Officer Aeneas Coffey, and modern improved stills embodying the same principles, are used. This still has two columns – the analyser and the rectifier: at the foot of the analyser steam is introduced, and at the top of the rectifier cold wash. By a

1 Name the four stages of production of malt whisky.
2 Why is peat used in drying malt?
3 What is a Saladin box?

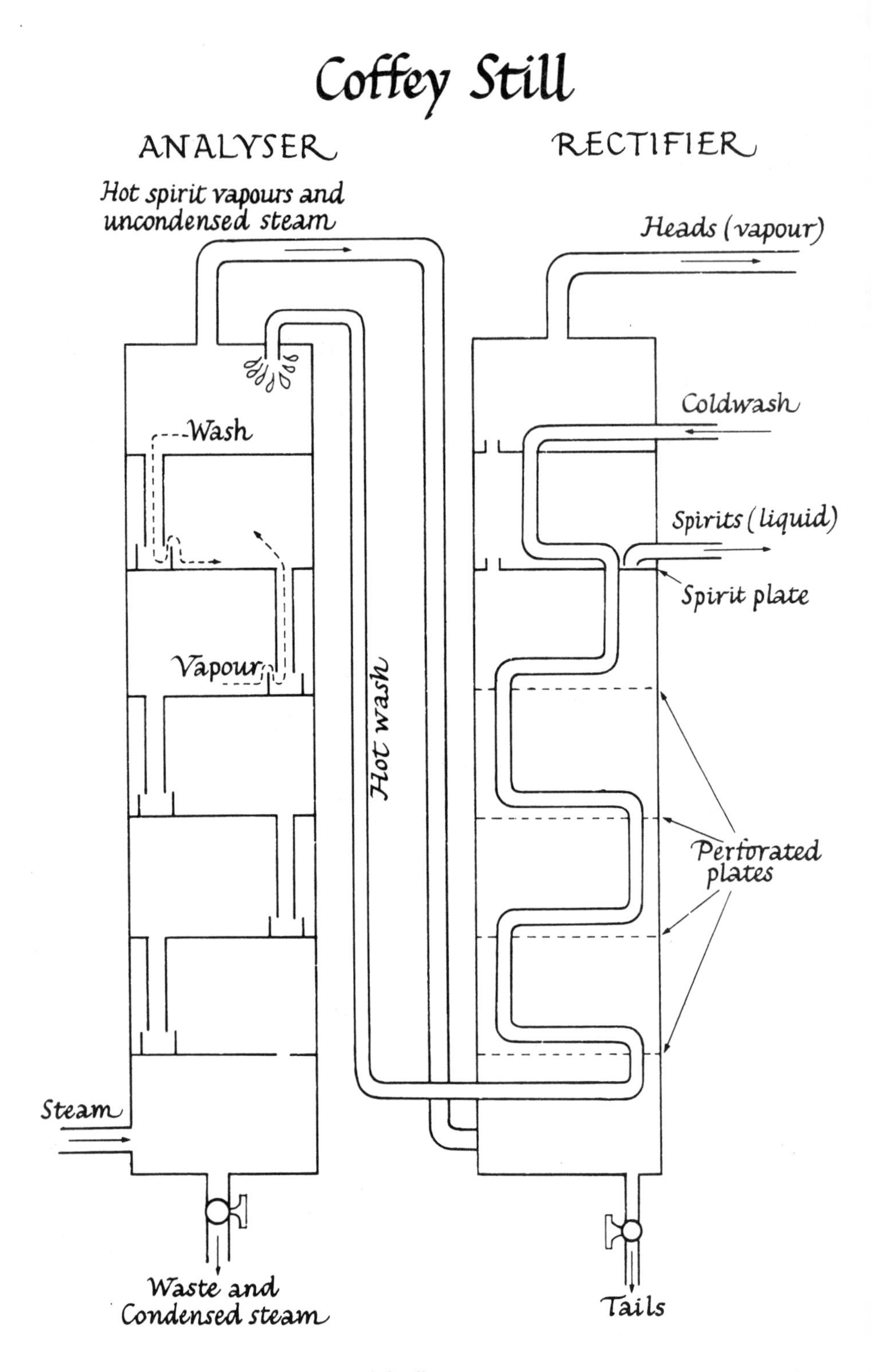

1 Malting, mashing, fermentation, and distillation.
2 To give flavour.
3 Machinery used in modern centralized maltings.

system of heat exchange the wash is already hot when pumped from the foot of the rectifier to the head of the analyser, and as it descends in the analyser the alcohol, some water, and other volatile fractions vaporize. The action of the analyser equates to the (first) distillation in a wash still; pot ale flows out at the bottom, while the vapours escape at the top. These vapours are not condensed to low wines, but pass directly to the bottom of the rectifier. The rectifier is hot at the bottom and cool at the top; through it, the vapours rise and condense successively as the temperatures of their dew-points are reached. The condensates either fall to the bottom as feints, or are collected – at the point where the temperature is 78.4°C – as spirit; foreshot vapours, passing out of the head of the still, are condensed and, with the feints, are added to the next lot of wash for redistillation.

This is the essence of the continuous still process, illustrated on page 320, which produces alcohol at a higher strength than does the pot still. Although purer than pot still malt spirit, this contains sufficient deleterious matter also to require three years' maturation in wood before it can be called 'whisky' and sold for consumption.

Although the spirit safe is of a different type, the stillman can observe the strength by controlling the rate at which spirit is taken from the still.

MATURATION

The water-white British Plain Spirits, whether from pot or patent still, is put into cask at a strength of 70% vol. Commonly, used sherry butts are employed; the wood and the sherry residues impart colour to the spirit. Sometimes, to achieve consistency of a light colour, maturation is carried out in new wood which gives less colour to the spirit.

During maturation most of the volatile impurities (which might have been foreshots) will evaporate through the pores of the wood; at the same time, the cool damp air of Scotland will enter in to replace them. Some good alcohol will also evaporate, so that the spirit in the cask gradually loses strength, although it does not lose much in bulk.

1 (a) How long does spirit have to be matured before it may be called whisky? (b) What is it called meanwhile?
2 Name the geographical groups of malt whisky distilleries.
3 How is maize treated before mashing, and why?
4 What are the two main parts of the Coffey still?

The original fermented components of the wash – water, alcohols, acids, aldehydes, ketones, and other esters – have been concentrated about seven times by the process of distillation, and the worst of them removed. The remainder, thrown closer together, tend to combine and in so doing form other compounds, drawing oxygen from the air through the pores of the casks. These compounds give flavours which will originate from the type of malt used, the peat used in malting and the extent to which it is used, the type of water used in malting and in mashing, the type of still used, the craft of the stillman, and the precise location of the warehouse where maturation took place.

Malt whiskies may be left to mature for as long as fifteen years, having many highly flavoured constituents to digest; grain whiskies do not improve perceptibly after three years.

RECTIFIED SPIRITS

It should be noted that a significant proportion of continuous still production of grain spirits (not necessarily all from Scotland) is re-rectified to about 80 per cent purity and distributed in tankers as 'ethanol' to producers of vodka, gin, and other flavoured spirit.

SCOTCH WHISKY BLENDING

There are between 100 and 120 malt distilleries in Scotland, each producing between one-third of a million and two million gallons of spirit annually, and each has its particular characteristics. In contrast, there are only fourteen grain distilleries, which each produce several million gallons annually; their product varies little, and a large proportion is further purified for the manufacture of gin and vodka.

1 (a) A minimum of three years. (b) British Plain Spirits.
2 Highland, Lowland, Islay, and Campbeltown.
3 It is milled, then cooked under steam pressure for $3\frac{1}{2}$ hours in a converter, to burst the starch cells.
4 The analyser and the rectifier.

Grain whisky on its own is quite drinkable, and a 'single grain' has recently been marketed. Malt whiskies of a single distillery (single malts) are more popular, though the products of only a few appeal; vatted malts – blends of single malts – are rare. Most branded whiskies, therefore, are malts blended with grain.

A blend may have as many as fifty single malts in its formula with possibly twice their total volume of grain whisky, and the proportions are the closely guarded secret of the blending company. Character and consistency are the essentials of good blending, which must result in a spirit recognizable for colour, strength, and flavour: the Scotch drinker can be as much of a connoisseur as any wine drinker. After vatting or blending takes place the mixture must be left, usually in wood, to marry for a period of months before bottling.

STRENGTH AND STORAGE

At bottling, the whiskies are reduced to shipping strength with pure Glasgow or Edinburgh tap water. The general strength for home sales is 40% vol. although some brands are stronger. This also applies to the continent, but the United States requires whisky at a strength of 85° US proof (42.5% vol.).

When whisky is distilled the gallonage filled into each cask is called the original gallonage. The gallonage remaining on maturation or after storage is called the regauge gallonage, and the difference which is lost in the intervening period is ullage. It is reckoned that in a normal year some ten million gallons of spirit are lost in ullage from the stocks maturing in warehouse – as much as is consumed of the final product in the UK during the same period.

Bottled whisky should not be stored in very cold conditions as it may become cloudy. Although the cloudiness will usually disappear when the spirit is restored to a normal temperature, under long storage the cloudiness may precipitate, and some of the flavour may be lost on filtering the sediment from the whisky.

1 As vapours rise in the rectifier, at what temperature do they condense?
2 What happens to the strength and bulk of whisky during maturation in Scotland?

SCOTCH WHISKY AND THE LAW

The description 'Scotch Whisky' is protected by law, which in essence states:

(a) The expression 'whisky' shall mean spirits which have been distilled from a mash of cereals which have been:
  (i) saccharified by the diastase of malt contained therein with or without other natural diastases approved for the purpose by the Commissioners; and
  (ii) fermented by the action of yeast; and
  distilled at less than 94.8% vol. in such a way that the distillate has an aroma and flavour derived from the materials used, and which have been matured in wooden casks in warehouse for a period of at least three years;

(b) The expression 'Scotch Whisky' shall mean whisky which has been distilled and matured in Scotland;

(c) The expression 'blended whisky' or 'blended Scotch Whisky' shall mean a blend of a number of distillates each of which separately is entitled to the description whisky or Scotch Whisky as the case may be;

(d) The period for which any blended whisky or blended Scotch Whisky shall be treated as having been matured as mentioned in sub-paragraph (a) of this paragraph shall be taken to be that applicable in the case of the most recently distilled of the spirits contained in the blend.

Although distillation of Scotch Whisky must take place in Scotland the law until 1980 allowed the spirit to undergo its maturation elsewhere, which had the undesirable effect of allowing immature spirits to be sold as Scotch Whisky abroad. However, matured Scotch Whisky can be and is exported in bulk and bottled abroad; but a blend of single Scotch malt whiskies with foreign, locally-produced grain spirit cannot be termed Scotch, because the whole was not distilled in Scotland. In 1977 this practice was common abroad, particularly in Japan.

While the minimum period of maturation is three years, whiskies are usually aged for a much longer time, although they will take on an unacceptable flavour of oak after about thirty to forty years. If the

1 At the temperature of their dew-points.
2 Strength is lost by evaporation, but bulk is only slightly reduced.

label on a blend shows an age, as '15-year-old', *all* the constituent whiskies of the blend must have been aged in wood for at least that length of time – including the grain whisky, even though this does not improve after three years in cask.

Although there are over 110 different distilleries, and a number of firms carrying out blending and bottling, the greater number of these are gathered into large corporations (which also include firms manufacturing gin and vodka). Firms have a large degree of autonomy in the production and marketing of their individual brands, but the larger resources of groups offer economic stability.

With so much money tied up in stocks, the industry faces severe financial problems, aggravated by the extremely high rate of duty which has to be paid immediately Scotch leaves the bonded warehouse. The industry needs to make firm and united representations to the Government on taxation and to be ever-watchful to protect the name of Scotch. The Scotch Whisky Association, which includes the principal firms, large and small, in the industry, performs this admirably, and, in addition, gives unstinting help to the cause of trade education and promotion.

WHISKEYS

Besides Scotch Whisky, some others spelt Whiskey merit attention.

Irish Whiskey

This is a pot-still whiskey made from a mash of malted barley with unmalted cereals – usually rye, barley, and wheat – and distilled three times in a pot still, the first and last runnings (heads and tails) being extracted from both the second and third distillations. It is matured in wood for a minimum of five years, and the best up to fifteen. It is made in the Republic of Eire and in Northern Ireland, and has a distinctive smoky aroma, originating from the use of rye and not from peat-smoke, for the Irish malt is kilned without smoke.

1 How many malt distilleries are there in Scotland? How many grain?
2 (a) In a blended Scotch whisky what generally is the proportion of malt whiskies? (b) How many malt and grain varieties might be in a blend?
3 What qualities does the blender aim for in blended whisky?
4 (a) Why may whisky become cloudy during storage? (b) Can this harm the whisky?

Rye Whiskey

This is made extensively in the United States and Canada, but lacks the rye smokiness because it is distilled in a continuous still to a higher degree of purity. It is based on the same mixture of grains as Irish Whiskey with the addition of maize. Canadian rye whiskey must be produced from cereal grain only, but American blended rye may have neutral spirit from other sources added. American straight rye must be distilled from a fermented mash of grain, of which not less than 51 per cent must be rye. It is commonly consumed with mixers, as gin is in England, but has a better social image across the Atlantic than gin.

American Whiskeys

These had their origins in 'moonshine', the illicitly-distilled spirit brewed by the hillbillies of Kentucky, Tennessee, and Arkansas. No doubt this dangerous potion still exists, as poteen does in Ireland, but it should be avoided at all costs. Moonshine and religion go together, as readers of Tennessee Williams know – religion needs sin, and moonshine provides it! The most famous legal spirit is Bourbon, which may be made from a sour mash of grains of which at least 51 per cent is maize (corn); malted barley is used for saccharification. Originally distilled in Bourbon County, Kentucky, it may now be made anywhere in America, and must be matured for not less than two years in new oak casks charred on the inside. A sour mash whiskey is one where the fermentation is carried out by the addition of 'spent beer' – the residue from the previous distillation. Single-distillery whiskeys are called 'straight' and are heavy, dry, mellow, and full-bodied; blended whiskeys, which must contain at least 20 per cent straight whiskey with 80 per cent plain spirit, are also common. Blended whiskeys are lighter than straight whiskeys, and, being blended with patent still spirit, are cheaper.

Although figuring as the base for many cocktails, good straight Bourbon is usually taken with branch-water, that is to say water drawn from a branch stream above the cow pastures and kids' bathing pools – but it usually comes out of a tap!

1 Between 100 and 120 malt, and 14 grain.
2 (a) About 33% malt whiskies. (b) Up to fifty malts but only one or two grain whiskies.
3 Consistency of colour, strength, and flavour, all contributing to character.
4 (a) If stored in very cold conditions. (b) Yes, if stored too cold for too long, when the cloudiness may precipitate causing loss of flavour.

Corn Whiskey

This is different from Bourbon in that it must be made from at least 80 per cent corn and may be aged (if it is aged at all!) in used or uncharred barrels. Fiery young 'corn likker' or 'white lightning' is popular in the areas which used to produce moonshine.

Other Whiskeys

As mentioned earlier, many foreign whiskey-producers blend imported Scotch malt whisky with local neutral spirits, but the Japanese have now set up distilleries to produce entirely Japanese whiskey in the Scottish style.

1 Which constituent of blended whisky determines the labelled age?
2 Can single malt whiskies from Scotland, blended with grain spirit manufactured abroad, be called Scotch whisky?
3 What method is used for distilling Irish Whiskey?

16
Vegetable Spirits

RUM

Rum originally came to Britain via the Navy, who found it in the West Indies. The spirit was so fiery, and its effects so fierce, that it was nicknamed Rumbullion or Rumbustion because it made the drinker so unruly. It was, and is, made from sugar-cane, or rather from the sweet black glutinous liquid called molasses which remains after all the marketable sugar has been extracted from the sap of the sugar-cane. For the UK, rum was defined in 1909 as 'the spirit produced by distillation of fermented sugar-cane products, distilled in sugar-cane producing countries'.

1 The most recently distilled of the spirits contained in the blend.
2 No: all parts must have been distilled in Scotland for a blend to be called Scotch whisky.
3 Threefold distillation in a pot still.

Although originating in the West Indies, which remains the principal zone, rum is produced extensively in North and South America, Indonesia, and Australia, in all of which sugar-cane is grown. In Egypt it is known as Tafia, and in Indonesia, it is Batavian Arrack.

Sugar-cane is a giant type of grass, and two or three crops a year can be obtained from a plot of land. It is simply propagated by planting a short length of cane in the ploughed soil; this cane then grows over a period of three to four months to over 2 m in height, with each stalk or cane about 10 cm thick. The fields of cane are often separated by canals which form a means of transport for the cut cane and an important fire-break; for, when ripe, the cane field is set on fire. This gets rid of the leaves and gives potash to the soil, besides chasing the snakes out; it is a wonderful sight. The cane is then cut by hand, using sharp knives – cutlasses in Jamaica, machetes in Guyana – and taken to the mill where it is crushed to extract the sweet sap.

The sugar is extracted from the sap by successive boilings to evaporate the water, followed by centrifuging to separate the crystals; further extracts are made of golden syrup and black treacle, until only molasses, incapable of further refinement, remains.

To make rum, the molasses is diluted with water, and fermented with yeast to give a wash of about 10% vol. strength; in some distilleries, the yeast is added in the form of 'dunder' – the residue from the previous distillation – but most modern distilleries use cultivated yeasts. The flavour of the rum can be affected by the length of fermentation – a long fermentation will produce a more highly flavoured rum, as will the addition of dunder.

Originally all rum was made by two distillations in a pot-still, heads and tails being separated during the second distillation; this produced a heavily flavoured spirit which was too pungent for popularity. In recent years, more and more rum has been made in patent continuous stills, which gives spirit with a lighter flavour; and some are passed through the still twice, to produce almost neutral cane spirit.

There are just as many variations in pot-still rums as there are in Scotch malt whiskies. The climate and soil of individual canefields will vary, as will the product from different crops in a year; the water

1 (a) Does Bourbon have to be made in Bourbon County? (b) Where is Bourbon County? (c) What proportion of its grain mash must be maize?
2 What are 'straight' Bourbon whiskeys?

varies from distillery to distillery, as do the stills and the stillmen. Blending is a much-practised art.

All rum is water-white when it comes from the still, whether this be pot or patent. The colour of golden rums comes from maturation in wooden casks, usually charred on the inside; but dark rums receive their colour from the addition of caramel, which does not affect the taste. Thus the darkness of colour is no guide to the pungency of a rum, although rums bleached by filtering through charcoal (like white Bacardi from Puerto Rico, and Santigo from Guyana) are the least flavoured. The most pungent are probably the golden rums of Jamaica, Barbados, and French Martinique, the last having a completely different flavour, rather like that of a chocolate truffle. All these are pure pot-still rums. With these exceptions, rums are made of a blend of pot- and patent-still spirits, giving a subtle gradation of flavour to suit the market.

To avoid too much evaporation, rum is often sent to temperate climates to mature; the period of maturation varies. A lesser legal maturation requirement reflects the absence of the more harmful congeners found in grain spirits at distillation. Rum may also be matured in glass.

In Germany a product called Rhum Verschnitt (rum sandwich) is made by blending a concentrate of pot-still rums with local corn spirit.

TEQUILA

The other drink of the Caribbean and Gulf of Mexico which is growing in popularity is Mexican tequilla. Like rum, it is of vegetable origin, but all resemblance ceases there. It is made from the juice of the agave, or century plant, which is a distant relation of 'mother-in-law's tongue'. Spiky leaves grow from a tuberous ball of roots which sits partly above, partly below ground.

After six to eight years, when the root has grown to the size of a beach-ball, it is dug up and the outer leaves are sliced off. The white centre is chopped up and steam-cooked in an autoclave to extract the sweet juice. When this juice is fermented, it is known as *pulque*, and

1 (a) No, it may now be made anywhere in America. (b) Kentucky. (c) At least 51%.
2 Single-distillery whiskeys.

this beery liquid is widely consumed on its own. When distilled, it becomes *mezcal*; tequila is mezcal from Tequila County, 300 km to the north-west of the capital.

Older methods of distillation produced a very rough tequila, which was traditionally taken neat in one shuddering gulp, after the drinker had taken a lick of salt sprinkled on the back of his lime-juice-moistened hand. The taste was described as unpleasant. Mexico now has several very modern distilleries producing spirit of high quality and pleasant, if rather neutral, flavour.

ARRACK

This name of Arabic origin, covers as many different spirits as the words eau-de-vie or vodka, and is found generally in the Near and Far East. It is distilled from such various fermented bases as rice, palm sap, yams, and dates. It is a spirit of the people and should be treated with great reserve by westerners.

VODKA

In Poland, Russia, and the Baltic states, agricultural crops vary greatly, with occasional large surpluses of one crop or another, and it has been traditional to convert these surpluses into potable spirit – *vodka* or 'little water'.

The vodka popular in the UK is without flavour and has a growing market. It is costly to produce: to lessen the cost, it is usually sold at a strength of 37.4% vol. rather than 40%, which reduces the duty.

It is made from rectified grain- or cane-spirit, further purified to 96 per cent pure alcohol: for final purification it is filtered through activated charcoal which removes every trace of colour and smell. This charcoal (which is also used in gasmasks) is of vegetable origin, often coconut shell (copra) which has been heated to over 1000°C in a vacuum, canned, and sealed. After a short period of use, the charcoal

1 What is molasses?
2 Where is rum made?
3 What is dunder?
4 How is most rum now distilled?

must be reactivated, and it is this process which is so expensive. The high-strength vodka is broken down to selling strength with demineralized water.

Colourless, odourless, tasteless vodkas are also made abroad; in Germany they may be known as Korn, Doppelkorn, or Alter Doppelkorn.

1 The sweet black glutinous residue after marketable sugar has been extracted from sugar cane.
2 In sugar cane-producing countries.
3 The residue of a previous fermentation, used instead of fresh yeast to produce more highly-flavoured rum.
4 In continuous stills, sometimes being passed through twice to produce neutral cane spirit.

17
Flavoured Spirits

Camomile, wormwood, hyssop, quinine, angelica, aloes, gentian, citrus peels, coriander, cinnamon, juniper, orris root, arrowroot, mint, vanilla, marjoram, centaury, cardamom, sage, cloves, rosemary, ginger, aniseed, fennel, liquorice, caraway, cassia. This may sound like an extract from a medieval pharmacopoeia, and may indeed have been so: in present days, it is a list of the common herbs and spices which are used to flavour spirits. Spirits are flavoured for two historical reasons: first, that the medicinal properties of herbs were most easily extracted by, and administered in, alcohol; and second, that strong spices disguised the taste of poor spirit. Nowadays, flavoured spirits are still taken for their curative effects, and are

1 (a) Is colour in rum a guide to pungency? (b) How do rums get their colour?
2 What are mezcal and Tequila and from what are they made?
3 Name a 'spirit of the people', to be treated with caution by westerners.
4 (a) What does 'vodka' mean? (b) At what per cent vol. is it usually sold, and why?
5 Why does vodka lack colour or smell?

also taken because they taste nice. Today however, the spirit is free of impurity.

The monastic brethren of the Middle Ages were great herbalists and great pharmacists. They perfected the arts of extracting flavours from plants, using cold or hot water, or fats, or spirits – hot, cold, or vaporized; thereby producing many curative potions which, at any rate, stimulated hunger. They originated a range of bitters, of which Angostura, Fernet Branca, and Underberg are present-day examples. The same techniques were and are applied to the extraction of fragrances for perfumes. The pungency of the bitter herbal extractions was alleviated by sweetening, although it was not until sugar-cane was developed in the Caribbean that liqueurs – sweetened and flavoured spirits – became popular. Only a few of the most commonly-found flavoured spirits will be mentioned – their variety, like their sweetened brethren, the liqueurs, is only limited by man's taste and ingenuity. A full chapter is devoted to liqueurs in *The New Wine Companion.*

The spirit used is generally re-rectified grain or cane spirit of such purity that the flavoured spirits made from it legally need no maturation. They are termed 'Immature Spirits' by H M Customs and Excise.

GIN

The northern countries of Europe had no grapes to make wine for brandy; but from very early times had produced rough spirits from grain-based alcoholic washes. The Dutch produced such a spirit from malted barley and rye and disguised its coarseness by flavouring it with the pungent berries of the juniper bush, which had the added advantage of benefiting the kidneys. The Dutch name for *juniperus communis* is Genever, and this name passed to the spirit flavoured with it. The English shortened the name to Gin and, during the eighteenth and nineteenth centuries, abused the name by applying it to any and every sort of foul compounded spirit which could be made cheaply and sold to the poorest, who bought it and drank it to achieve oblivion and escape from the social misery in

1 (a) No. (b) Generally by addition of caramel.
2 Mezcal is the spirit distilled from pulque; Tequila is mezcal from Tequila County.
3 Arrack.
4 (a) 'Little Water'. (b) 37.4% vol. to lessen the rate of duty.
5 Because it has been filtered through activated charcoal.

which they existed. It should never be forgotten that it is social conditions which cause misuse of alcohol, not the reverse. Gladstone knew this, and although he tried to introduce legislation to control the production and sale of alcohol (and to a large degree succeeded), his most important contribution to moderation was social reform. Contented people do not misuse alcohol, but use it to enrich their lives.

Modern Production

Nowadays, gin is perhaps the purest spirit sold. The compounders buy twice-rectified spirits of a purity of 70% vol. from patent-still distillers; they redistil this through fractionating columns which extract all the harmful fractions and leave a spirit which is 96 per cent pure ethyl alcohol. The base material may be grain – much comes from grain distilleries in Scotland; or it may be molasses from sugar-cane, which is also used for making rum in sugar-cane producing countries. Whatever the source, the product is a practically odourless, tasteless, pure spirit, and is itself the base material for gin, vodka, and liqueurs.

To make the best gin, pure spirit is poured into a pot-still and powdered herbs are added; after soaking (maceration) for some time, the still is heated, often by steam coils immersed in the liquid. The first 'runnings' contain too much pungency from the flavouring materials, as do the last; the pure alcohol is recovered from them by re-rectification. Only the heart is taken for gin, and only needs breaking down with pure water before sale.

Types

There are various types of gin on sale: Dutch Genever has been mentioned, and is of two main types: *Oude Genever* (old Geneva) is based on a pot-still distillate of a wash made from malted barley and rye; it retains a smoky pungency, is oily in consistency, and does not

1 Give the German names for Vodka.
2 Give two historical reasons why flavoured spirits are produced.

appeal to everyone. It is often put up in stone bottles. *Yonge Genever* (young Geneva) is patent-distilled as described for grain or cane spirit, and is generally indistinguishable from London Gin.

'London Gins' vary enormously; they can be made anywhere in the world, the word 'London' denoting generally a lightly flavoured, dry, gin; Plymouth Gin is very similar. Different brands have different herbs to flavour them, in different concentrations; some are flavoured only with juniper and coriander, others may have as many as twenty herbs. A sweetened gin, 'Old Tom', used to be popular in the UK but is now seldom seen. It is still popular on the continent, as are lemon and orange gins; these, however, are completely dry and are flavoured with fresh citrus peel by maceration in the cold spirit. This process also imparts colour to them.

AKVAVIT

Further north, in Scandinavia and the Baltic states, the almost universal flavouring is cumin, from caraway seed (*carum carvi*), an umbelliferous plant like cow-parsley or fennel. The spirit goes by the universal name of akvavit or 'snaps' meaning 'a mouthful'. Most, but not all, are water-white, flavoured only with cumin, and served like vodka, as cold as they can be got. Danish Jubileumsakvavit, also flavoured with dill and created to celebrate the Jubilee of the State Distillery, deserves separate mention; as does Export Akvavit, flavoured with Madeira, and Akeleje with 'a spectrum of fragrances that rivals the colours of the columbine'. These may be taken at room temperature. Snaps is always taken with food, traditionally pickled fish, pea-soup, or strong cheese. The basic spirit is made during the winter from potatoes, which are well cooked to break down the tough starch granules; or in the summer from maize flour from which the corn oil has previously been extracted. (This extraction provides a valuable by-product, and obviates excessive foaming in the multi-column continuous stills.) The starches, from either source, are saccharified by synthetic diastase, and fermented using baker's yeast.

1 Korn, Doppelkorn, or Alter Doppelkorn.
2 Because the medicinal properties of herbs were easily extracted and administered in alcohol; the herb flavours disguised the taste of poor spirit.

Akvavits of other Scandinavian countries are Norwegian Linie, so-called because it is sent across the equator and back in ships (each bottle being numbered), Swedish Skåne, and Finnish Kuskenkorvo and Finlandia.

FLAVOURED VODKA

Until modern methods of fractional distillation were perfected, the spirits made from root vegetables and from rye carried a rank flavour that needed to be disguised. Peach- and cherry-flavoured vodkas are found, but the most famous is the Polish Zubrowska (Buffalo), which is coloured green and flavoured with the coarse grass on which the buffalo feed; a blade of the grass is often placed in the bottle.

1 Give examples of herb bitters on the UK market.
2 Define a liqueur.
3 What are 'Immature Spirits'?
4 From what did gin originally take its name, and why?
5 (a) When making gin by distilling with herbs, what fraction of the distillate is used? (b) How is the alcohol in the other fractions recovered?

1 Angostura, Fernet Branca, Underberg.
2 A spirit which has been sweetened and flavoured.
3 Spirits so purified by distillation that they are not required by HM C & E to undergo maturation before sale.
4 From the Dutch name Genever for the juniper berry whose flavour was used to moderate the coarse taste of the rye spirit 'moutwijn'.
5 (a) The heart. (b) By further rectification.
(Questions continued on page 341.)

PART FIVE

Legal Aspects

18

Effects on the British Trade of National and EEC Legislation

'In France, everything is permitted except that which is specifically forbidden; in Germany, everything is forbidden except that which is specifically permitted; in the EEC, everything is permitted, even when it is specifically forbidden.'

An anonymous cynic

'If I make water in public, or whisky in private, I shall be liable to prosecution. That is the law.'

A Lord Chief Justice of England

1 Distinguish between London Gin and Old Tom Gin.
2 What does maceration consist of?
3 What is cumin and what spirits does it flavour?
4 What is 'Skåne'?
5 What is Zubrowska vodka?

The law, to some people, means police, prosecution, and punishment – but happily, to most it means peace and protection. Every citizen is expected, nay required, to know the law and to obey it; ignorance of the law is no defence although, in Great Britain, a person is generally held to be innocent until judged guilty.

What then is this law – deemed by Dickens, through Mr Bumble, to be 'a h'Ass'? How, and why, does it affect the Wine and Spirit Trade? It is no more nor less than a set of rules governing the relationship of persons with one another, and with the community which is the sum of all persons. But just as people are complicated and devious, so are the laws that govern their relationships; however, they can be categorized and simplified.

Think of the law as a piece of cloth in which the warp is composed of the threads of civil and criminal law, regulating the behaviour of one individual towards another individual, and towards society. Misbehaviour of a person (and companies are counted as individual persons too) towards another person is a wrong, or tort, which upon proof may be redressed by damages for loss suffered, or sometimes by enforcement of contract. Misbehaviour of a person towards society is a crime, guilt of which may be punished by society with reprimands, supervision, fines, or loss of liberty.

Threads of the weft of our piece of cloth, interwoven across the civil and criminal law, are the objects of law which, for our purposes, are broadly:

Protection of the community
Protection in trade, for all who buy or sell
Protection in employment
Levying of taxes to finance Government.

These threads are reinforced by different types of law, which in Great Britain are usually categorized as follows:

Treaty law – agreed internationally, ratified by Parliament, and superseding national laws, which are amended to accord.
Common law – the old, traditional, unwritten, law, as defined and refined by judge and jury through the centuries, to become case law; this has increasingly been codified into

1 London Gin is dry, Old Tom sweet.
2 Soaking flavouring agents in alcohol or water to extract the flavour.
3 Caraway seed from the plant *carum carvi*; used for flavouring Akvavit (Snaps) in Scandanavia.
4 A type of Swedish Akvavit.
5 Vodka flavoured with Zubrowska (Buffalo) grass.

Statute law – the main body of the law today, consisting of Acts of Parliament: some, like the Finance Acts which levy the taxes for a particular year, are annual (although the particular provisions of each remain in force until repealed): most others are 'permanent', until revised, added to, or repealed. Many Acts, as for example the Food & Drugs Act, contain in one or more clauses the words '... the Minister may, by order, direct that.........' and these words give rise to an increasing mass of
Administrative law – such as the Labelling of Food Regulations. These Statutory Instruments (SIs) have to be laid before Parliament and may be debated and voted upon; but they take effect from the date decreed by the Minister until nullified by Parliament. Similarly under Treaty Law, the 'Treaty of Rome', to which Great Britain acceded on 1 January 1973, has its own body of Administrative Law known as
EEC Regulations – issued either by the Commission (the permanent 'Civil Service' of the Community) or by the Council of Ministers of the participating States. These are binding on all Member States, who may be fined heavily by the Court of Justice if they fail to implement them; however, 'derogations' (delays and exceptions) too numerous to catalogue excuse performance of the Regulations. (These are numbered 'serial/last two numerals of year'.) See Appendix No 1 for a list of some relevant Regulations.
EEC Directives, like Regulations, are binding under the Treaty, but are directed towards individual Member States, or towards individual companies or persons in those states. (These are numbered 'last two numerals of year/serial'.) EEC Notices (numbered 'C last two numerals of year/month/serial') list such matters as approved qwpsr names, approved grape varieties, etc.

So the law can be likened to a fabric or, as some might have it, a net to catch the unwary. Laws have to be enforced to be of any use; those that enforce them are Government Departments and Local Authorities. But whereas in the past it was possible in Great Britain to group the various laws under the Government Departments which had responsibility for them, recent Governments have tended to shift responsibilities from ministry to ministry. It is easier, therefore, to group laws according to their objects; the Departments quoted in the following notes are those to which the administrative function was attributed in 1987.

Protective legislation is broad and far-reaching, covering every aspect of the individual way of life. There are laws which protect the buyer and seller, and the employer and employee; many of them are interwoven and complex. Here, the basic ideas have been put in the simplest forms. It is hard to think of anybody who at some time or another does not fall into one or other of these categories.

To treat this subject fully would require a book of its own; the ensuing sections, each devoted to one of the general objects that have been listed, give only the most relevant Acts, Orders and Regulations in current operation. The basic purposes of each section are summarized in note form; these notes are expanded a little in places where the Wine and Spirit Trade is especially concerned but, even for British students, they should only be taken as a guide indicating the particular law(s) to study. Governments are always making new laws, and the present government of the UK is no exception. The 'licensing laws' regulating the sale of 'alcoholic beverages' differ between the provinces of the realm – and those of England are about to be amended.

It hardly needs emphasizing that these laws are so complicated that no one should *act* on them without consulting a qualified lawyer – in England this will be a solicitor.

For readers outside the United Kingdom, the notes are intended as a guide to the sort of things that are regulated by law, and why; the countries in which such readers work will have parallel laws, and they must search for these for their own purposes. For other nationals, the examples given should at least instruct them in the curious habits of these islands!

PROTECTION OF THE COMMUNITY

This heading covers individuals in bulk – the general public – who may or may not be members of the Wine and Spirit Trade or its customers.

Licensing Acts

These regulate the sale of alcoholic beverages for consumption on or off the licensed premises in quantities of two gallons or less. No Licence is now required for sales of alcoholic beverages in quantities exceeding two gallons (equivalent to one case of wine). Licences are granted by the local Bench of Justices. 'Off' licences allow sales throughout the day; 'on' licences are restricted to certain hours which vary from place to place and between day and day, and are also restricted to particular rooms on the premises. Licensees must not permit, among other things, any drunkenness on the premises. 'On' licences may be extended in time for special occasions by an 'Extension' licence, or may be removed to normally unlicensed premises (such as a community centre) by an 'Occasional' licence. It should be noted that only an existing licensee can obtain Extension licences. In the case of an Occasional licence the licensee may be an officer of a non-licensed Club or Society, such as a local Branch of a political Association, or a village Dramatic Society. But no one may sell alcohol for consumption *without* a licence.

On licensed premises in the UK, no person under the age of 18 may serve in a bar, although they may serve in a restaurant or work (but not sell) in an off-licence. Persons under the age of 14 may not enter a bar, neither may anyone buy alcoholic drinks for persons under 18 in a bar. Over 16, however, young people may buy and consume beer or cider (but not wine) in a restaurant with a substantial meal.

Other types of Justices' licences are the 'Restaurant' licence, where alcohol may only be sold for consumption with a meal; the 'Residential' licence, where alcohol may only be served to residents of the hotel or guest-house for consumption by themselves or their guests; and the 'Club' licence, where alcohol may only be served to members in accordance with the rules of the club as approved by the licensing Justices; this extends to guests, whose numbers will have been limited by club rules, and for golf clubs, to those 'day-members' who have paid a 'green-fee'. Provided that alcoholic drinks (defined as containing more than 1.5% vol. alcohol) are not *sold* for *consumption* no licence is needed. Firms may conduct free tastings in private premises or on public premises if local bye-laws so permit, but *not* on the Queen's Highway. There are many more

1 What offence is committed in misbehaviour of (a) one person towards another; and (b) a person towards society? (c) What branches of law deal with these offences?
2 What are 'derogations'?

complications to this Act, and different rules apply in Scotland, Wales, and Northern Ireland.

Legislative Department: Home Office
Enforcing Authority: Police

Customs and Excise Management Act

It is in the public interest that manufacturers of alcoholic beverages should be licensed, ostensibly so that the public may not be poisoned, but really so that no such beverage should reach the consumer without tax being paid on it. A brewer requires a 'Brewers' licence: so does the manufacturer of British Wine (made-wine *q.v.*), and the vineyard owner making English wine for sale. The ordinary citizen may make his own beer, wine, or cider at home for consumption by his family, employees, and guests; but if he sells it, he must be licensed, and must pay tax. Distillers need a 'Distillers' licence for their apparatus, and this applies whether or not they are distilling alcohol; the difference is that the producer of distilled water pays no tax on his product, but the distiller of alcohol does – heavily. Lastly, the maker of gin, bitters, or liqueurs requires a 'Compounders' licence.

Legislative Department: Treasury
Enforcing Authority: HM Customs & Excise

Road Traffic and Transport Acts

The 1984 Act, amending that of 1972 makes it an offence to be in charge of a motor vehicle while having a breath-alcohol content of more than 35 micrograms (μg) of alcohol/100 ml measured by an approved testing apparatus; or for refusing to take a test when requested by a police officer. This equates to the previous measure of 80 mg/100 ml of blood-alcohol content. Most of the old loopholes in this contentious law have been closed by the new Act and by case law.

Heavy people have more blood in their systems than light: and men more than women; hence the legal capacity varies from person

1 (a) A tort. (b) A crime. (c) Civil and criminal law respectively.
2 Delays and exceptions which excuse performance of EEC regulations. They are the oil which prevents the wheels of the Community grinding to a halt!

to person. It also varies with food intake, which slows passage of alcohol into the bloodstream. And as the human body uses alcohol as a fuel, 'burning' up to 4 μ (micrograms) per hour – but half that when asleep, it wastes from the body. Therefore, although the rate of absorption and wastage of alcohol from the system varies also with the age and health of the subject, people ought to be able to work out for themselves a safe and sensible level of consumption – and if they can't they should not drive if they drink. Those who are unaccustomed to alcohol may yet be unfit to drive with a blood-alcohol level well below the legal limit; conversely, those with a high tolerance to alcohol may find that, after a late party at home the night before, they risk driving to work in the morning with more than the permitted blood-alcohol limit, for the body does not 'burn' so much alcohol when asleep.

Legislative Department: Department of Transport
Enforcing Authority: Police

FACTORIES ACT
OFFICES, SHOPS, AND RAILWAY PREMISES ACT
HEALTH AND SAFETY AT WORK ACT ETC

Although these acts protect individuals, they also protect the public, for the owner of premises is bound by them to provide a safe place to carry on business. A shopkeeper who leaves a cellar entrance unguarded may find himself liable, as may a publican who has a rickety bar-stool – to say nothing of serving drinks in cracked or chipped glasses!

Department of Employment
Principal Legislative Department: Department of Trade

PUBLIC HEALTH ACT
WATER ACT
CLEAN AIR ACTS
DISPOSAL OF POISONOUS WASTES ACT
CONTROL OF POLLUTION ACT
HEALTH AND SAFETY AT WORK ACT
LOCAL BYE-LAWS

1 Who grants licences for the sale of alcoholic beverages for consumption on or off licensed premises?
2 (a) Can the organizer of a social event himself apply for an 'Occasional' licence? (b) Can he sell alcohol for consumption without a licence?
3 Name three other types of Justices' licence.

It was an offence under Common Law to allow the dissemination of any dangerous thing, but these laws go further, with detailed specification. They are very relevant to the Wine and Spirit Trade – the possibility of pollution of streams by the discharge of pot-ale has already been mentioned in Chapter 15.

Legislative Departments: Department of the Environment
Department of Employment
Department of Health and Social Security

PARTNERSHIP ACT
COMPANIES ACTS
INSOLVENCY ACTS

These Acts govern the relationships between firms and their owners, suppliers, and customers. Furthermore, limited companies are legally 'individuals', as are their shareholders; in many cases the latter are so numerous as to constitute 'the general public', who should not be defrauded any more than customers or suppliers. The Companies Acts define, *inter alia*, the powers and duties of Directors, requiring them to file reports and accounts annually. One important provision is that a company that cannot discharge its liabilities must cease trading.

Legislative Department: Department of Trade

PROTECTION IN TRADE

Here, the individual comes in four modes – as wholesale or retail buyer or seller: the law protects them all, and managers in the Wine and Spirit Trade and allied trades must know how it does so. First, consider the protection of the retail buyer, or consumer.

FOOD AND DRUGS ACT
LABELLING OF FOOD ORDERS

1 Local Bench of Justices.
2 (a) Yes. (b) No.
3 'Restaurant' licence, 'Residential' licence, and 'Club' licence.

These laws generally control what may be sold as food; what additives are permitted and in what quantities, for the flavouring or preserving of foods and drinks; and how the products shall and may be labelled. The list of permitted additives is contained originally in a Schedule to the Food and Drugs Act, and this has been amended by various Orders. The rules concerning labelling have been absorbed or superseded by EEC Regulations as far as unsophisticated wines are concerned, but they still apply to all other alcoholic beverages.

Under the regulations, labels *must* contain certain information: here are the main provisions.

(1) *An 'appropriate description'*, which may be, for instance, 'Bordeaux'.
(2) *The country or countries of origin*, which may be deduced from the wording ('Mis au Chateau' has been construed as implying French origin); or specific, as 'Scotch Whisky', or 'Produce of Yugoslavia and Hungary', or 'Produce of the countries of the EEC'.
(3) *The name and address of the responsible importer or bottler* (which may in certain cases be coded).

Besides this essential basic information, labels for some special categories require additional items of information as follows (beer, cider, and perry are not considered):

(a) For light wines of whatever origin, the nominal volume of contents. If this volume is one of those approved for use within the EEC, the 'e' mark must be included in the same field of vision; this applies whether the wine is of EEC origin or not.
(b) Wines which are not made entirely from grapes must show *the strength*. These include wines with additives like aromatized wines and aperitif wines. Also within the group are tonic wines which must show not only the strength but also the exact amounts of the additives which justify the claim.
(c) Wines not made from freshly gathered grapes (commonly called fruit wines but including 'made wines' *q.v.*). These must include *the fruit or non-fruit basis*, plus *the strength*.

1 Name three licences under the Customs and Excise Management Act.
2 Give two amendments to previous legislation enforced by the 1984 Road Traffic and Transport Act.
3 Does consumption of a given quantity of alcohol affect the blood-alcohol of all consumers equally?

(d) Fortified wines: other outside influences creep into the labelling regulations for some of these wines.
Port and Madeira: Under the Anglo-Portuguese Trade Treaty Acts of 1914 and 1916, these two wines have particular protection in that they can only bear these names if they come from Portugal accompanied by appropriate Certificates of Origin. Nor may wines be labelled 'Port-type' or 'Port-style' as often happened in the past.
Sherry: Because of a Civil High Court judgement in what has become known as 'the Sherry case', Sherry comes only from the Jerez district of Spain. However, the judgement also allowed certain generic Sherries to be labelled, e.g. 'British Sherry', provided that the qualifier immediately preceded the word 'Sherry' and was in exactly the same size and type style. 'Cyprus Cream Sherry' is therefore inadmissible because the word 'cream' intervenes. Since 1987, only 'British', 'Irish' and 'Cyprus' sherry are permitted.
(e) Spirits and liqueurs, in addition to the basic requirements (1), (2), and (3) above, need the alcoholic strength on the label. If whisky, gin, rum, vodka, or brandy is under 37.2% vol., a statement of dilution must appear on the label, except for 'old landed' brandies which may have dropped naturally below this strength.
(f) Old wines without adhesive labels, like vintage Port, still need a label with the appropriate requirements. This may be a luggage-type label tied around the neck of the bottle.

Not only are the contents of the label subject to legislation; the manner of showing the information is also regulated. The minimum size of type, the relative sizes of type for different words, and even the juxtaposition of words (as illustrated by 'Sherries', above) are controlled. Further information on these matters may be gained from the works of other authors mentioned in the bibliography or, where any doubt exists, from a legal practitioner.

Legislative Department: Ministry of Agriculture,
Fisheries, and Food

1 Brewers', Distillers', and Compounders'.
2 Breath-alcohol limit of 35 micrograms of alcohol/100 ml; and offence of refusing to take a breath test.
3 No: it depends on food intake, weight, sex, age, and state of health of the individual consumer.

Enforcing Authorities: Local Government: Trading Standards Officers
Wine Standards Board

SALE OF GOODS ACT

Though old and to a certain extent superseded, this Act still gives protection for buyer and seller. Even older law, the common law, operated in one famous case where a Scottish lady was treated by her friend to a bottle of ginger beer, put up in a stone bottle: the first glass was fine, but the refill disgorged 'a slug in an advanced state of decomposition', which caused this lady to suffer 'severe illness'. The Sale of Goods Act could not help her, as she hadn't bought the ginger beer; nor could it help her companion, who had suffered no damage. But the common law laid a duty on the bottler not to be negligent with regard to dangerous things and the lady was compensated for her illness. This case, *Donaghue v Stevenson*, is now overtaken by more recent legislation, but the common law on negligence applies equally to dangerous premises; and various Acts have laid down rules which apply to both customers and staff.

TRADE MARKS ACT
COPYRIGHT ACT
SALE OF GOODS ACT

These Acts and Articles give the trader a number of rights which *inter alia* protect him from false information when purchasing a business; from wrongful use of his trademarks or copyrights without due user agreements; and from non-performance of contracts of sale. Regulations and Directives of the Common Market protect him from agreements which effectively prevent, restrict, or distort competition, or which, by the exercise of a 'dominant position', discriminate against individual traders or groups of traders. But the trader should be warned that these Acts, Regulations and Directives are in some respects double-edged. For instance, rights in a trademark may easily be lost by not defending them in court; and a contract of sale may

1 What UK laws control the permitted addition of flavourings and preservatives to food and drink?
2 Name the main labelling requirements for alcoholic beverages under the Labelling of Food Orders.
3 Show by example how the description of a beverage not totally derived from grapes must be shown on the label.

have to be completed without payment if the contract does not state specifically the date and other terms of payment.

Supply of Goods and Services Act
Trade Descriptions Act
Fair Trading Act

Once again these Acts operate primarily to protect the consumer, but the trader as buyer is generally deemed also to be a consumer. The Supply of Goods and Services Act calls for 'exclusion clauses' to be eliminated so that concealed conditions do not apply to the goods supplied and also gives powers against fraudulent or spurious guarantees and warranties. It also provides that a 'person' receiving such goods or services may regard them as a gift after a lapse of six months, and a 'business' receiving the same may so regard them after six years. If the addressee has not indicated his willingness to buy, the sender may be fined £200 for sending an invoice and £400 if he threatens legal action.

The Trade Descriptions Act makes it a criminal offence to apply a false description to goods, and here 'false' includes 'misleading' descriptions. The Fair Trading Act gives power to the Director General of Fair Trading to stop trading abuses, and he may require written undertakings from the traders to this end. The Director General works with the Consumer Protection Advisory Committee, and sometimes he may make Statutory Instruments – usually where the application is general. Offenders may be taken to the Restrictive Practices Court.

Legislative Department: Department of Trade
Enforcing Authorities: Local Government: Trading Standards Officers
Restrictive Practices Court

PROTECTION IN EMPLOYMENT

There are various Acts with which every employer in the Wine and Spirit Trade or any other trade should be thoroughly conversant,

1 Food and Drugs Act and Orders (e.g. Labelling of Food Orders) made under it.
2 (a) An 'appropriate description'. (b) The country or countries of origin. (c) Name and address of the importer or bottler.
3 For example, fruit basis, exclusively pear juice.

such Acts relating to the safety, health, welfare, and terms of service of employees, and also to their engagement and termination of employment.

LEVYING OF TAXES TO FINANCE GOVERNMENT

One of the greatest problems of rulers is the raising of money. Armies have to be paid, ships have to be built – and, in Charles II's time, mistresses had to be supported. Pepys' diaries tell of his problems in this regard, and the rolls of the Customs and Excise Department in King's Beam House museum give the proof. Customs and Excise duties on goods imported from abroad or produced at home provided the money for the King's Privy Purse – from which all had to be paid; even though, by then, Parliament had to sanction both the expenditure and the levy.

Why 'Customs *and* Excise'? What is the difference? Customs duties are protective tariffs levied on imported goods: tariff might be levied on brandy to protect or promote whisky, or on the goods of one country less than those of another. Excise duties are 'fiscal' in that they confiscate the money of the citizen for the intended good of the state; they apply to all liquors 'Released to Home Use'.

Within the EEC there are no longer (in theory) any discriminating customs duties between members. Common Customs Tariffs (CCTs) are, however, levied on goods imported from 'third' (non-EEC) countries. These are complicated not only by their variation between different products, but also by variations for the same or similar product between more- and less-favoured nations or groups of nations, and in that a specific product (Sherry, for instance) may be allowed a quota volume at a lower rate of customs duty or even none at all. There are, additionally 'Compensatory levies' raised against wine imported into the Community from countries which have not concluded treaties with the Community; the treaties promise not to undercut the price of the Community wines. The levy is made whether or not the prices *do* undercut Community wines, and the computations are complicated.

1 Name the enforcing authorities of the Food and Drugs Act 1955 and the Labelling of Food Orders.
2 Name (a) an Act; and (b) a branch of the Law, giving protection to buyer and seller.
3 How may third parties utilize duly registered Trade Marks and Copyrights?

Of course, all taxes are complicated, and they change rapidly with the wind of politics; traders can consult the tables of actual rates each year, but need to understand the basis on which they are levied, so that they can apply the rates and forecast the effect of changes on their businesses.

Duties on wines, spirits, and liqueurs are levied in bands according to alcoholic strengths. Within each band (as appropriate) they are levied in increasing order of severity on products supplied in bulk, in bottle, and when sparkling. Why? Bulk is preferred to bottle to protect the British glass trade; sparkling wine is fun and, as such, is taxed more heavily!

The bands are as follows:

Wines	
less than 13% vol. (British wine only)	flat rate
less than 15% vol.	flat rate
15% to less than 18% vol.	flat rate
18% to less than 22% vol.	flat rate
22% vol. upwards.	according to strength but expressed as a flat rate 'per litre of pure alcohol'
Spirits, all	
Liqueurs, all	

As the most usual strength of spirits in the UK is 40% vol. and the most usual bottle size 75 cl, the duty on a bottle of spirits can be calculated as 30 per cent of the 'pure spirit litre' rate.

As if all this were not enough, the trader in wines and spirits suffers, as other traders do, the effects of Corporation Tax and Value Added Tax.

Legislative Department: Treasury
Enforcing Authorities: Commissioners of Inland Revenue
Commissioners of Customs and Excise

1 Local Health Authorities, Trading Standards Officers; the Wine Standards Board detects breaches, which the Min. A F F enforce.
2 (a) Sale of Goods Act. (b) Common Law.
3 Through duly registered user agreements.

APPENDIX 1

NOTES ON EEC DIRECTIVES AND LEGISLATION

337/79:★ Sets up a common organization of the market in wine and lays down general rules for the market; the rules include provision for an annual return by wine-growers of areas under vines, and their varieties; and by vinegrowers *and* merchants of their stocks of wines – in great detail. This Regulation also provides machinery for controlling the minimum price of imported wines, and for levying compensatory duties on cheaper imported wines. It regulates the amount of sugar that may be added to musts to improve them in different areas; and it defines wines of various types: '*Wine*' is the product of fermentation of freshly gathered grapes, anywhere in the world; '*Table Wine*' is wine from the EEC; '*Liqueur Wine*' is wine which has been fortified with spirit during fermentation. The Regulation also controls blending. This Regulation replaces Regulations 24/62, 816/70, 2506/70, and the many amending Regulations such as Regulation 1160/76.

338/79:† Lays down general rules for '*Quality Wines Produced in Specific Regions*' (qwpsr – or, in French, vqprd).

★ Amended by Regulations 2961/79; 453, 454, 459 and 3456/80; 3577/81; 2144 and 3082/82; 1595/83; and 1208/84.

† Amended by Regulations 454, 459 and 3456/80; 3578/81; and 2145/82.

1 Under which Act is a buyer for resale deemed to be a consumer?
2 Who has power under the Fair Trading Act to control false and misleading descriptions? Name the Committee of that office.
3 Distinguish between (a) Customs Duties and (b) Excise Duties.

	These are wines of the EEC, produced in France, Italy, Germany, and Luxembourg.
2005/70:	Lists the permitted grape varieties for the production of table wine and qwpsr.
1769/72 amended by 374/74:	Allows coded closures to replace VA Forms.
2247/73 and Notice C73/1/76:	Lists the qwpsr by country and region.
355/79:	Lays down general rules for the description of wines and grape musts, on labels and in promotional literature, including wine lists, and rules for container sizes.
358/79:	Lays down the definitions and general rules for sparkling wines including qwpsr.
1153/75:	Lays down rules for documents to accompany wines, and for the keeping of registers by merchants and vinegrowers; in many cases, the branding of containers with a registered number can obviate the need for full accompanying documents. (R/1769/72.)
997/81:*	Lays down detailed rules for the description of wines and grape musts; the terms which may be used are laid down for each country, and for third countries (non-EEC), specifying the regional and district names, and the names of grape varieties, that may appear on the label. Certain other label information is compulsory (most of which has been

* Amended by Regulations 2618/81; 1224/83 and 1071/84.

1 Supply of Goods and Services Act.
2 Director General of Fair Trading: Consumer Protection Advisory Committee.
3 (a) Duties levied on Imports. (b) Duty levied on imports and home production for home use or consumption.

required for some time under existing UK legislation). All other information is optional, within laid down limits, and must be authorized by the country where the wine is sold or the advertisement appears. It has been obligatory since May 1988 throughout the EEC for the alcoholic strength to be shown on the label.

2984/78: Lays down rules for the laboratory testing of wines, and for documentary evidence of these tests to accompany all imported wines. (This Regulation repeals Regulation Nos. 1599/71 and 1770/72.)

997/81: Lays down permitted sizes and containers for musts and wines between 0.05 and 5 litres.

1475/77: Amends 2113/74 requiring the 'e' mark to be put on all containers of permitted size.

The above is a digest of the most important measures only, at the time of going to press.

DESCRIPTION AND PRESENTATION OF WINES AND GRAPE MUSTS – LABELLING PROVISIONS

The table overleaf is for guidance only. It should not be regarded as an authoritative interpretation, and should be read in conjunction with the text of current Regulations.

The indications in the table which are marked 'E' (essential) *must* appear on the label, and those marked 'O' (optional) *may* appear on the label as far as permitted by the laws of individual member countries. Indications other than these *shall not* appear on the label.

Table of Labelling Provisions overleaf

1 (a) How are duties on wines levied? (b) What other taxes do wines and spirits attract?
2 How is duty on spirits and liqueurs levied?
3 What would the rate of duty be on a bottle of spirits of 40% vol. alcohol and capacity 75 cl?

Table of labelling provisions

Indications appearing on any label attached to the bottle	*Types of wine*				
	EEC wine			*Imported wine*	
	1	2	3	4	5
	Table Wine	Table Wine permitted to indicate Geographical Origin (e.g. Vin de Pays)	Quality Wine psr	Imported Wine	Imported Wine permitted to indicate Geographical Origin
1 The words 'table wine'	E	E			
2a The words 'quality wine psr' or equivalent expression (e.g. Appellation Contrôlée)			E		
2b Superior Quality Description				O	
3 The word 'wine'				E	O
4 The specified region of origin in the EEC (or geographical unit in third countries)			E	E	
5 Country of origin (but see 5a and 5b below) when circulated outside producer state)	E	E	E	E	E

1 (a) In bands according to alcoholic strength. (b) Value Added Tax.
2 According to strength expressed as a flat rate 'per litre of pure alcohol'.
3 30% of the 'pure spirit litre' rate.
(End of Question and Answer footnotes.)

	1	2	3	4	5
5a In the case of EEC wines which have been *coupaged*, the words 'Wine from different countries of the European Community'	E				
5b In the case of EEC wines not turned into wine in the Member State in which the grapes were grown, an indication of the countries in which the grapes were grown and the wine made.	E				
6 The name and address of the bottler/consignor (EEC wines)/ importer (imported wines) as appropriate. May have to be coded.	E	E	E	E*	E*
7 The nominal volume of the contents	E	E	E	E	E
8 The 'e' mark	O	O	O	O	O
9 Indication as to whether the wine is red, rosé, or white	O	O	O	O	O
10 The actual and/or total alcoholic strength	E	E	E	E	E

*If wines are bottled in country of origin, name and address of both bottler and importer must be stated.

	1	2	3	4	5
11a A recommendation to the consumer as to the use of the wine (e.g. 'serve chilled')	O	O	O	O	O
11b Sweet/Dry description	O	O	O		O
12 Indication of one or two vine varieties		O	O		O
13 Indication of vintage year		O	O		O
14 Vineyard name		O	O		O
15 Indication of a geographical unit other than a specified region, or a wine producing region in third countries		O	O		O
16 Traditional special details		O	O		O
17 Quality control number			O*		O
18 Serial number of container			O		
19a Indication of method of production/type of product/special colour characteristics	O	O	O		

* E for Germany—see page 142.

	1	2	3	4	5
19b Indication of a recognized award to the wine			O		O
20 Indication of bottling on the premises of production (e.g. château-bottled)		O	O		O
21 Use of a brand name	O	O	O	O	O
22 Distributor's name and address	O	O	O	O	O
23 A citation awarded to the distributor/ wholesaler/ retailer (e.g. 'By Appointment to . . .')	O	O	O	O	O
24 History/Ageing		O	O	O	O

APPENDIX 2

THE ANNUAL CYCLE OF THE VINE AND VINEYARD WORK

The Annual Cycle of the Vine and Vineyard Work

For a vineyard in the Northern *Temperate Zone with a heat-sum of about 1200°C/days*

	(Year begins) October	*November*	*December*	*January*
Weather	Fine at beginning of month, then rains start. Average rainfall 80 mm Average temperature 12°C	Rain. Temperature falls. Average rainfall 70 mm Average temperature 8°C	Rain, perhaps some snow. Frost at night. Average rainfall 60 mm Average temperature 4°C	Generally cold and wet; some snow and heavy frosts at night. Average rainfall 50 mm Average temperature 4°C
Vine	Grapes fully ripe; leaves start to fall.	Leaf-fall complete; vine moves into dormant state.	Vine dormant; sap retreats from canes.	Vine dormant.
Work	Harvesting until about 15th. Equipment cleaned and put away.	Leaves and rubbish cleared from vineyard. Some early pruning may take place. Continue clearing vineyards for rest and eventual replanting.	Soil ploughed up to roots, to protect them, and to provide trough to drain winter rains.	Renewing posts and wires. Bench grafting in nurseries.

	February	*March*	*April*	*May*
Weather	Much the same as January. Average rainfall 50 mm Average temperature 4°C	Slightly brighter weather, warming up slightly. Windy. Average rainfall 60 mm Average temperature 8°C	Warmer, with showers, some heavy; risk of hail in showers, and of frost at night. Average rainfall 60 mm Average temperature 11°C	Warmer again, with showers and risk of hail; still some frost risk early in month. Calm later. Average rainfall 60 mm Average temperature 15°C
Vine	Vine dormant.	Sap starts to rise; buds become prominent towards the end of the month.	Shoots appear towards end of month.	Shoots grow rapidly. Moths and beetles start depredations
Work	Pruning. Scions taken for later field grafting. Bench grafting continues in nurseries.	Soil ploughed away from roots, leaving hollow by stems to collect spring rain. (Débuttage.) Pruning continues (main month).	Frost precautions. Field grafting. Bench grafts planted out in nurseries. One-year-old grafted shoots from nurseries, and rootstocks for next year's field grafting, planted out in vineyard. Fertilizing.	Shoots trained to wires (Main work). Cutting out suckers and unwanted watershoots. First spraying. Ploughing/hoeing to aerate soil and keep down weeds.

	June	*July*	*August*	(Year ends) *September*
Weather	Much brighter and calmer weather, with occasional showers. Some risk of hail. Average rainfall 50 mm Average temperature 16°C	Much warmer, with occasional rain. Average rainfall 50 mm Average temperature 18°C	Hot weather, with some thunder and risk of hail. Average rainfall 50 mm Average temperature 19°C	Much the same as August, but cooler towards end of month with morning mists. Still some risk of hail. Average rainfall 50 mm Average temperature 16°C
Vine	Shoot and leaf growth continues. Flowering and pollination at beginning of month, berries form by end.	Berries, hard and green, now can be recognized as grapes. Vine growth continues.	Grapes swell and change colour. Leaf growth slows down.	Leaves change colour. Grapes near maturity, reaching it towards end of month.
Work	After flowering, spraying continues and is repeated after rain. Tying up, desuckering, and weeding continue.	Spraying. Hoeing weeds. Cutting back excessive shoot growth and cutting out grape bunches considered too numerous for vine to bring to maturity.	Spraying when necessary, i.e., in particularly humid conditions. Trimming foliage to expose grapes to sun. Preparing equipment for harvest.	Testing grapes for sugar with refractometer. Harvest starts from 15th onwards when sugar readings show grapes are mature.

APPENDIX 3

MALADIES OF THE VINE

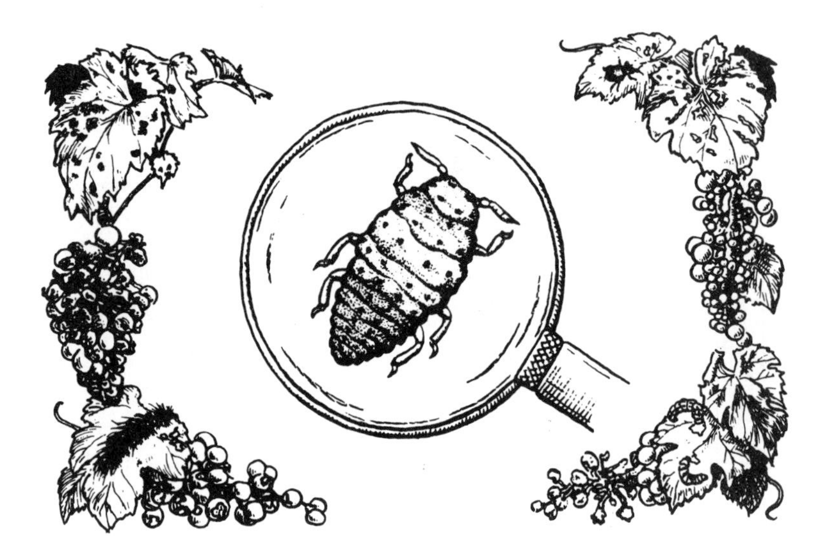

CLIMATIC

Malady	*Symptoms*	*Treatment*
Winter frost (below −16°C)	Trunk and roots split, vine dries up and dies.	Bury whole vine.
Spring frost affecting emergent shoots.	Shoots wither and die.	High training style. Local heating; spray water on vines before frost.
Bad weather at time of flowering.	Sparse bunches = Coulure Undeveloping (shot) berries = Millerandage	
Storms (hail, wind, and torrential rain).	Hail and wind break shoots, tear leaves; Hail holes ripe grapes, which rot; Rain washes soil from roots.	Seed storm-clouds to drop their load elsewhere. Prune away damage and replace soil.
Excessive heat (43°C +)	Leaves scorch, change colour: grapes shrivel.	
PARASITIC		
Phylloxera Vastatrix Feeds on leaves, stems, and roots. Wounds do not heal readily.	Holes in leaves and stems. Vine withers and dies.	Grub up and burn affected vines, sterilize soil, replant with resistant rootstocks. Impossible to eradicate entirely.
Nematodes (Eelworms) Attack roots.	Vine withers and dies.	

PARASITIC (*contd*)

Malady	*Symptoms*	*Treatment*
CATERPILLARS of: ***Pyralis*** (Meal Moth)	Eat shoots and young leaves	Spray before bud-burst and flowering with approved insecticide, and dress soil round stems with insecticide.
Cochylis (Grape-Berry ***Eudemis*** Moths)	Eat flowers and berries - which then rot.	
Melalontha (May-Bug)	Gnaw surface roots; vine 'does poorly'.	
BEETLES **Altise** (Leaf-Beetle)	Eat leaf to lacy state.	Spray with insecticide.
Hanneton (Root- **Vespere** Beetles)	Vine 'does poorly'	Dress soil round stems with suphur dust.
OTHER PARASITES **Snails**	Eat leaves.	Physical removal.
Birds	Peck and eat fruit.	Netting, bird-scares, children with gongs – keeps *their* hands busy!
Wasps	Puncture skins, let rot get in.	Wasp traps.
Wild Herbivores	Eat young shoots.	Fencing – 2.5 m high for deer, buried 0.5 m for rabbits. Repellent sprays. Wax-paper tubes for new plantings.

FUNGAL (CRYPTOGAMIC)

Malady	*Symptoms*	*Treatment*
MILDEWS ***Oidium Tuckerii*** (Powdery Mildew)	Floury white dust on shoots, flowers, grape. Flowers fall off, grape splits, rots and dries.	Sulphur spraying, powder or liquid, from time of bud-burst to after flowering.
Peronospera (Downy Mildew) Thrives in humidity. Two-week cycle.	Oily transparent stains on leaves, followed by white deposit. Leaves and grapes shrivel and fall.	Spraying with copper sulphate (Bordeaux mixture) or proprietary products. Dead leaves (carriers) cleared away.
ROTS **Black Rot** Thrives in humidity.	Black spots on leaves; grapes shrivel.	As for ***Peronospera***.
Anthracnose Thrives in humidity.	Stains on leaves, later becoming holes.	As for ***Peronospera***.
Botrytis Cinerea (grey Rot) Warm, damp weather. Attacks particularly vines weakened by other pests and diseases.	Grey mould on leaves and grapes. Affects pulp of grape, giving nasty taste to wine.	Good vineyard hygiene; constant spraying with copper sulphate in humid weather. New expensive sprays are effective to some extent.
VIRAL		
Court-Noué (Fan-Leaf) (Akin to 'die-back' found in blackcurrants)	Leaves misshapen and discoloured: shoots grow laterally.	Grub out affected vines and replant with virus-free stock after sterilizing soil.
PHYSIOLOGICAL		
Chlorosis Intolerance to lime or boron in soil prevents assimilation of iron.	Leaves turn golden yellow; lack of chlorophyll prevents photosynthesis.	Replant with tolerant rootstocks. Treatment of isolated cases with ferrous sulphate may be successful.

APPENDIX 4

MALADIES OF WINE, THEIR SYMPTOMS, PREVENTION AND CURE

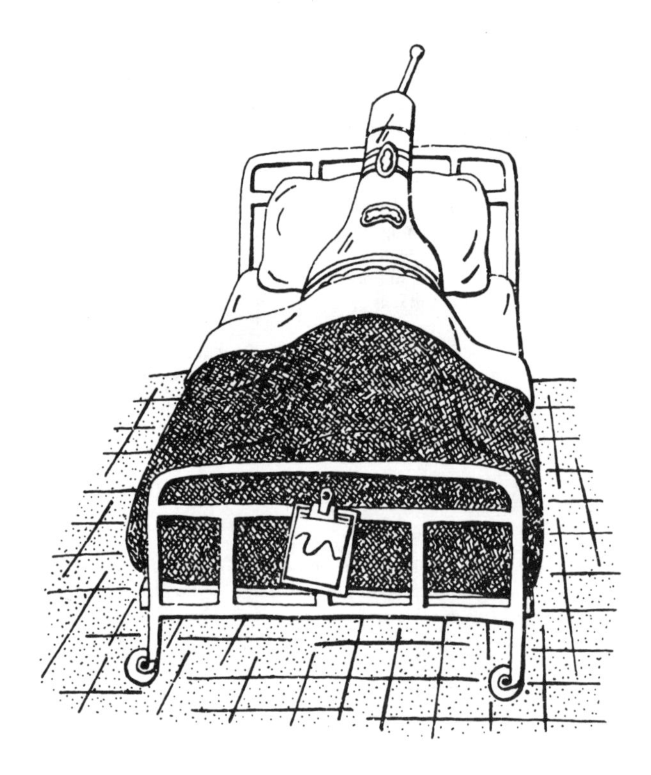

(1) Maladies of chemical origin
(Wines low in acidity are more susceptible to these)

1	*2*	*3*	*4*
Malady	*Symptoms*	*Prevention*	*Cure*
Iron 'Casse' (modification of appearance and colour)	Grey deposit: smell of bad eggs: astringent after-taste	Good cellar hygiene: lower pH: when legal to use citric acid there are additional advantages	(a) Blue fining with potassium ferrocyanide: danger that overdose may leave poisonous product in wine (b) Fining with calcium phytate
Copper Casse	Oxidation: orange/brown deposit: musty smell: bitter taste	As above	(a) Blue fining, as above (b) Fining with rubeanic acid
Oxidic Casse (or oxidation)	Orange/brown colour and tannin: more common in wines with metal content	SO_2: sparging with nitrogen during bottling: pasteurization: addition of ascorbic acid	No certain cure: fining with PVP (polyvinylpyrrolidone), followed by filtration
Protein Casse	Opaque grey haze: no off-flavour: common in fortified wines (and Rhine wines in some years)	Fining with Bentonite during manufacture	Fining with Bentonite

1	2	3	4
Malady	*Symptoms*	*Prevention*	*Cure*
Tartrates	Noticeable in many fortified wines after some months' shelf-life	Refrigeration—less effective for calcium tartrates	
(a) Potassium	Sugar-like crystals, soluble when heated		
(b) Calcium	Sugar-like crystals, not affected by temperature		
Colour Deposit	Colour pigments (mainly in red wines) causing cloudiness	Refrigeration: limited use of SO_2 crystals: over-treatment may cause colour-loss and/or SO_2 flavour	
	(2) MALADIES OF BIOLOGICAL ORIGIN		
Wine Flower	White, 'flor'-like skin floating on surface: slight haziness and considerable deposit: objectionable oxidized 'nose': although Wine Flower and acetic bacteria are distinct, wine with Flower can become further contaminated with acetic bacteria at a later stage	Sterile filtration: addition of SO_2: restriction of exposure to air	None

1	2	3	4
Malady	*Symptoms*	*Prevention*	*Cure*
Acetic Fermentation	Acetic Bacteria: smell of vinegar: can spread from one cask to another (see note above)	Sterile filtration: addition of SO_2: restriction of exposure to air	None
Second Fermentation (Malolactic or yeast)	Bubbles and cloudiness: smell of apples/yeast: common in sugared (improved) white wines and/or low strength wines	Sorbic acid in limited quantities: complete fermentation: sterile bottling: thermal bottling (warm/hot bottling) at 45°C	Sterile filtration: refrigeration and kieselguhr filtration
Rod bacteria infections	Uncommon nowadays with improvement in cellar hygiene	SO_2 filtration: thermal bottling	Re-filter: sterile bottling
(a) *Tourne* (tartaric fermentation)	Dull appearance: 'mousy' smell: flat and acetic		
(b) *Graisse*	Flat, oily, and viscous: in extreme cases forms threads		
(c) *Amertume*	Bitterness and loss of colour		
(d) *Lactic taint*	Turbid, with mousy smell and taint: flat and acetic		

APPENDIX 5
BIBLIOGRAPHY

The first books listed are those of general interest to the reader; these may include references to the regions or other subjects treated separately by chapter in *Wine Regions of the World*. Then follows a list of books appropriate for further reading on individual chapter subjects. It is suggested that the wine and spirit trade press should always be consulted for current information.

GENERAL LIST

Allen, H. Warner. *A History of Wine* (Faber and Faber Ltd, 1961).
Austin, Cedric. *The Science of Wine* (University of London Press, 1968).
Bezzant, Norman, *The Book of Wine* (Ward Lock, 1988).
Born, Wina. *Concise Atlas of Wine* (Ward Lock, 1974).
Broadbent, J. M. *Wine Tasting* (Christies Wine Publications, 1977).
Burroughs, D. and Bezzant, N. *The New Wine Companion* (Heinemann, 1988).
Don, R. S. *Teach Yourself Wine* (Hodder & Stoughton, 1977).
Gold, E. A. *Wines and Spirits of the World* (Virtue, 1973).
Hallgarten, Peter A. *Spirits and Liqueurs* (Faber & Faber, 1979).
Hogg, Anthony. *Off the Shelf* (Gilbey Vintners, 1973).
Holland, Tim. *Behind the Label* (Wine and Spirit Education Trust, 1975).
Johnson, Hugh. *Wine* (Nelson, 1973).
Johnson, Hugh. *World Atlas of Wine* (Mitchell Beazley, 1985).
Lichine, Alexis. *Encyclopaedia of Wines and Spirits* (Cassell, 1982).
Marrison, L. W. *Wines and Spirits* (Penguin, 1973).
Massel, A. *Applied Wine Chemistry and Technology* (Heidelberg Publications, 1969).
Ordish, George. *The Great Wine Blight* (Dent, 1972).
Robinson, Jances. *Vines, Grapes, and Vines* (Mitchell Beazley, 1986).
Sichel, Allan and Sichel, Peter. *The Penguin Book of Wines* (Penguin, 1974).
Vandyke Price, Pamela. *A Dictionary of Wines and Spirits* (Northwood Publications, 1980).

LIST FOR FURTHER READING BY CHAPTERS

CHAPTER 3, *General*

Jacquelin, Louis and Poulin, René. *The Wines and Vineyards of France* (Paul Hamlyn, 1966).
Vandyke Price, Pamela. *Eating and Drinking in France Today* (Tom Stacey, 1972).

Woon, Basil. *The Big Little Wines of France*, vols. 1 and 2 (Wine and Spirit Publications, 1976).

CHAPTER 3, *Bordeaux*

Féret, Ed. *Bordeaux et ses Vins* (Féret et Fils, 1982).
Faith, Nicholas. *The Winemasters* (Hamish Hamilton, 1978).
Penning-Rowsell, E. *The Wines of Bordeaux* (Penguin, 1985).
Peppercorn, D. *Bordeaux* (Faber and Faber, 1982).
Vandyke Price, Pamela. *Guide to the Wines of Bordeaux* (Pitman, 1977).

CHAPTER 3, *Burgundy*

Arlott and Fielden. *Burgundy Wines and Vines* (Davis-Poynter, 1976).
Chidgey, Graham. *Guide to the Wines of Burgundy* (Pitman, 1977).
Hanson, A. *Burgundy* (Faber and Faber, 1982).
Poupon, P. et Forgeot, P. (Ott, trans.). *The Wines of Burgundy* (Universitaires de France, 1979).
Yoxall, H. W. *The Wines of Burgundy* (Penguin, 1976).

CHAPTER 3, *Northern France*

Bréjoux, Pierre. *Les Vins de Loire* (Cuisine et Vin de France, 1956).
Forbes, Patrick. *Champagne, the Wine, Land, and People* (Gollancz, 1972).
Hallgarten, S. F. *Alsace and its Wine Gardens* (Deutsch, 1957).

CHAPTER 4

Ambrosi, H. (trans. Gavin). *German Wine Atlas and Dictionary* (Ceres-Verlag, 1976).
Hallgarten, S. F. *Rhineland Wineland* (Arlington Books, 1951).
Hallgarten, S. F. *German Wines* (Faber & Faber, 1976).
Fowler, trans. Stabilisierungsfond für Wein. *German Wine Atlas and Vineyard Register* (Davis-Poynter, 1976).

CHAPTER 5

Anderson, B. *Vino* (Papermac, 1980).
Belfrage, N. *Life Beyond Lambrusco* (Sidgwick and Jackson, 1985).
Dallas, Philip. *Italian Wines* (Faber & Faber, 1983).

Panarella, G. Carlo. *Italian Wine and Brandy Buyers Guide* (Rome, 1975).
Paronetti, Lamberto (trans. Roncarati). *Chianti* (Wine and Spirit Publications, 1970).
Roncarati, Bruno. *Viva Vino* (Wine and Spirit Publications, 1985).

CHAPTERS 6 AND 7
Croft-Cooke, Rupert. *Madeira* (Putnam, 1961).
Jeffs, Julian. *Sherry* (Faber & Faber, 1982).
Perfeito, Valente. *Let's Talk About Port* (Istituto do Vinho do Porto, 1978).
Read, Jan. *Guide to the Wines of Spain and Portugal* (Pitman, 1977).
Read, Jan. *The Wines of Spain and Portugal* (Faber & Faber, 1973).

CHAPTER 8
Ordish, George. *Vineyards in England and Wales* (Faber & Faber, 1977).

CHAPTER 10
Gunyon, R. E. H. *The Wines of Central and S. E. Europe* (Duckworth, 1971).

CHAPTER 12
Melville and Morgan. *Guide to Californian Wines* (E. P. Dutton, 1976).
Toplos and Dobson. *California Wineries*, vol. 1 (Vintage Image, 1975).

CHAPTER 13
Evans, Len. *Australia and New Zealand Complete Book of Wine* (Lansdowne, 1985).
Knox, Graham. *Estate Wines of South Africa* (David Philip, 1976).
Simon, André. *The Wines, Vineyards and Vignerons of Australia* (Paul Hamlyn, 1967).

CHAPTER 14
Bradford, Sarah. *The Englishman's Wine* (Macmillan, 1979).
Ray, Cyril. *Cognac* (Peter Davies, 1973).

CHAPTER 15
Wilson, Ross. *Scotch Made Easy* (Hutchinson, 1959).
Publications of the Scotch Whisky Association.

CHAPTER 17
Hallgarten, Peter A. *Spirits and Liqueurs* (Faber & Faber, 1979).

CHAPTER 18
The ABC of Licensing Laws (National Union of Licensed Victuallers, 1977).

Glossary

Adega (P) – Cellar (*q.v.*).
Alcohol – In wine terms, this is ethyl alcohol, C_2H_5OH, formed by conversion of grape sugar.
Actual alcohol is the % vol. actually present in wine: **Potential alcohol** is the % vol. of alcohol that could be produced by complete fermentation of the sugars remaining in the wine: **Total alcohol** is the sum of the **Actual alcohol** and the **Potential alcohol**.
Anreichern – Improvement of musts (*q.v.*).
Areometer – Saccharometer (*q.v.*).
Assemblage – Blending (*q.v.*) of wines of the same age from the same region, district, commune, or vineyard of origin.
Baixo Corgo – The district of the Alto Douro region of Portugal being downstream of the Corgo (*q.v.*) river.
Balling, Brix – US scales of sugar density of grape musts. *See also* Saccharometer.
Ban de vendange – Literally, prohibition of the harvest – actually, proclamation of the date when the harvest may start, particularly in Champagne and Bordeaux.
Beaumé – Scale of sugar density of grape musts. Name of French scientist who devised it. *See also* Saccharometer.
Blending – Mixing different wines, or similar wines of different years or vineyards, in order to offset their differences: *see Vin de Cuvée, Coupage, Assemblage.*
Blue-fining – Fining wines affected by iron (ferric) casse or copper casse (*q.v.*) with potassium ferrocyanide to remove the metals. As over-dosing produces prussic acid, treatment must be carried out under the supervision of a qualified analyst. Called 'blue-fining' because iron salts are immediately deposited as Prussian blue. The

process is forbidden in several countries, notably France and South Africa. (*See* Appendix 4.)

Bora – A violent cold wind of Northern Italy and Western Yugolavia. *See* Katabatic Winds.

Bordeaux mixture – Copper sulphate and lime, used for spraying vines against pests and diseases.

Botrytis cinerea – A fungus that feeds on grapes. In cold, wet conditions it forms a grey mould, rotting the grapes and giving an off-flavour to the wine. This is known as **Grey rot** or **Pourriture Grise**. In warm, humid conditions, it sucks the water out of the grapes, concentrating the sugar, bleaching black grapes (undesirable), turning white grapes to rose then brown. It also increases the glycerine content. On white grapes known as **Noble Rot** (*q.v.*).

Bound SO_2 – Sulphur dioxide added to wine which has combined with sugars or aldehydes, or has converted to sulphuric acid. *See* Sulphur.

British Wines – 'Made-wines' (*q.v.*) elaborated in the UK. Distinguish from English Wines, which are made in England from English-grown grapes.

Burgundy mixture – Copper sulphate and calcium carbonate (washing soda) used for spraying against pests and diseases.

Cantina (It). – Cellar (*q.v.*).

Casse – Unhealthy haze or deposit in wines. (*See* Appendix 4.)

Cellar –

(1) Any place for storing bottled wines: the best are underground, maintaining an even temperature of 11°C (52°F); they can be 'wet' (when water can run through them) or 'dry'.

(2) A building for vinification and manufacture of wines: in French, *chai*; in German, *Keller*; in Spanish, *bodega*; in Portuguese, *adega*; in Italian, *cantina*.

Chai (Fr) – Cellar (*q.v.*).

Chaptalisation – Process named after French scientist Jean Chaptal, Comte de Chanteloup (1756–1832). Improvements of musts. In the EEC the addition of sugar is strictly controlled for various areas (see map on page 27) and is expressed in degrees of potential alcohol. It varies from nil in the South to maximum 3° in the North, according to original must-weight, vine variety, and other factors in each area. Will shortly be barred entirely in favour of improvement with concentrated must, which is permitted and not so strictly controlled at present.

Claret – A term for light red wine originating from Bordeaux *Clairet* and adopted as a generic term. May now only be used for wines with Bordeaux AC, although 'Australian Claret' and 'South African Claret' may be permitted.

Clone selection – Establishment of a variety by successive grafting of scions from vines showing the desired characteristics, e.g. Gewürztraminer.

Consejo (Sp) – The official governing body of a wine region. Pronounced kon-seck-o. Distinguish from **Cosecho**, pronounced similarly, which means vintage or harvest.

Consorzio – An association of regional Italian wine-producers. Each Consorzio has its own rules which are additional to DOC laws; many have special labels.

Corgo – A river joining the river Douro at Regua, which divides the Port-producing region of the Alto Douro. **Cima Corgo**, producing the best wines, is east of the Corgo, up-river as far as the Douro is concerned; **Baixo** (lower) **Corgo** is west of the Corgo, down-Douro. Cima is a large area, stretching to the Spanish frontier, producing little that is not of superb quality.

Coupage – Blending (*q.v.*) of wines of different years or of different origins, or of both.

Courtier – Broker who acts as intermediary between vineyard owner and négociant (*q.v.*). Full title Courtier Piquet-en-Vin.

Cross – A new variety obtained from cross-pollination of two varieties of *vitis vinifera*, e.g. Riesling × Silvaner (Müller-Thurgau).

Cuvée – Selection; contents of a vat (*Curve*: Fr) – *see Vin de Cuvée.*

Débourbage – The process of standing must in a tank after pressing to allow soil particles and other solids to settle.

Dégorgement – The action of removing sediment from bottles of sparkling wine.

—*à la volée*: hand extraction of the sediment without freezing.

—*automatique*: extraction by machine after the sediment has been frozen.

Dosage – Adding sweetened wine (*liqueur d'expédition*) to disgorged bottles of sparkling wine. Mixed by *poignettage* – gentle shaking of

the bottles by hand or machine.
Edelfäule (Ge) – Literally 'noble rot' (*q.v.*).
Federation Internationale des Marchands du Vin, Spiritueux, Eaux-de-vie, et Liqueurs (FIVS) – A voluntary grouping of national associations concerned with the production and marketing of wines and spirits, who consider problems common to the trade – labelling regulations, education, promotion, social aspects, etc. There are various sub-committees and an executive committee, which meet during the year; an annual Congress is attended by all participating nations. Great Britain is represented.
Féret – The definitive listing of Bordeaux vineyards, edited and republished every four or five years. (Used to be 'Cocks et Féret'.)
Fixed acids – The principal fixed (non-volatile) acids to be found in wine are tartaric, malic, succinic, and lactic.
Free SO_2 – Unbound sulphur dioxide active in wine as a bactericide. *See Sulphur.*
Goût de Terroir – Applied to wine, either (1) an earthy taste; or (2) identification of the soil and climate of the region, in the wine.
Grey rot – A disease of the grape caused by *botrytis cinerea* (*q.v.*) in its maleficent form.
Hybrid – A variety obtained by cross-pollination of *v. vinifera* with a 'wild' vine, such as *v. riparia* or *v. labrusca,* e.g. Cinsaut × Noah (=BACO 22A).
Hydrometer – Instrument for measuring specific gravities of liquors, e.g. Sikes Hydrometer for gauging the strength of wines and spirits in the UK for excise purposes. *See also* Saccharometer.
Improvement of Musts – Known as **Chaptalisation** (Fr) (*q.v*) **Anreichern** (Ge) (*q.v.*). Not permitted in Italy. Sugar may be added to a maximum of 3% vol. potential alcohol, depending on the region, provided the original must has at least 5.5% vol., measured by an instrument known as a **Saccharometer** (*q.v.*).
INAO – Institut National des Appellations d'Origine. The organization which is responsible in France for laying down the rules for admission of wines to AC, VDQS, or Vins du Pays. Infractions of its decrees are dealt with by the Service de Répression des Fraudes.
Katabatic winds – These occur in regions close to mountains holding glacier snows. Heavy cold air flows down the valleys, attaining speeds of 60–80 kph, continuing until the prevailing wind

is strong enough to hold it in the hills. Typical examples are the **Mistral** (*q.v.*) of Southern France and the **Bora** (*q.v.*) of Northern Italy and Yugoslavia.

Liqueur de Tirage – A mixture of sweetened wine and yeast for inducing a second fermentation.

'Made-wines' – These are not entitled to the name 'wine' because they are not made by the fermentation of freshly gathered grapes in their district of origin, but from preserved musts or dried raisins. **British wines** (*q.v.*) come into this category and, despite the implied slur, are consumed by many with enjoyment.

Mistelle (Fr), **Mistella** (It), **Mistela** (Sp and P) – Grape juice that has been 'silenced' – prevented from fermenting by the addition of alcohol. Distinguish from **Süsswein** (Ge), where the yeasts are removed or killed without alcohol, and further fermentation is possible. Used for sweetening. **Pineau des Charentes** is a mistelle produced in Cognac for drinking.

Mistral – A violent cold dry wind of the Rhône Valley. See Katabatic Winds.

Muffa nobile (It) – Literally 'noble rot' (*q.v.*).

Mustimètre – Saccharometer (*q.v.*).

Négoçiant (FR) – A shipper: a merchant who buys wine in bulk, treats it and often engages in blending (*q.v.*) to produce a saleable wine.

Noble rot – The beneficent form of ***Botrytis*** (*q.v.*). In France it is called **Pourriture Noble**, giving the luscious wines of Sauternes and the Coteaux du Layon; in Germany, **Edelfäule**, giving the Beerenauslese and Trockenbeerenauslese wines; in Italy, **Muffa Nobile**, giving sweetness to Orvieto wines; in Austria it gives **Ausbruch** wines; in Hungary, it is known as aszú, and gives Tokay Eszencia and Aszú. It sometimes occurs in other parts of the world, particularly California.

Oechsle – German scale of sugar density in grape musts. See also Saccharometer.

Oïdium Tuckerii – A mildew which settles on vines. Can be extinguished by spraying with Bordeaux or Burgundy mixture (*q.v.*).

Palus – Rich alluvial land unsuitable for growing grapes for fine wines.

Phylloxera – This vine louse, native to the eastern states of North America, is only countered by grafting on resistant stocks from the same regions, by rigorous quarantine, or by planting in sand. Cyprus, Chile, Australia, Colares in Portugal, and the Rhône Delta sands are free of it; so also are some vineyards in Champagne, the Upper Douro in Portugal, and elsewhere – for unexplained reasons.
Plafond Limite de Classement (plc) – Ceiling yield.
Poignettage – Gentle shaking of bottles of sparkling wine which have received dosage (*q.v.*) to mix the contents.
Pourriture (Fr) – Rot, rottenness. Pourriture gris is Grey Rot (*q.v.*). Pourriture noble is Noble Rot (*q.v.*).
Rebêche – Juice from final intense pressing of marc in hydraulic or screw press. Contains bitter constituents of stalk and pip, and is unsuitable for wine.
Remuage – Shaking sediment from a second fermentation in bottle onto the cork.
Rendement de base – Basic yield.
Riesling
1 The first syllable is pronounced -ee-, as in 'relieve', 'field'.
2 The classic grape of Germany and Alsace; in other countries known as Rhine-Riesling or Johannisberger (-Riesling) to distinguish it from:
3 Wälsch-Rizling, also called Italian Rizling, Laski Rizling, or Olasz-Rizling. Although Wälsch is pronounced Welsh, Wales has nothing to do with it. The spelling 'Rizling' has recently been introduced to distinguish.
Saccharometer – Also known as an **Areometer, Hydrometer**, or **Mustimètre**, is used for calculating the potential alcohol of must. Calibration may be in specific gravity (distilled water = 1000), **°Beaumé, °Oechsle, °Balling**, or **°Brix** (distilled water = 0). The scales relate as shown on pages 19 and 20.
Sulphur – In the form of sulphur dioxide, SO_2, this is the only permitted antiseptic and de-oxidant treatment for wines permitted in many countries. It can be applied in compressed liquid form, or as alkaline salts which react with the wine acids to release SO_2. However applied, some SO_2 binds itself to sugars and aldehydes in the wine, and some is converted to sulphuric acid; together, these

fractions are referred to as 'bound' sulphur. The remainder, as sulphurous acid, is the protective element and is known as 'free' sulphur: more than a minute quantity smells unpleasant. Total sulphur (= 'bound' + 'free' sulphur) is therefore limited to 450 parts per million (ppm), but free SO_2 should not exceed 60 ppm.

Tailles – Must from later pressings after the marc has been dug over or 'cut' (*taillé*) – Champagne.

Tastevin – Silver tasting cup. Common in Burgundy.

Total acidity – The sum of fixed acids (*q.v.*) and volatile acids (*q.v.*) usually expressed in terms of g/litre sulphuric acid.

Tufa – Rock of vulcanized chalk, easy to work for caves, full of minerals, porous, and water-retentive. Found in Touraine district of Loire, and parts of southern Italy.

Valinch – Pipette or 'thief tube' for drawing sample from a cask.

Varietal – A wine named after the grape variety from which it is made.

Vin de cuvée – Must of the first pressing (free-run juice from the central zone of the grape); particularly, Champagne.

Vin Doux Naturel (VDN),
Vin de Liqueur (VdL)
These differ in that VDN must have a higher degree of potential alcohol before addition of spirit than VdL. All VDN are AC; only some VdL are.

Volatile acid – Acetic acid (vinegar). A normal by-product of alcoholic fermentation, acceptable up to 0.6–0.8 g/litre (600–800 ppm).

Index

Principal entries are numbered in **bold type**, reference to maps, charts and diagrams in *italic type*, and other entries in plain type.

Entries with appended ★ (in general grape varieties and soil types) include only principal and extraordinary references.

Abbocado, 199
Abboccato, 166, 184
Abruzzo, **186**
Absinthe, 239
AC, AOC, 45
Acetaldehyde, 33
Acetic fermentation, 33
Acetobacter, 24, 28, 109, 207
Acetylization, 307
Acid:
  citric, 20
  malic, 20, 225
  succinic, 33
  tartaric, 26, 36
Acidification, of musts, 26
Acidity:
  fixed, **33**
  volatile, **33**
Acids:
  fruit, 32
  in grapes, 19, 20
  in wine, 32
  measurement, 20
Aconcagua, River, Valley, 295–6
Adamado, 227
Adda, River, 172
Adelaide Metropolitan, 285
Administrative Law, 343
Aeration, of musts, 25, 26
Afames, 262
Agave, Century plant, 330
Agiorgitiko grape, 230
Aglianico del Vulture DOC, 187
Aglianico grape, 186
Agly, River, 131
Aguardiente, 303
  de Orujos, 314
Agueda, 213
Ahr, 159
Aigle, 247
Airén grape, 201
Aix-en-Provence (Coteaux-d'), 123
Akeleje, 336
Akvavit, **336**
  production, 336
  types, 336
Alavesa, 195
Alba, 169
Albariza, 205
Albarizona, 205
Albarolo grape, 172
Albi, 69
Albumen, for fining wine, 35
Alcohol, sensible use of, 347
Alcoholic drinks, definition, 345
Aldehydes, 33
Aleatico di Puglia DOC, 187
Alella, **199**
Aleppo, 63
Alexander Valley AVA, 275
Alexandria Muscat grape, 261, 266, 289
Alföld, 254, **255**
Algarve DO, 212, **213**
Algeria, **266**
Alicante, 202
Alicante-Bouschet grape, 266–7
Aligoté grape, 95, 105, 263

Alios, 9, 54
Almansa, 202
Almude, 223
Aloes, 333
Aloxe-Corton, 103
Alsace, **88**, *90*
  ACs, **90**
  Comité Régionale d'Experts, 90
  Crémant, 91
  Grand Cru AC, 91
  history, 88
  'Sélection des Grains Nobles', 91
  soil and climate, **89**
  spirits, 312
  vines and wine names, **89**
  wine characteristics, **91**
Alter Doppelkorn, 332
Alto Adige, 168, **175**
Alto Douro: *see* also Port, *221*, 223
  soil and climate, **220**
  vines, **220**
Alvarelhão grape, 227
Alvarinho grape, 226
Amarante, 226
Amarone, 166, 174
American vines 3, 269
American Whiskeys, **326**
Amontillado, 207, 210
Ampelidacae, 3
Ampurdán, **198**
Analyser, 319
Anatolia, 263
Andalucia, **203**
Anderson Valley AVA, 274
Andes, range, 295–6
Angelica, 333
Angels' share, 307
Angevin, 51
Anglo-Portuguese Trade Treaty Acts, 350
Angostura, 334
Angoulême, 304
Aniseed, 333
Anjou/Saumur, 80, **81**, *84*
  production, 81
  soils, 81
Anjou-Coteaux-de-la-Loire AC, 83
Anjou-Gamay AC, 83
Annata, 166
Annual Yield, 46
Anreichern, 140
Anthocyanins, 28, 32, 34
Anthracnose, 282
Añada, 208
Añina, 205
Apennines, Mts, 163, 168, 172
Appellation Réglementée (AR), 312
Appellation Simple (AS), 45
Appellation (d'Origine) Contrôlée (AC), 42, 45
Apple wine, 236
Appropriate Description, 349
Approved Viticultural Areas (AVAs), **274**
Aqua Vitae, 302, 315
Aramon grape, 127, 267
Arbanne grape, 74
Arbois, Mousseux, AC, 109
Arbois-Pupillin, 109
Arena, 205
Argas, 257
Argentina, **296**
Arinto grape, 213, 215
Armagnac, **309**
  climate, 310
  distillation, **310**
  districts, 309–10
  grape varieties, **310**
  maturation, **311**
  continuous still, 310
  oak, **311**
Aromatized wines, **238**
  flavouring, 239
Arràbida, Sierra, 227
Arrack, **331**
Arrope, 203
Arrowroot, 333
Arroyo Seco AVA, 275
Artemisia absinthium, 239
Aspect, **7**
Assimilation, 6, 12
Assisi, 184
Assyrtico grape, 230
Asti, 171
  Barbera d', 171
  Moscato d', DOC, 171
  Spumante DOC, **171**
Aszú, 256
Atacama, desert, 295
Attemperation, *see* Refrigeration

Attica, 231
Auch, 311
Auckland, 288
Aude, Département, 127, 133
Aurore hybrid, 277
Ausbruch, 250
Auslese, **143**
Ausone, Ch, 63
Australia, 10, **282**, *284*
  climate and soils, **283**
  exports, 283
  grape varieties, **283**
  history, 282
  spirits, 329
  vineyard regions, **284**
  wine types, 283
Austria, **248**, *249*
  grape varieties, 249–50
  regions and production, 249
Autolysis, of yeasts, 31
Autovinification, 29, 222
Auvergne, 70
Auxerrois grape, 231, 234
Auxey-Duresses, 104
Azal grape, 226
Azeitao, 227
APNr, 143

Babeasca grape, 257
Babíc, 254
Bacardi, 330
Bacchus grape, 137, 235
Bacharach, Bereich, 160
BACO 22A hybrid (Baco Noir), 310
Bacteria, acetic, 28
Bacteria, **20**, 23
Bad Dürkheim, 150
Bad Kreuznach, 153
Badacsony, 255
Baden, **157**
  Bereiche, **157**
Bädische Bergstrasse Kraichgau, Bereich, 157
Bädische Frankenland, Bereich, 157
Bagaçeira, 314
Bag-in-box, 37, 286
Bairrada, 212, **213**
Balaton, Lake, 254–5
Balatonfüred-Csopak, 255
Balbaina, 205
Balling scale, 19
Ban de Vendange, 63
Banat, 252, 258
Bandol AC, 125
Banyuls AC, **131**
Banyuls Rancio, Grand Cru, AC, **131**
Barbados, 330
Barbaresco DOCG, 169
Barbera:
  d'Asti, 171
  grape, 169, 175, 253, 297
Barco do Rabelo, 222
Bardolino DOC, 173
Barley, Malted barley, 317, 319, 325
Barolo, DOCG, 169
  maturation, 34
Baronnies, Vin de Pays de Coteaux de, 122
Barossa Valley, 285
Barrica, 195
Barrique, 54
Barro-s, 205
Barsac AC, 58
Bas Beaujolais, 108
Basalt, 150, 178
Basic Yield, 46
Basilicata, **187**
Bastardo grape, 217, 227
Basto, 226
Bas-Armagnac, 309
Batavian Arrack, 329
Baux-de-Provence, Coteaux des, 126
Bayern, 141
Béarn, Vins de, 68
Béarnaise sauce, 68
Beaujolais, *106*, **107**
  -Beaujeu AC, 108
  crus, 107
  Supèrieur ACs, 107, 108
  -Vaux AC, 108
  Villages AC, 108
Beaumé:
  degrees, 57, 91
  scale, 19
Beaumes-de-Venise, Muscat de, AC, 120
Beaune, Côte de, 95, *101*, **103**
Beerenauslese, **143**, 146
Beira Alta, 227
Beka'a Valley, 264

Bel Air, Ch, 64
Belemnitic chalk, 73
Bellegarde AC, Clairette de, 128
Bellet AC, 126
Bench grafting, 95
Bench of Justices, 345
Bennett, William, 279
Bentonite, 36, 54, 240
Bereich, 141, 143
Bergerac, **65**
  ACs, 66
Bergkelder, 292
Bermet, 253
Bernkastel, Bereich, 155
Bessards, les, 117
Bianca della Lega, 183
Bianco d'Alessano grape, 187
Bianco, 166
Biferno de Molise, 186
Bingen, Bereich, 149
Bio-Bio, River, 296
Birch Wine, 236
Birngeist, 312
Bitters, 334
Black Sea, 263
Blagny, 104
Blanc de Noirs, 31, 109
Blanc Fumé AC, 86
Blanchots, Les, 99
Blanco, 193
Blanquette de Limoux AC, **132**
Blanquette grape (Mauzac), 132
Blaufrankish (Limberger) grape, 250
Blaye AC, 65
Blended Whiskeys (US), 326
Blending*, 34, 77, 78, 97, 124, 141, 209, 223, 253, 266, 316
Bligh, Governor, 282
Blonde, Côte, 116
Bloom, 20
Blush wines, 31
Bobal grape, 197, 203
Boberg region (Cape), 290
Bocksbeutel, 156
Bodega-s, 206
Bodensee, Bereich, 158
Bois Ordinaires et Communs, 305
Bolsena, Lake, 184
Bombino Bianco grape, 187
Bon Blanc (Chasselas) grape, 111
Bonarda grape, 297
Bonne chauffe, 306
Bonnes-Mares, 102
Bonnezeaux AC, 83
Bons Bois, 305
Bora, (katabatic wind), 176
Bordeaux, **48**, *49, 55, 59, 67,*
  AC, 52, 56, 58, 60, 61, 62
  -Côtes-de-Castillon, 63
  grape varieties, **53**
  history, **50**
  pattern of trade, **51**
  topography and soil*, **49**, 54, 56, 57, 60, 63, 64
  types of wines and spirits, **52**
  mixture, 53
  Mousseux, AC, 52
  Supérieur AC, 62
  vinification, **54**
  viticulture, **53**
Borderies, 305
Borracal grape, 226
Bosco grape, 172
Bosnia, 254
Botanical Gardens, Sydney, 282
Botrytis, 52, 58, 96, 146, 184
Bouchet (Cabernet Franc) grape, 63
Bougros, 99
Bound SO_2, 32
Bouquet garni, 241
Bourbon Whiskey, 326
Bourboulenc grape, 122
Bourg AC, 65,
Bourg & Blaye, **65**
Bourgeois, 56
Bourgogne:
  AC, 98, 105, 107
  Aligoté AC, 98
  Aligoté Bouzeron AC, 98
  Clairet AC, 99
  Grand Ordinaire AC, 98, 107
  Hautes Côtes de Beaune AC, 100
  Hautes Côtes de Nuits AC, 100
  -Irançy AC, 99
  Passe-tout-Grain AC, 98
Bourgueil AC, 85
Bourret grape, 240
Bouvier grape, 253
Bouzy, 78
Braga, 226

Branch-water, 326
Branco, 227
Brandewijn, brandy, 302
Brazil, **297**
Breath-alcohol content, 346
Breede River Valley region (Cape), 290, **293**
Breisgau, Bereich, 157
Breisgau, wine co-operative, 157
Bresse, Poulets de, AC, 110
Breton (Cabernet Franc) grape, 85
Brewer's Licence, 346
British Columbia, **279**
British Imperial Preference, 282, 294
British Legislation for the Wine Trade, **341**
British Plain Spirits, 318, 321
British Sherry, 350
British Sweets, 236
British Wines, **236**
Brix scale, 19
Brochon, 100
Brouillis, 306
Brouilly AC, 107
Bruce-Page Government (AUS), 282
Brune, Côte, 116
Brunello di Montalcino DOCG, 179, **182**
Brut, 79
Bruto, 200
Bual, 217–18
Bucelas, 212, **213**
Bucher-Guyon press, 22
Bud grafts, *11*
Buena Vista (Society), 270
Buffalo grass, 337
Bugey VDQS, 110
Bulgaria, **258**
Bulls' Blood of Eger, 257
Burgengau, Bereich, 159
Burgenland, **250**
Burgundy, **93**, *97*, *98*, *101*, *106*
  climate, **94**
  districts and quality wines, **98**
  grapes, **95**
  history, **93**
  production, 94
  soils, **94**
  vinification, **96**
  viticulture, **95**
Burying vines, 5
Busby, James, 282
Bush training, 15, *16*, 114
Butt, Sherry, 207
Buttage, 53
Buzbag, 263

Cabardès VDQS, 130
Cabernet d'Anjou Rosé AC, 83
Cabernet Franc grape*, 53, 56, 63, 81, 85, 296
Cabernet, Cabernet Sauvignon, grape*, 5, 53, 56, 65, 85, 127, 175, 197, 230, 257–8, 263, 266, 275, 280, 285, 296
Cabernet-de-Saumur Rosé AC, 83
Cadiz, 204
Cahors AC, 69
Cailloux, 56
Cairanne, 119
Calabria, **187**
Calcareous soil*, 54, 155, 210
California, **271**, *272*
  frost precautions, **273**
  regions, AVAs, **274**
  soils, **274**
  topography and climate, **271**
  vines, **274**
  viticulture, 273
  wines, **274**
Calvados:
  AR, 312
  -de-la-Vallée-de-l'Orne AR, 312
  -du-Calvados AR, 312
  -du-Pays-d'Auge AC, 312
Calvi, 133
Camomile, 239, 333
Campagne, 304
Campania, **186**
Campbeltown Malts, 319
Campo de Borja, **197**
Canada, **278,** 278–9
Canaiolo Nero grape, 181
Cane spirit, 216
Canes, 9, 13, 15
Cannonau di Sardegna DOC, 189
Cantabrian Mts, 195
Cantina Sociale, 165
Cantine, 165
Cap of skins, 28

Cape, The, **288**
  climate and soil, **289**
  estate wines of origin, **290**
  grape harvest yield, 289
  history, **288**
  marketing of wine, 294
  vine varieties, **289**
  wine labelling, **289**
Caramany, 131
Caramel, 241, 330
Caraway, 333, 336
Carbon dioxide (CO_2), 17, 21, **24**, 140
Carbonic maceration, 28
Carcavelos, 212, **213**
Cardamom, 333
Carignan grape*, 119, 123, 127, 200, 266–7, 275
Cariñena, **197**
Carmel Valley AVA, 275
Carmenère grape, 56
Carmignano DOC, 183
Carneros AVA, 275
Carpano, Antonio, 241
Carpathian Mts, 257
Carrascal, 205
Cartaxo, 227
Carum carvi, 336
Casa do Douro, 225
Casa Vinicola, 166
Casablanca, 268
Cascade system, 45
Casein, 36, 124
Casks:
  Bordeaux, 34
  Burgundy, 34, 104
  charred, 326
  Piemontese, 34
Caspian Sea, 263
Cassia, 333
Cassis, Crème de, 124, 241
Cassis AC, 124
Castillon-la-Bataille, 63
Cataluña, **198,** *198*
Catarratto grape, 188
Catawba (grape), 269, 277
Caucasus Mts, 263
Cava, CAVA DO, **199**
Ceiling Yield, 46
Cencibel grape, 201
Cenicero, 195
Centaury, 239, 333
Central Coast Region, California, 275
Central Europe, *249*
Central Valley Region, California, **276**
Central Vineyard Region (Loire), 80, **86,** *87*
Central and South-Eastern Europe, **245,** *246*
Centrifuge-ing, 23, 27, 29
Ceramic tiles, 23
Cerdon VDQS, Vin de Bugey Mousseux, 110
Cérons, 54, **60**
Cervantes, 192
Cesanese grape, 185
Cèvennes Mts, 80, 126
Chablais, 247
Chablis:
  AC, *98,* **99**
  Grand Cru, 99
  Petit, AC, 99
  premier cru, 99
Chai, 57
Chalk, as de-acidifying agent, 26
Chalk Hill AVA, 275
Chalk soil*, 86, 133, 150, 197, 231, 265, 304, 309
Chalone AVA, 275
Chambertin, Le, 100
Chambertin-Clos-de-Bèze, 100
Chambéry, 111, 240
Chambolle-Musigny, 102
Champagne, **71,** *75*
  climate, **73**
  corks, **79**
  districts, 74,
  grape varieties, **74**
  growers, 72
  history, 71
  makers, 'Houses', 72
  regulation, **72**, 76
  soil, **73**
  vineyard grading, 73
  vinification, **76**
  viticulture, **76**
Chante Alouette, 117
Chanteau, 307
Chapman, Joseph, 270
Chaptalisation, 26, *27,* 57, 140, 380
Charcoal, activated, 331

Charcoal, 330
Chardonnay grape*, 5, 74, 83, 95, 99, 103, 132, 235, 257–8
Charentais, 303
Charentais still, process, 305, 313
Charente, River, 304
Charente Maritime, Département, 304
Charlemagne, Emperor, 136
Charmat, Eugène, 52
Chassagne-Montrachet, 95, 104
Chasselas (many synonyms) grape, 87, 90, 111
Château Grillet AC, 116
Château, 52
Châteaumeillant VDQS, 88
Châteauneuf-du-Pape AC, 120
Château-bottled, 52
Château-Chalon AC, 109
Châtillon-en-Diois AC, 117
Chaudière, 306
Cheilly-les-Maranges, 105
Chénas AC, 107
Chenin Blanc grape*, 5, 70, **81**, 132, 275, 289
Chenove, 100
Cher, River, 80
Chevalier de Sterimbourg, 117
Cheval-Blanc, Ch, 63
Cheverny VDQS, 87
Chianti, 179
  Classico DOCG, 181
Chiaretto del Garda DOC, 173
Chiavennasca (Nebbiolo) grape, 172
Chichée, 99
Chiclana, 205
Chile, **295**
  climate, 5, 295
  fortified wines, 295
  labelling terms, 296
  vines, 10
Chilling, *see* Refrigeration
Chinon AC, 79, 85
Chiroubles AC, 107
Chlorophyll, 6
Choapa, River, 295
Chorey-lès-Beaune, 103
Christchurch, 288
Chusclan, 119
Ch de Nozet, 87
Ch de Tracy, 87
Ch d'Epire, 83
Cinnamon, 239, 333
Cinqueterre DOC, 172
Cinsaut grape, 5, 119, 266–7, 289
Circulatory vat, 222
Cirial grape, 201
Ciro DOC, 188
Citric acid, 20
Citrus peel, 239, 333
CIVC, **72**
Civil law, 342
Clairette:
  de Bellegarde AC, **128**
  de Die AC, 117
  de Languedoc AC, **128**
  grape, 117, 124, 240, 266–7
Claret, 51
Clarete, 227
Clare/Watervale, 285
Clarksburg AVA, 276
Classico:
  definition, 165
  districts, 173, 175, 183
Classification:
  of 1855, Gironde, **56**
  of 1954, St Émilion, 63
  of 1959, Graves, **57**
Clay, Kimmeridgian, 74
Clear Lake AVA, 274
Clicquot, Veuve, 72
Climat, 136
Climate, **5**
Clochemerle, 108
Clones, **4**, 137
Clos, Les, 99
Clos de Chêne Marchand, 87
Closures:
  crown cork, 36
  screw-top, 36
Clos-de-Tart, 102
Clos-de-Vougeot, 96
Clos-du-Papillon, 83
Clos-Saint-Denis, 102
Cloves, 239, 333
Club Licence, 345
Coastal region (Cape), **290**
Coffey, Aeneas, 319
Cognac, **303**, *303*
  ACs, **308**
  Certificate of Age, 309

climate and soil, **304**
distillation, **305**
districts, 304
early landed late bottled, 309
grape varieties, 305
maturation, **306**
strength, 306, 309
Colares, 10, 212, **214**
Cole Ranch AVA, 274
Colheita, 227
Colli Aretini, Chianti dei, 181
Colli Euganei, 175
Colli Fiorentini, Chianti dei, 181
Colli Orientali del Friuli DOC, 177
Colli Pistoiese, Chianti, dei, 181
Colli Senesi, Chianti dei, 181
Colline Pisane, Chianti dei, 181
Collio Goriziano DOC, 176
Collioure AC, 131
Colombard grape, 65, 275, 289, 305, 310
Color, 209
Columbia, basin, 277
Comblanchien, 102
Comissão do Região dos Vinhos Verdes, 212
Commandaria, 261
Commissioners:
of Customs & Excise, 354
of Inland Revenue, 354
Common Customs Tariff (CCT), 353
Common Dry, 218
Common Law, 342
Common Rich, 218
Community, (European Economic), 41
Como, Lake, 172
Companies Acts, 348
Company of Musketeers, 312
Compartimenti, 169
Compensatory Levies, 353
Completer grape, 248
Compounder's Licence, 346
Comte Tolosan, Vin de Table du, 45
Concannon, James, 279
Concord (grape), 269, 277
Condenser, 306, 311, 317
Condom, 311
Condrieu AC, 116
Congeners, 319
Consejo, 193
Consorzio-i, **166**
Constantia, 288, **293**
Constantine, 267
Consumption habits, 136
Continuous press, 22, 305
Continuous stills, 310, **319**, *320*, 326
Converter, 319
Cooperative Winegrowers Association (Cape), 288
Copra, 331
Corbières, 128, **129**
Cordoba (Arg), 297
Córdoba (Sp), 210
Cordon Bleu, 309
Cordon de Royat, *14*, 76, 96
Cordon training, 15
Corfu, 230
Corgo:
Baixo, 219, 223
Cima, 219, 224
Corgoloin, 102
Coriander, 239, 333
Cornas AC, 117
Corowa, 285
Corporation Tax, 354
Corsica, ACs, **133**
Corsica, **133**
Corton:
AC, 103
Charlemagne AC, 103
hill of, 95
Corvina grape, 173
Corvo, 188
Costières du Gard AC, **128**
Cot (Malbec) grape, 66, 69, 83
Côte Blonde, 116
Côte Brune, 116
Côte Chalonnaise, **105**
Côte de Beaune, 95, *101,* **103**
Côte de Beaune AC, 103
Côte de Beaune Villages AC, 103, 105
Côte de Brouilly AC, 107
Côte de Dijon, 100
Côte de Nuits Villages AC, 103
Côte de Nuits, 94, **100**, *101*
Côte des Blancs, 74
Côte du Jura AC, 109
Côte du Jura Mousseux AC, 109
Côte d'Or, 94, **100**, *101*
Côte Mâconnaise, **105**, *106*

Côte Rôtie AC, 116
Coteaux Champenois, 78
Coteaux de Cap Corse AC, 133
Coteaux de Pierrevert VDQS, 124
Coteaux de Tricastin AC, 119
Coteaux du Languedoc VDQS, **129**
Coteaux du Loir AC, 85
Coteaux du Lyonnais AC, 109
Coteaux du Vendômois, 86
Coteaux d'Aix-en-Provence AC, 123
Coteaux d'Ancenis VDQS, 81
Coteaux-de-l'Aubance AC, 83
Coteaux des Baux-de-Provence AC, 123
Coteaux-du-Layon AC, 83
Côtes de Bourg AC, 65
Côtes de Brulhois VDQS, 66
Côtes de Buzet AC, 66
Côtes de Cabardès et de l'Orbiel VDQS, 130
Côtes de Canon Fronsac AC, 64
Côtes de Duras, **66**
Côtes de Forez VDQS, 88
Côtes de Fronsac AC, 64
Côtes de la Malapère VDQS, 130
Côtes de Marmandais VDQS, 66
Côtes de Montravel AC, 65
Côtes de Provence AC, 123
Côtes de St Mont VDQS, 68
Côtes de Toul VDQS, 92
Côtes de Ventoux AC, 120
Côtes de Vivarais VDQS, 119
Côtes du Frontonnais AC, 69
Côtes du Rhône, **112**
  climate, **113**
  North, *115,* **116**
  production, 113
  soil, **113**
  South, *118,* **119**
  vinification, **114**
  viticulture, **114**
Côtes du Roussillon AC, 131
Côtes du Roussillon Villages AC, 131
Côtes d'Auvergne VDQS, 88
Côtes Roannaises VDQS, 88
Côtes-du-Lubéron AC, 123
Côtes-du-Rhône-Villages AC, 119
Cotnarí, 257
Coulée-de-Serrant, 83
Country of Origin, 349
Courtier, 52
Cow-parsley, 336
Crackling, 278
Crémant de Bourgogne AC, 99
Crémant de la Loire AC, 80
Crépy AC, 111
Crete, 230
Criadera, 208
Criminal law, 342
Croatia, **253**
Croatina grape, 169
Crop control, 13, **45**, 76, 137
Crossing, **4**
Crozes-Hermitage AC, 117
Crushing, 21, 54, 108, 123
Cuero, 193
Cultivar, 290
Cumin, 336
Cunningham hybrid, 217
Customs & Excise Management Act, **346**
Cutlasses, 329
Cyprus sherry, 262, 350
Cyprus, 10, **260**, *261*
Czechoslovakia:
  districts, 251
  soil, population, 251

Daimiel, 202
Dalmatia, **253**
Damascus, 263
Danube, 245–6, 258
Dão, **214**
Dark grains, 318
Dated Ports, 222–3
De Rolin, Nicholas, 94
De Rothschild, Baron Edmund, 264
De Vogüe, Comte Robert-Jean, 72
Dealul-Mare, 257
Debina grape, 230
Débourbage, 23, 77
Dégorgement, 78
Deidesheim, 150
Dejuicer, 22
Delicias, 280
Denominação do Origem, 42, 227
Denominación de Origen (DO), 42, *192,* 193
Denominazione di Origine Controllata (DOC), 164, **165,** 167, 169
Desize-les-Maranges, 105

Deutscher Tafelwein, 141
Deutscher Weintor, 150
Deux Sèvres, Département, 304
Deuxièmes Tailles, 77
Dew-points, 321
De-acidification, of musts, 26
De-stalking,-stemming, 21, 54, 108, 123
Diastase, 317, 319, 336
Diatomaceous earth, 36
Die, Clairette de, 117
Digestifs, 309
Diglucosides, 4
Dill, 336
Dimiat grape, 258
Direct producers, **4**
Directives, EEC, 41, **343**
Distillation, Irish Whiskey, 325
Distillation, principles, 301
Distillers Licence, 346
Doce, 227
DOCG, 42, **165**
Dolce, 166
Dôle, 247
Dolomites, Mts, 163
Dom Pérignon, 72
Dominant Position, 351
Don Quixote, 193
Donnaz, 172
Doppelkorn, 332
Dordogne, River, 50, 62, 66
Dorin (Chasselas) grape, 247
Dosage, 78
Dourado grape, 226
Douro Branco Grape, 215
Douro Valley, 226
Downey, John G., 270
Draff, 317–18
Dragasani, 258
Drakenstein, range, 290
Drava, River, 253
Dreimännerwein, 155
Drosophila, 24
Drum malting, 317
Dry Creek Valley AVA, 275
Dubrovnik, 254
Duero (Douro), River, 191, 197
Dulce, 200, 209
Dunder, 329
Durbanville, 292
Dutchess hybrid, 277, 297
D'Artagnan, 309, 312
D'Yquem, Château, 58

Early Landed Late-Bottled Cognacs, 309
Éauze, 311
Eau-de-vie-de-marc, 313
  -de-Champagne, 77, 314
  -originaire-d'Aquitaine, 53, 314
Eau-de-vie-de-poiré, 312
Eau-de-vie-de-vin, 302
  -de-Bourgogne, 312
  -de-Faugères, 312
  -originaire-d'Aquitaine, 53, 312
Ebro, River, 191, 194
Échézaux, 102
Edelfäule, 146
Edelzwicker, Vin d'Alsace AC, 91
Edna Valley AVA, 275
EEC:
  Comité Consultatif Viti-Vinicole, 41
  commission, 41
  control of:
    advertising, 42
    bottle sizes, 42
    minimum prices, 42
    vinestock type, 41
    vineyard planting, 41
    wine labelling, 42
    wine quality, 42
  Council of Ministers, 41
  Court of Justice, 343
  derogations, 343
  Directives, 41, **343**
  legislation, 41, **341**
  Notices, **343**
  objectives, 41
  Regulations, 41, **343**
  Wine Management Committee, 41
  Wine and Spirit Importers Group (of FIVS), 41
  wine-zones, 26, *27*
Eerste, River, 292
Eger, 256
Egg-whites, 35, 54, 209
Égouttoir, 22
Egri Bikaver, 256–7
Egypt, **265**
Ehrenfelser grape, 137

Einzellage-n, 136
Eisenberg, 250
Eiswein, **143**
EK filter, 36
El Dorado County AVA, 277
Elba Bianco DOC, 183
Elbling grape, 155, 231
Elixir, 302
Emilia-Romagna, **179**
Enfer d'Arvier, 172
Engarrafado na origem, 227
English wines, **232**, *233*
  climate and soils, **232**
  grape varieties, **234**
  vinification, **235**
  viticulture, **235**
English Vineyards Association, 234
Enology, Departments of:
  Davis, 274
  Fresno, 274
Ensenada, 280
Entraygues-et-du-Fel VDQS, 70
Entre Rios, 297
Entre-deux-Mers AC, **61**
Enzymes, **24**
Épernay, 72
Epirus, 230
Épluchage, 21
Epoxy resin, 23
Espadeiro grape, 226
Espalier training, 15
Espumante, 227
Espumoso, vino, 199
Estaing VDQS, 70
Estoril, 213
Estramadura, 227
Estufa,-gem,-do, 217, **218**
Est! Est!! Est!!! di Montefiascone, 185
Eszencia, Tokay, 256
Ethanol, 322
Ethyl alcohol, **24**, 32
Etna, 188
Étoile Mousseux AC, 109
European (Economic) Community, EC, EEC, **41**
Euvitis, 3
Export Akvavit, 336
Extension Licence, 345
Ezerjó grape, 255
Faber grape, 137
Faible, 307
Falerno, 186
Falkenstein, 250
Faro DOC, 188
Faugères AC, 129
Federação dos Vinicultores do Dão, 212, 214
Federal Viticultural Council, 282
Feints, 318, 321
Fendant (Chasselas) grape, and wine, 247
Fennel, 333, 336
Fer grape, 70
Fermentation, **24**
  acetic, 33
  alcoholic, 24, 32, 317
  anaerobic, 25
  effect of aeration, 25
  effect of pressure, 25
  factors affecting, **25**
  malolactic, 26, 32, **33**, 72, 225
  permitted treatments, **25**
  temperature, 25, 60
  temperature, red wines, 28
Fernet Branca, 334
Feteasca grape, 257, 263
Feu nu, 306
Fez, 268
Fiddletown AVA, 277
Fiefs Vendéens VDQS, 81
Figari, 133
Filter, filtration, 23, 27, 36, 241, 330
Filtrado, 179
Fine, Marsala, 188
Fine Champagne Cognac, 308
Fine Marne, 312
Finger Lakes, 269, 277
Fining, **35**, 54, 60, 124, 240
Finlandia, 337
Finos, 207
Fins Bois, 305
Fish albumen, 35
Fitou AC, **129**
Five Kings, The, 261
FIVS, 41
Fixed acidity, **33**
Fixin, 100

Flagey-Échézaux, 102
Flavoured Spirits, **333**
Flavours, extraction, 334
Fleurie AC, 107
Fleys, 99
Flor, 109, 207, 262
Flowers,-ing, 9, 17
Flûte d'Alsace, 90
Focsani, 257
Folle Blanche (Picpoul) grape, 81, 305, 310
Food and Drugs Act, 343, **349**
Forbes, 285
Foreshots, 318, 321
Forst, 150
Fortification, 26, 207
Fortified Wine, 26
Fraise, 313
Framboise, 313
France, **43**, *44*
  Central and Eastern, **93**
  Central Southern, *125*
  consumption habits, 136
  Mediterranean, 70, **112**
  Northern, **71**
  South-West, **48**
  wine production, 45
  wine regions, 46
Franken/Franconia, Bereiche, **156**
Franken/Franconia, soils, 156
Franschhoek, 292
Frascati DOC, 185
Free SO_2, 32
Freisa grape, 171
Friuli-Venezia Giulia, 166, **176**
Frizzante, 171
Fronsac, Fronsadais, **64**
Frontignan, Muscat de, ACs, 132
Frontonnais AC, Côte du, 69
Frost, 10, 95, 273
Fructose, 24
Fruit acids, 32
Fruit Spirits, **302**
Fruit wines, **236**
Fruska Gora, 252
Fuenmayor, 195
Fugger, Bishop, 185
Fumet, fish, 242
Furmint grape, 250, 256
Fusel Oils, 319
Fyé, 99

Gaglioppo grape, 187
Gaillac AC, 69
  Doux AC, 69
  Mousseux AC, 69
Galestro, 183
Galicia, **194**
Galium odoratum, 242
Gamay grape*, 81, 83, 85, 95, 107, 247, 250
Gamza (Kadarka) grape, 258
Gard, Costières de, **128**
Gard, Département, 127, 133
Garda, Lake, 163, 173
Garganega grape, 175
Garnacho-a, grape*, **196–9,** 201
Garonne, river, 50, 66
Garrafeira, 227
Garrigues, 128
Gattinara DOC, 169
Gay-Lussac, 24
Geisenheim (Weininstitut), 146, 235
Geneva, canton, 247
Geneva, Lake, *see* Léman; Lac
Geneva Double Curtain, *14*, 176, 235
Genever, 334, 335–6
Gentian, 239, 333
Georgia, 263
Gerace, 187
Germination, 317
Gers, Département, 309
Gevrey-Chambertin, 100
Gewürztraminer grape, 5, 89, 158, 199
Gianacis, Nestor, 265
Gigondas AC, 119
Gin, **334**
  production, **335**
  types, 335
Ginger, 236, 333
Gironde, Département, 48, 66
Gironde, River, 50
Giropalettes, 78
Givry, 105
Gladstone, 288, 335
Glass, 23
Glucose, 24

Glühwein, 159
Glycerol, 26
Gobelet training, 15, *16*, 96, 107
Gönc,-i, 256
Gonçalves, Captain João (Zarco), 215
Gondolas, 17, 21
Goulburn Valley, 286
Goût de terroir, 57
Governo all'uso Chianti, 181
Gozo, 265
Graciano (Morrastel) grape, 197
Grafting, **10**, 95
  bench, 95
  bud, *11*
  field, 12
  hand, *11*
  machine, *11*, 12
Grain Spirits, **315**
  history, 315
Grain Whisky, **319**
  maturation, 321
Granite*, 81, 95, 107, 113, 122, 178, 212, 214, 220, 226, 274, 289, 291, 319
Grape:
  constituents, **19**
  pulp, 18
  seeds, 18
  skin, 19
  structure and composition, **18**
  zones, **19**, 22, 31
Grape sugar, 19
  concentration, 24
  measurement, 19
Grappa, 314
Grasa grape, 257
Gravel*, 50, 54, 57, 62, 86, 127, 131, 274
Graves:
  district, 51, 54, **57**, *59*
  grape varieties, 58
  Supérieure AC, 60
Graves-de-Vayres AC, **62**
Great Plain, districts and grape varieties, 255
Great Western, 286
Grecchetto grape, 184
Greco di Bianco DOC, 187
Greco di Tufo DOC, 186
Greco grape, 185
Greece, **229**
  history, 229
  island vineyards, **230**
  regions, 229–30
  soils, 229
Green malt, 319
Green Valley-Solano AVA, 275
Green Valley-Sonoma AVA, 275
Greffieux, les, 117
Gremio-s, 212, 225
Grenache grape*, 119, 127, 128, 131, 276, 280; *see also* Garnacha/o
Grenouilles, Les, 99
Grevenmacher, 231
Grey rot, 96
Grillet AC, Château, 116
Grillo grape, 188
Grinzing, 251
Groslot, Grolleau grape, 83
Grosslage, 143, 149
Gros-Plant-du-Pays-Nantais VDQS, 81
Grüner Veltliner grape, 249–51
Grumello, 172
Guadalquivir, River, 191, 208
Guadiana, River, 191
Guenox Valley AVA, 274
Gumpoldskirchen, 250
Gun-flint, aroma, 86
Gutedel, (Chasselas) grape, 90, 158
Guyot training system:
  Double, *14*, 15, 61, 76, 95, 235
  Simple, *14*, 15, 53, 63, 76, 205
Gypsum, 26, 206

Haardt mountains, 150
Hail, 50, 58
Halbtrocken, 141
Halkidiki, 230
Hand grafts, *11*
Hanepoot (Muscat) grape, 293
Haraszthy, Agoston, 270
Hardy, 304
Haro, 194–5
Hárslevelü grape, 256
Harvest-ing, **17**, 57, 58, 76, 130, 281
Haut Médoc AC, 56
Hautes-Côtes-de-Beaune, Bourgogne, AC, 100
Hautes-Côtes-de-Nuits, Bourgogne, AC, 100

Hautes-Pyrénées, Département, 67
Hautvillers, 71
Haut-Armagnac, 309
Haut-Brion, Ch, 58
Haut-Comtat, 122
Haut-Montravel AC, 65
Haut-Rhin and Bas-Rhin, Départements, 89
Haut-Savoie, Département, 110
Hawkes Bay, 288
Haze, in wine, 35
Heads, 306
Health and Safety at Work Act, 347
Heart, 306, 335
Heat-sum, **6**, 232
Heidelberg, 157
Heidsieck, 72
Hennessy, 304
Hérault, Département, 127, 133
Herbs, aromatizing, **239**, 333
Hercegovina, **254**
Hermitage, AC, 117
Hermitage, grape name, 284, 289
Hermosillo, 280
Hessische Bergstrasse, **158**
Heurige-n, 251
Highland Line, 318
Highland Malts, 318
Himbeergeist, 313
Hine, 304
Hippocrates, 238
Hochheim, 147
Hock, 147
Homs, 263
Hospices de Beaune, 94, 103, 314
Howell Mountain AVA, 275
Hudson, River, 277
Huguenots, 51
Humagne grape, 247
Humboldt Current, 295
Hungary, *249*, **254**
  regions, 10, 254
  soil and production, 254
Hunsrück, 152
Hunter River Riesling (Sémillon) grape, 283
Hunter River Valley, 285
Huxelrebe grape, 137, 235
Hybrids, **4**, 217–18
Hydrogen sulphide (H_2S), 31
Hydrolysis, 307
Hydrometer (saccharometer), 19
Hyssop, 239, 333
H M Customs & Excise, 346, 353

Iberian Peninsula, 190, *191*
  climate, **190**
Idaho State, USA, 277
Ile de Ré, 305
Ile d'Oléron, 305
Imbottigliato, 166
Immature Spirits, 334
Imotski, 254
INAO, 43
India, 7
Indonesia, 329
Indre, river, 80
Inferno, 172
Infiascato, 166
Infusion, 240
Inoculation, 27, 31, 207, 262
Institut National des Appellations d'Origine, 43
Instituto do Vinho do Madeira, 212, 219
Instituto do Vinho do Porto, 212, 225
International Federation of Wine Merchants, 41
Inzolia grape, 188
Irouléguy, 68
Irrigation, 217, 273, 279, 293, 296–7
Isabella hybrid, 217, 297
Ischia, 186
Isinglass, 54
Islay Malts, 318
Israel, **264**
Istanbul, 263
Istrian Peninsula, **253**
Italy, **161**, *162*
  Central and Southern, **178**, *180*
    soils and climate, **178**
  climate, **162**
  DOC law, **164**
  history, **161**
  Northern, **168**, *170*, *174*
  vines, **163**
  viticulture and production, **164**
  wine labelling terms, 165

Jacquet hybrid, 217
Jalle de Blanquefort, 56
Jamaica, 330
Jardin de la France, Vin de Table du, 45
Jasnières AC, 85
Jefferson, Thomas, 51
Jerez de la Frontera, 204
Jerez/Xeres/Sherry, **203**
Johannisberg, 140
Jubilæumsakvavit, 336
Juliénas AC, 107
Jumilla, 202
Juniper, 333, 334
Juniperus communis, 334
Junta Naçional do Vinho, 212
Jura, **108**
  soil, 109
Jurançon AC, 67, **68**
Jurançon grape, 310
Justices' Licences, 345

Kadarka (Cadarca) grape, 253, 255–6, 258
Kaiserstuhl/Tuniberg, Bereich, 157
Kallstadt, 150
Kalterersee, 176
Kanzler grape, 235
Kanzlerberg, 91
Katabatic wind, 113, 176
Kékfrankos (Limberger) grape, 255–6
Kéknelyu grape, 255
Kelowna, 279
Kerner grape, 137
Ketones, 33
Kieselguhr, 36
Kiln, 317
Kimmeridgian clay, 74, 86, 94
King's Beam House Museum, 353
Kir, 124
Kirsch Fantaisie, 313
Kirschwasser, 313
Kisalföld (Small Plain), 254, **255**
Klevner (Pinot), 91, 248
Klöch-East Styria, 251
Kloeckera apiculata, 20
Kloster-n, 136
Klosterneuburg, 250
Knights Templar, 261
Knights Valley AVA, 275
Knipperlé, 90
Kokkineli, 230, 262
Korn, 332
Kosher wine, 264
Kosmet, 252
Krems, 250
Kuskenkorvo, 337
Kvarner, 253
KWV, 288, 291

La Cienega AVA, 275
La Côte, 247
La Mancha, **201**
Labelling disciplines, 349–50
Labelling of Food Orders, **349**
Lacryma Christi del Vesuvio, 186
Lactic acid, 33
Ladoix-Serrigny, 103
Lafite, Ch, 96
Lafões, 227
Lagar-es, 206, 218, 221
Lago di Caldaro, 176
Lalande-de-Pomerol, 64
Lambrusco grape, 179
Lamego, 226
Landes:
  Département, 68, 309
  pine forests, 54, 304
Landwein-gebiet, 141
Langeberg, range, 293
Langenlois, 250
Langhorne Creek, 285
Languedoc, 62, 126, **128**
  Clairette du, AC, 128
  Coteaux de, AC, 127
  origin of name, 129
  vermouths, 240
Languedoc & Roussillon:
  districts, *125*, **128**
  soil and climate, **127**
  vines, 127
Lanson, 72
Late Bottled Vintage Port, 223
Latisana, 177
Latour de France, 131
Laudun, 119
Lausanne, 248
Lavaux, 247
Law, objects of, 342
Layon, river, 80

Lazio, **185**
Leavening, of musts, 27
Lebanon, **264**
Lebrija, 209
Lees, maturation on, 81
Legal Aspects, **339**
Léman, Lac, 110–1, 247
Lemon Gin, 336
Length, of maturation, 34, 35
Lenz Moser, 235, 249
Les Monts Damnés, 86
L'Étoile AC, 109
Leuconostoc, 33
Levant, The, **260**
Levante (Sp), **202**, *202*, 205
Levkas, 230
Leyburn, Roger de, 62
Libourne, 62
Licensing Acts, **345**
Licences:
  Brewer's, 346
  Compounder's, 346
  Distiller's, 346
  Extension, 345
  Justices', 345
  Occasional, 345
  'Off', 345
  'On', 345
  Residential, 345
  Restaurant, 345
Liebfrauenkirche, 149
Liebfraumilch, *143*, **149**
Lignin, 34
Liguria, 168, **172**
Lima, 226
Limassol, 262
Limekiln AVA, 275
Limestone soil*, 58, 61, 64, 69, 81, 85, 89, 94–5, 105, 108, 113, 122, 127–8, 132, 156, 199, 227, 229, 248, 285, 296
Limnio grape, 230
Limoges, 306
Limousin, 306
Limoux, 70, 132
Lindos, 230
Linie Akvavit, 337
Liqueur Cognacs, 309
Liqueur de tirage, 78
Liqueurs, 334
Liquorice, 333
Lirac AC, 122
Listrac AC, 56
Litoral, 297
Little Karoo, 294
Livermore AVA, 275
Loam*, 89, 147, 157
Locorotondo DOC, 187
Lodge, Port, 222
Loess*, 84, 95, 146, 156, 157, 159, 251
Logroño, 195
Loir, River, 85
Loire, The, **79**, 80, *82*, *84*, *87*
  districts, **80**
  history, 79
  production, 80
  soils, 81
Lombardia, 168, **172**
London Gin, 336
London Market, 218
London Particular, 218
Lorraine, 92, 312
Lote-s, 219, 223
Lot-et-Garonne, Département, 66
Loupiac AC, 61
Low Wines, 318
Low Wines Still, 318
Lowland Malts, 318
Lubéron Côtes de, AC, 123
Lunel AC, Muscat de, 132
Luxembourg:
  country, 155, **231**
  Marque National, 42
Lyne arm, 317

Macabeo grape, 196, 198–200, 267
Macedonia, 230, 252
Macération Carbonique, 28, 61
Maceration, 28, 130, 240, 335
Macharnudo, 205
Machetes, 329
Machines:
  bottling, 23
  cultivating, 137
  grafting, *11*, 12
  harvesting, **17**
Mâconnaise, Côte, **105**, *106*
Mâcon Villages AC, 105, 107
Macon (Supérieure) AC, 105

Mâcon-Lugny AC, 107
Madeira, **215**, *216*, 336
  fortification, 218
  history, **215**
  maturation and styles, **218**
  vines and wines, **217**
  vinification, **218**
  viticulture, **217**
Madeira Wine Company, 219
Madeleine Angevine grape, 235
Madera County AVA, 276
Maderisé, 240
Made-wines, **236**
Madiran AC, **67**
Maduro, 227
MAFF, 41
Maindreieck, Bereich, 156
Mainviereck, Bereich, 156
Maipo Valley, 296
Maiwein, 242
Maize, 319, 336
Makheras mts, 261
Malaga, **210**
Malapère, Côtes de la, VDQS, 130
Malbec grape, 56, 65, 66, 68, 69, 263, 296–7
Malic acid, 20, 225
Malmsey, 217–18
Malolactic fermentation, 26, 32, **33**, 72, 225
Malt Whiskies, **316**
  districts, 318–19
Malta, **265**
Maltose, 317
Malvasia Preta grape, 220
Malvasia (Malvoisie) grape*, 81, 176, 181, 185, 188, 199, 215, 220, 297
Manchuela, 201
Manseng grape, 68
Manta, 222
Manzanares, 202
Manzanilla, **208**
Maps, isothermal, 6
Maraschino, 254
Marc, 77
Marche, **183**
Marcillac VDQS, 70
Maréchal Foch hybrid, 278
Margaux AC, 56
Mariout vineyard, 265
Marjoram, 239, 333
Markgräferland, Bereich, 158
Marl-y soil*, 9, 65, 95, 130, 149, 156, 181, 231
Marlborough, 288
Marque National de Luxembourg, 42, 232
Marque, 77, 308
Marsala, **188**
Marsanne grape, 117
Martell, 304
Martinique, 330
Mash tun, 317
Mashing and fermentation, **317**
Maturation in bottle, **35**
Maturation of wines, **33**, 57
Maturation on lees, 81
Maule, River, 295–6
Maures, Massif de, 123
Maury, ACs, **132**
Mauzac, 69, 70, 132
Mavrodaphne grape, 230
Mavron grape, 261–2
Mavrud grape, 258
Mazuelo (Carignan) grape, 197
McArthur, Captain John, 282
McDowell Valley AVA, 274
Mead, 238
Méal, le, 117
Mechanical Harvesting, **17**, 130
Mecsek, 255
Médoc:
  AC, 56
  district, 54, *55*, **56**
  grape varieties, 56
Meknès, 268
Melnik grape, 258
Mendocino County AVA, 274
Mendoza, 296
Menetou-Salon AC, 87
Mentrida, 201
Mercaptans, 31
Mercurey, 105
Meritza, River, 258
Merlot grape*, 53, 56, 58, 61, 63, 65, 175, 177, 248, 258, 263, 280, 296, 297
Merrit Island AVA, 276
Metabisulphite, of Potassium, 23
Method of study, xiii

Méthode ancienne, 123
Méthode Champenoise, 52, 99, 109, 117, 132
Méthode rurale, 69, 132
Methuen Treaty, 219
Meursault, 95, 104
Mexico, wine regions, 280
Mexico, 5, **279**
Mezcal, 331
Mezesfeher grape, 255
Mezzogiorno, 186
Microclimates, **7**, 58, 201
Micropore filter, 36
Mijo, 226
Mikveh, 264
Milawa, 286
Mildew, 53
Millet, 319
Milly, 99
Minas Gerais, 297
Minerals, 6, 8, 19
Minervois AC, 127
Ministry of Agriculture (MAFF), 41
Mint, 333
Mirabelle de Lorraine AR, 312
Mirabelle, 232, 312
Mireval AC, Muscat de, 132
Misket grape, 258
Mission, 270
Mistela, 200
Mistelle, 131, 240
Mistral, 113
Mittelhaardt, Bereich, 150
Mittelrhein, **159**
Modena, 179
Moët et Chandon, 72
Molasses, 328
Moldavia, 257, 263
Molette grape, 111
Molinara grape, 173
Molise, **186**
Monastrell grape, 203
Monbazillac AC, 66
Monçao, 226
Mondego, River, 214
Moniacke Castle wines, 236
Monimpex, 254
Montagne de Reims, 74
Montagny, 105
Montalbano, Chianti dei, 181
Montefiascone, 185
Montepulciano, Vino Nobile de, DOCG, 181, **182**
Montepulciano d'Abruzzo DOC, 186
Montepulciano grape, 184, 187
Monthélie, 104
Montilla y Morilés, **210**
Montlouis AC, 80, 85
Montmorillonite, 36, 54
Montrachet, 104, 314
Montravel AC, 65
Moonshine, 326
Moors, 203
Mór, 255
Morey-Saint-Denis AC, 102
Morgon AC, 107
Morio-Muskat grape, 137
Morocco, **267**
Moscatel grape, 205, 209, 210
Moscato d'Asti, 171, 241
Moscato grape, 169
Moselle, River, 231
Moseltor, Bereich, 155
Mosel-Saar-Ruwer, **153**, *154*
  soil and grapes, 155
Mosto, 207
Mother of wine, 208
Mother-in-law's Tongue, 330
Moulin-à-Vent AC, 107
Moulis AC, 56
Mountain (Malaga), 210
Mourisco Semente grape, 220, 227
Mourvèdre grape, 119, 128, 267
Mousseux, 80
Mudgee, 285
Müller-Thurgau grape*, 90, 137, *138*, 231, 234, 249
Muffa nobile, Il, 184
Mura, River, 253
Mûre, 313
Murfatlar Hills, 257
Murray, River, 286
Musar, Ch, 264
Muscadelle, 53, 58
Muscadet, 81
Muscadet-des-Coteaux-de-la-Loire AC, 81
Muscadet-de-Sèvre-et-Maine AC, 81
Muscat, Muscatel grape*, 120, 131, 188, 240, 257, 261, 280, 295

Muscat:
  de Beaumes-de-Venise AC, 120
  de Frontignan AC, 132
  de Lunel AC, 132
  de Mireval AC, 132
  de Rivesaltes AC, 132
  de St Jean de Minervois AC, 132
Muscat d'Alsace, grape, AC, 89, 199
Musigny, Les, 102
Muskat Ottonel grape, 250
Musketeers, Company of, 312
Must, 23
  improvement of, 26, 305
  regulation of temperature, 54, 96

Nahe, **152**, *152*
Naked flame, 306
Nama, 260
Nantais, district, 80, **81**, *82*
Nantais, Gros Plant du Pays, VDQS, 81
Naoussa, 230
Napa Valley AVA, 275
Napoléon Bonaparte, 71
Napoléon brandy, 307
Navarra, **196**
Néac, 64
Nebbiolo grape, 169
Neckar, 141
Nederburg, 291
Négoçiant, 35, 52, 96
Négoçiant-éleveur, 34, 52
Negrette, 69
Nematodes, 3
Nemean Valley, 230
Neuburger grape, 250
Neuchâtel, **248**
Neusiedlersee, 250
Neustadt, 150
Nevers, 80
New South Wales, **285**
New Zealand, *285*, **286**
Niagara, peninsula, 278
Niagara (grape), 277, 297
Nicoresti, 257
Nicosia, 261
Niederösterreich:
  districts, **250**
  soil, grapes, and wines, **250**
Niellecio grape, 133
Nierstein, Bereich, 149
Nile Delta, 265
Nitrogen, 8, 76, 130
Noah (grape), 269, 310
Noble rot, 58
Noble vine varieties, 89
Nocera grape, 266
Normandy, 312
Norte, 297
North America, **269**
  history, 269–70
North Coast Region, California, 274
Northern Massif (H), 255, **256**
Northern Sonoma AVA, 275
North-Eastern States, USA, **277**
North-West Africa, **265**
  history and economics, 265
North-West States, USA, **277**
Nuits, Côte de, 94, **100**, *101*
Nuits-St Georges, 102, 314
Nuragus di Cagliari DOC, 189
Nussberger grape, 251
Nutmeg, 239

Obermosel, Bereich, 155
Oberrhein, 141
Occasional Licence, 345
Occidente, 297
Oechsle, 140
Oechsle scale, 19
Oeillade grape, 124
Oeuil-de-Perdrix, 248
Off Licence, 345
Ohio, River, 269, 277
Oidium tuckerii, 51, 217
Oja, Rio, 194
Ojo de Gallo, 197
Okanaga, River, 279
Okinagan Desert, 269
Olaszrizling (Wälsch-) grape, 255
Old Brown Cognac, 308
Old Castille, **194**
Old Tom Gin, 336
Oléron, Ile de, 305
Olive, still head, 306
Oloroso, 208–9
Oltrepó Pavese, 172
On Licence, 345
Onion-skin (pelure d'oignon), colour, 109, **122**
Ontario, 269, **278**

Opol, 254
Optima grape, 137
Oran, 267
Orange Gin, 336
Oregon State, USA, 277
Orris root, 239, 333
Ortenau, Bereich, 157
Orujo, 209
Orvieto DOC, 184
Otago, 288
Oude Genever, 335
Oued Medjerda, 266
Oued Miliane, 266
Oujda, 268
Ox blood fining, 60
Oxidation:
  of juice, 17, 21, 235
  of spirit, 307
  of wine, 207

Pachérenc-du-Vic-Bilh AC, 68
Packaging, 36
Padthaway/Keppoch, 285
Paicines AVA, 275
País grape, 296
Palette AC, 124
Palissage, 89
Palmela, 227
Palomino grape, **205**, 276, 289, 293
Palus, 50
Pamid grape, 258
Pantelleria, Passito di, 189
Pape Clément, 120
Pardillo grape, 201
Parellada grape, 199
Parramatta, 282
Paso Robles AVA, 275
Passito-i, 166, 181
Pasteur, 108
Pasteurization, of musts, 27
Pasteurization, of wines, 36, 240–1
Patrimonio AC, 133
Pau, 67
Pauillac AC, 56
Pays d'Oc, Vin de Table du, 45
Pays Nantais, Gros Plant du, VDQS, 81
Peat, 317, 319
Pécharmant AC, 65
Pech-Langlade (Côtes du Frontonnais), 70
Pectins, 19
Pedicel, 18
Pedro Ximénes (PX) grape, 200, **205**, 209, 210, 266
Peleponnese, 230
Pelure d'oignon (onion skin) colour, 109, **122**
Penafiel, 226
Peñafiel, 197
Penedés, **199**
Pergole Trentine training, 176
Periquita, 227
Perlan, 247
Perlwein, 232
Pernand-Vergelesses, 103
Perth, 283
Peru, 5
Pétillant, 80, 110
Petit Chablis AC, 99
Petit Meslier, 74
Petit Verdot, 56, 296
Petit Champagne Cognac, 304, 308
Petit-Rouge grape, 172
Petrus, Ch, 64
Phillips, Captain Arthur, 282
Photosynthesis, 6
*Phylloxera vastatrix**, 3, 10, 51, 128, 195, 214, 217, 252, 261, 271, 295, 297
Picolit, 177
Picpoul de Pinet VDQS, 130
Picpoul grape*, 68, 119, 128, 198, 240, 310
Piemonte, 168, **169**
Pierrevert VDQS, Coteaux de, 126
Pineau de la Loire (Chenin) **81**
Pineau d'Aunis, 86
Pinhão, 220
Pinhel, 227
Pinot grape, 5, 263
Pinot Beurot grape, 81
Pinot Blanc grape, wine, 90, 171, 176, 231
Pinot Chardonnay Mâcon AC, 107
Pinot Gris, Grigio – grape, wine, 89, 91
Pinot Meunier grape, 74
Pinot Noir grape*, 74, 86, 95, 103, 253, 275
Pinotage grape, 289

Pipe, 219, 222
Pisco brandy, 295
Plafond Limite de Classement (plc), 46
Plavina, 254
Plemenka, 253
Plymouth Gin, 336
Po, River, 163
Poinchy, 99
Pointe de Grave, 56
Poire Williams, 312
Polar currents, 5, 273, 295
Polish Zubrowska, 337
Pollino DOC, 187
Pomace, 34, 77
  spirits, **313**
Port, **219**, *221; see also* Alto Douro
  bottling, 223
  definition, 219
  fermentation and fortification, **221–2**
  labelling, 223
  Late-Bottled Vintage, 223
  maturation, **222**
  production, 225
  Ruby, 223
  styles, 223
  vineyard authorization, **223**, *224*
  vineyard classes, 224
  vinification, **221**
  Vintage, 223
  White, 223
Port of Indicated Age, 223
Port with Date of Harvest, 223
Porto Vecchio, 133
Portugal, **211**
  climate and soil, **212**
  demarcated regions, **212**
  government control, **212**
  history, **211**
  labelling terms, **227**
Portugieser, Blauer Portugieser, grape, 150, 250
Posip, 254
Post training, 15, 137
Pot Ale, 318, 321
Potassium metabisulphite, 23
Potatoes, 336
Poteen, 326
Potter Valley AVA, 274
Pot-still, 305, 317–18, 329
Pouilly Fumé AC, Blanc Fumé AC, 86
Pouilly Fumé de Pouilly AC, 86
Pouilly-Fuissé AC, 86, 107
Pouilly-Loché AC, 107
Pouilly-sur-Loire AC, 87
Pouilly-sur-Loire, 80, **86**
Pouilly-Vinzelles AC, 107
Poulsard grape, 109
Pourriture noble 83; *see also* Noble Rot
Poverty Bay, 288
Prädikat-en, **143**, *143*
Premières Côtes de Blaye AC, 65
Premières Côtes de Bordeaux AC, *59*, **61**
Premières Côtes de Gaillac AC, 69
Premières Tailles, 77
Press, **21**, 23
  basket, 21
  Bucher-Guyon, 22
  continuous screw, 22, 305
  Vaslin, 22
  Willmes, 22
Press-wine, 34, 124
Preto Mortágua grape, 215
Preuses, Les, 99
Pre-heater, 306
Primitivo di Manduria DOC, 187
Primitivo grape, 186
Priorato, **200**
Prissey, 102
Procanico (Trebbiano) grape, 183
Produttori, 166
Prohibition:
  in Canada, 278
  in USA, 271
Provence, *121*, **123**
  AC districts, 123
  Côtes-de-, 123
  soils, **122**
  VDQS districts, **126**
  viticulture, **123**
Provinci, 169
Prûfungsnummer, **143**
Prugnolo Gentile (Sangiovese) grape, 182
Pruning, **12**
Puerto de Santa Maria, 204
Puerto Rica, 330
Puglia, **187**, 241
Puligny-Montrachet, 95, 104

Pulque, 330
Pump-s, 23
Purification (distillation), 301
Putto, Consorzio, 181
Puttonyo-i, 256
Pyrenees, mts, 126
Pyrénées,-Atlantiques, Département, 67
Pyrénées-Orientales, Département, 127, 133

Qualitätswein bestimmte Anbaugebiet (QbA), 42, **141**, 176
Qualitätswein mit Prädikat (QmP), 42, 140, **143**
Quarts-de-Chaume AC, 83
Quartzite, 146, 149
Queen Victoria, 147
Queensland, 283
Quetsch, 232, 313
Quincy AC, 86
Quinine, 239, 333
Quinta, 221
Qwpsr, vqprd, 141, 343

Rabat, 268
Rabelais, 79
Rabigato grape, 220
Racking, of wines, 32, 34, 35, 124
Rain, 6
  during harvest, 17, 57
Rainwater, 218
Ramisco grape, 214
Rancio, 120, 129, **131**
Ranina (Bouvier) grape, 253
Rasteau AC, 120
Rasteau, 119
Rayas, 207
Ré, Ile de, 305
Rebêche, 77
Recie, 174
Recioto, 166, 174, 175
Rectified spirits, **322**
Rectifier, 319
Red wines:
  fermentation, **28**, 54
  maturation, **34**, 54, 57
Reek, 317
Refosco, 177
Refractometer (Saccharometer), 17, 19
Refrigeration:
  of musts, 26
  of wines, **36**, 209, 240–1
Regaleali, 188
Regiãos Demarcadas, 212, 226
Région de Mercurey, 105
Régione Siciliano, Q, 188
Régua, 220
Regulations, EEC, 41, **343**
Rehovot, 264
Reichensteiner grape, 137, 235
Reims, 71
Remich, 231
Remuage, 78
Rendement de base, 46
Reserva, 227
Residential Licence, 345
Respiration, 6
Restaurant Licence, 345
Retsina, **231**, 238
Retz, 250
Reuilly AC, 86
Rheingau, **145**, *146*
Rheinhessen, **147**, *148*
Rheinpfalz, **150**, *151*
Rhein-Mosel, 141
Rhodes (Rhodos), 230
Rhône, Côtes du, **112**, *115*, *118*
Rhône Delta, 10
Rhône sands, 128
Rhum Verschnitt, 330
Ribeiro, 194
Ribero del Duero, **197**
Ribolla, 177
Riesling (Rhine-, Johannisberger-, White-) grape, 89, **137**, 145, 149, 159, 171, 231, 251, 263, 275, 283, 296
Rio de Janeiro, 297
Rio Grande do Sul, 297
Rio Negro, 197
Rioja, **194**, *195*
  climate, **195**
  grape varieties, viticulture, **196**
  Alavesa, 194, 195
  Alta, 194, 195
  Baja, 194, 196
Ripaille, Ch, 111
Riserva, 166, 182, 187
Riserva speciale, 182

Rishon le Zion, 264
Riverina, 285
Riversonderend, range, 293
Rivesaltes ACs, **132**
Rkatsiteli grape, 258, 263
Road Traffic and Transport Acts, **346**
Roaix, 119
Robertson, 294
Rochefort, 304
Rochegude, 119
Roche-au-Moines, 83
Roederer, 72
Roma, 283
Romania, regions and production, **257**
Rondinella, grape, 173
Roots, 9
Rootstocks, 4, 10
Rooty Hill, 285
Roriz grape, 220
Rosado, 227
Rosé de Loire, 80
Rosé d'Anjou AC, 80
Rosé wines:
  fermentation, **31**
  maturation, 34
Rosemary, 333
Rosette AC, 66
Rosé-des-Riceys AC, 74, 78
Rosso Conero DOC, 184
Rosso, 166
Rotgipfler grape, 250
Rothschild, Baron Edmund de, 264
Rôtie AC, Côte, 116
Roussanne grape, 117
Roussette de Bugey VDQS, 111
Roussette grape, 110
Rousset-les-Vignes, 119
Roussillon, **130**
Rubesco, 184
Ruby Port, 223
Rueda, **197**
Rülander (Pinot Gris) grape*, 157, 158, 231, 251
Rufete grape, 220
Rufina, Chianti dei, 181
Ruinart, 72
Rully, 105
Rum, **328**
  countries of production, 329
  definition, 328
  distillation, 329
  maturation, 330
  sugarcane processing, 329
Rumbullion, 328
Rumbustion, 328
Russian River Valley AVA, 275
Rust, 250
Rutherglen, 286
Ruzica, 252
Rye Whiskey, Canadian, American, **326**
Rye, 325

Saar-Ruwer, Bereich, 155
Sables, Vins de, 130
Sablet, 119
Sabra, 264
Sabrosa, 226
Saccharometer, 19
Saccharomyces, 20, 207
  cerevisiae, 20, 25
  ellipsoidus, 20, **21**
  oviformis, 20
  uvarum, 20
Sacramento Valley, 276
Sage, 333
Saladin Box malting, 317
Sale of Goods Act, 351
Saltillo/Parras, 280
Salvagnin (Pinot Noir) grape, 247
Samos, 230
Sampigny-les-Maranges, 105
San Joaquin Valley, 276
San Juan del Rio, 280
San Pasquale AVA, 276
San Severo DOC, 187
Sancerre AC, **86**
Sand-s-y soil*, 66–7, 122, 128, 130, 212, 214, 227, 289, 309
Sandstone-soil*, 58, 67, 149, 153, 155–6, 289
Sangiovese grape*, 179, 181–2, 182, 184
Sanlùcar de Barrameda, 208
Santa Catarina, 297
Santa Clara county, 276
Santa Cruz Mountain AVA, 276
Santa Maddalena, 176
Santa Maria Valley AVA, 276

Santa Ynez AVA, 276
Santenay, 104
Santigo, 330
Santorini (Thira), 230
Sao Paulo, 297
Saperavi grape, 258, 263
Sardegna, **189**
Sartène, 133
Sassella, 172
Sassicaia, 183
Saumur, **83**
Saumur-Champigny AC, 85
Saussignac AC, 66
Sauternes, AC, 58
Sauternes and Barsac, 54, **58**, *59*
Sauvignon grape*, 53, 58, 60, 61, 65, 83, 85, 86, 188, 266, 275, 280, 296
Sauvignon-de-St Bris VDQS, 95, 98
Sava, River, 253
Savagnin grape, 109
Savatiano grape, 231
Savennières, 83
Savigny-lès-Beaune, 103
Savoie:
  Département, 110
  districts, **110**
Savuto DOC, 187
Scheurebe grape, 137
Schiave grape, 176
Schist-ous soil*, 81, 113, 128, 130–1, 146, 159, 212, 220, 274
Schloss, Schlösser, 136
Schloss Böckelheim, 153
Schönburger grape, 235
Sciacarello grape, 133
Scotch Whisky, Scotch, *316*, 325
Scotch Whisky Association, 325
Secco, 166
Seco, 200
Second fermentation, for sparkling wines, **78**
Séguret, 119
Selo de Origem, 226
Semidulce, 200
Sémillon grape*, 53, 58, 60, 65, 266, 283, 289, 296–7
Semiseco, 200
Serbia, **252**
Sercial, 217–18
Sérine (Syrah) grape, 116
Serpentine press, 22
Serpentine, 311
Serprina grape, 175
Settling, of must (débourbage), 23
Setúbal, Moscatel de, **225**
Setúbal, 213
Sèvre, River, 80
Seyssel AC, 111
Seyval hybrid, 235, 278
Sforzato (Sfursat), 173
Shale, 289
Shenandoah Valley AVA, 277
Sheris Sack, 203
Sherry, **203**, *204*
  classification, **206**
  climate, **204**
  fermentation, 206–7
  fortification, 207
  history, **203**
  preparation for sale, **209**
  soil and grapes, **205**
  vinification, **206**
  viticulture, **205**
Shipper, 34, 35
Shiraz (Syrah) grape, 283, 289
Sicilia, **188**, 241
Siebengebirge, Bereich, 159
Sierra Foothills Region, California, **277**
Sillery, 78
Silvaner grape, 137, 149, 231, 235; *see also* Sylvaner
Simone, Ch, 123
Simonsberg, range, 290
Single Malt Whiskies, **318**
Siphon vat, 29, 222
Sipon (Furmint) grape, 253
Sitges, Malvasia de, 199
Skåne Akvavit, 337
Slate*, 146, 149, 153, 155, 159, 200
Slovenia, **253**
Small Plain (Kisalföld), **255**
Smederevka, 252
Smith and Woodhouse, 188
Smudge-pots, 273
Snake, River, 277
Snaps, 336
Soave DOC, 173, **175**
Soil, **8**, 50, 54
Solera-s, 201, **208**, 219

Somló, 255
Somontana, **197**
Sonoma Mountain AVA, 275
Sonoma Valley AVA, 275
Sopron, 255
Sorbara, Lambrusco di, 179
Sour Mash, 326
Sousão grape, 220, 276, 289
South African Sherry, 289, 294, 350
South America, **295**
South Australia, **285**
  grape varieties, 285
  wine districts, 285
  wine production, 285
South Coast Region, California, **276**
Southern Hemisphere, **281**
Southern Vales, 285
Soviet Union, **263**
Spätburgunder (Pinot Noir) grape, 146, 157, 158
Spätlese, **143**
Spain, **190**, *191*
  central, **201**
  history, **192**
  quality wine regions, **194**
Spanish earth, 209,
Spanna (Nebbiolo) grape, 169
Sparging, 317
Sparkling wine styles, *79*
Speciale, Marsala, 189
Specific gravity, 19
Spent wash, 318
Spices, 333
Spirit safe, 318, 321
Spirits, **299**
  history, 301
  rectified, **322**
Spitzenweine, 248
Spraying, 273
Spumante, 166
Spur training, 13, *14*
Stainless steel, 23
Stara Planina, range, 258
Starch, 317, 319
Statute Law, 343
Statutory Instruments, 343
Steen grape, 289
Steigerwald, Bereich, 156
Steinweine, 156
Stellenbosch Farmers' Winery (SFW), 292
Stellenbosch, 292
Ste Croix-du-Mont AC, 61
Ste Foy-de-Bordeaux AC, **62**
Still:
  Charentais, 305
  low wines, 318
  spirit, 318
  wash, 317
Strengths of wine, 52, 81, 267
Stripping, distillation, 301
Styria, **251**
St Amour AC, 107
St Chinian VDQS, 129
St Émilion AC, **63**
St Émilion Côtes, 63
St Émilion des Charentes, 305
St Émilion Graves, 63
St Etienne-de-Baïgorry, 68
St Georges, Les, 102
St Gervais, 119
St Jean de Minervois, Muscat de, AC, 132
St Joseph AC, 116
St Juan, 296
St Julien AC, 56
St Macaire AC, 61
St Maurice-sur-Eygues, 119
St Nicholas-de-Bourgueil AC, 85
St Pantaléon-les-Vignes, 119
St Péray AC, 117
St Pourçain VDQS, 88
St Raphäel, 239
St Rémi, 71
St Romain, 104
St Véran AC, 107
St Vérand, 107
Sub-climates:
  continental, 5, 94, 113, 191
  maritime, 5, 190
  mediterranean, 113, 191
  upland, 5, 280, 295
Succinic acid, 33
Südliche Weinstrasse, Bereich, 150
Süd-Tirol, 175
Süsswein, Süssreserve, 25, 140
Sugar:
  addition to must, 25

effect on fermentation, 25
ugar-cane, 215, 329
uisun Valley AVA, 275
Sulphiting, 23, 25, **32**, 54, 60, 77
Sulphur, permitted levels, **32**
Sulphur candle, 23
Sulphur dioxide (SO_2), 23, **32**, 77
Sumoll grape, 199
Sunning, of grapes, 206
Sunshine, 6
Superiore, 166, 172, 174, 175, 186
Sur lie, 248
Suze, 239
Swartland, 291
Swellendam, 294
Switzerland, **246**
Sylvaner, 82, 90; *see also* Silvaner
Syrah, Petit Syrah, 5, 116, 127, 275
Syria, **263**

Tafelweingebiete, **141**
Tafia, 329
Tagus (Tejo, Tajo), River, 191, 213
Taille Chablis, 76
Tailles, Premières, 77
Tank method, 52, 200
Tannat, 67, 68
Tannins, 19, 20, 32, 34, 307
Tarragona, **200**
Tartaric acid, 26, 36
Tartrates 24, 36
precipitation, 36, 240–1
Tasmania, 283
Tastevin, 104
Taunus mts, 147
Taurasi DOC, 186
Tavel AC, 122
Tawny Port, 223
Taxes, **353**
Taza, 268
Te Kauwhata, 286
Tea, 239
Temecula County AVA, 276
Templeton AVA, 275
Tempranillo-a, grape, **196–7**, 297
Ténarèze, 309–10
Tent (Tinto), 193
Tenuta, 166
Tequila, **330**
method of production, 330–1
region, 331
Terlano, 176
Teroldego grape, 176
Terra Alta, **201**
Terraces, 112, 184, 217, 220, 226
Terrantez grape, 217
Terret grape, 120, 128
Têt, River, 131
Tetrapacks, 37
Thermotic extraction, 29, 31, 130
Thrace, 263
Thunderstorms, 50
Thurgau, canton, 248
Tiber, River, 184
Ticino, canton, 248
Tierçons, 307
Tignanello, 183
Tinajas, 201
Tinta das Baroccas grape, 289
Tinta Francisca grape, 220
Tinta Negra Mole grape, 217
Tinta Pinheira (Pinot Noir) grape, 215
Tirage, Liqueur de, 78
Tirnave, 257
Tischweine, 248
Tocai grape, 176
Tokaj (place), 256
Tokay:
Aszú, 256
Eszencia, 256
Szamorodni, 256
Tokay d'Alsace, 92
Tomelloso, 202
Tonel-es, 221
Tonic wines, 236
Topping-up, of casks, 34
Torgiano DOC, 184
Torreon, 280
Torres Vedras, 213, 227
Toscana, **179**
Total SO_2, 32
Touraine, 80, *84*, **85**
Touriga grape, 215, 220, 227, 276
Training, principles and systems, **13**, *14*, *16*, 61, 89, 114, 137
Trakya, 263
Traminer grape, 4, 176, 250–1, 263

Transdanubia, **255**
  districts and grape varieties, **255**
Treaty Law, 342
Treaty of Rome, 343
Treaty of Windsor, 211
Trebbiano d'Abruzzo, 186
Trebbiano (Ugni Blanc) grape*, 5, 175, 179, 181, 184, 289, 297
Trebujena, 205
Trellis training, 217, 225
Trentino, 168, **175**
Tresterbranntwein, 314
Tricastin AC, Coteaux de, 119
Trocken, 141
Trockenbeerenauslese, **143**, 146
Tronçais forest, oak, 102, 306
Troödos mts, 261
Trousseau grape, 109
Tufa, 85, 186
Tulbagh, 291
Tunisia, **266**
Turin (Torino), 241
Turkey, **262**
Turk's Cap, still head, 306
Tursan VDQS, 68

Uco Valley, 297
Ugni Blanc grape*, 5, 65, 124, 266–7, 280, 305, 310
Uisge Beatha, 315
Uiskie, 315
Ukraine, 263
Ullage, 323
Umbria, **184**
Underberg, 334
Ungstein, 150
United States of America, **270**
  history, 270–1
Unmalted cereal grains, 319, 325
Utiel-Requena, 202

Vaccarèse grape, 120
Vacqueyras, 119
Val de Loire, ACs, 80
Val d'Aosta, 168, **172**
Valais, **247**
Valbuena, 197
Valdeorras, 194
Valdepeñas, **201**
Valençay VDQS, 88
Valencia, 202
Valgella, 172
Valinch, 104
Vallée de la Marne, 74
Valmur, 99
Valpantena, 173
Valpolicella DOC, 173
Valréas, 119
Valtellina, 172
Value Added Tax, 354
Van der Stel, Simon, 288
Van Riebeeck, Jan, 288
Vanilla, 239, 333
Vanillin, 34
Vaslin press, 22
Vat-s, 23
Vaud, **247**
Vaudésirs, 99
VdL, 131
VDN, 120, 131
Vecchio, 166, 182, 187
Vega Sicilia, 197
Vendange Tardive, 91
Vendemmia, 166
Veneto, 168, **173**
Ventoux, Mont, 120
Ventoux AC, Côtes de, 120
Verbesserung, 140
Verdeca grape, 187
Verdejo Blanco grape, 197
Verdelho, grape and wine, 217–18
Verdello grape, 184
Verdicchio del Castelli di Jesi DOC, 183
Verdicchio grape, 183
Verduzzo, 177
Vergine, Marsala, 188
Vermentino di Gallura DOC, 189
Vermentino grape, 133, 172
Vermouth, 128, **239**
  Bianco, 239, 241
  Blonde, 239
  French, dry, **240**
  Italian, sweet, **241**
  manufacture, **240**
  uses, **241**
Vernaccia de San Gimignano DOC, 183
Vernaccia di Oristano, 189
Verona, 173

Vesuvius, mt, 186
Victoria, Queen, 147
Victoria, wine types, 286
Victoria, *284,* **286**
Vidal hybrid, 277
Vienna, region, 251
Vienne, River, 80
Vignes, Jean-Louis, 270
Vila Nova de Gaia, 219, 222
Vila Real, 226
Villafranca del Penedés, 199
Villages:
  Beaujolais, 108
  Côte de Beaune, 103
  Côte de Nuits, 103
  Côtes-du-Rhône, **119**
  Mâcon, 107
Villaudric VDQS, 69
Vin de Blanquette AC, 132
Vin de Consommation Courante (VCC), 45
Vin de Corse AC, 133
Vin de Cuvée, Champagne, 77
Vin de Haut-Poitou VDQS, 85
Vin de l'Orléanais VDQS, 88
Vin de Table, 45
Vin de Thouarsais VDQS, 85
Vin Délimité de Qualité Supérieure (VDQS), 42, 45
Vine, The, **9**
  constituent parts, 9
  life-span, 9
Vin fou, 109
Vinha, 227
Vinhão grape, 226
Vinho de consumo, 227
Vinho de mesa, 227
Vinhos Verdes, **225**
Vini dei Castelli Romani, 185
Vinification, **28**, 54
  red wines, 54
  white wines, 54
  en blanc, 248
Vin Nature de Champagne, 78
Vino Cotto, 188
Vino de tavola, vini de tavole, 164, 186, 188
Vino tipico, vini tipici, 164
Vins de Béarn, 68
Vins de Liqueur (VdL), **131**
Vins de Moselle VDQS, 92
Vins de paille, 109
Vins de Pays, 45
  Départementals, 45, 70, 88, 109, 133
  locales, 81, 122, 126
  Charentais, 45
  de Franche-Comté, 110
  de l'Aude, 45
  de l'Ile de la Beauté, 134
  des Balmes Dauphinoises, 111
  des Coteaux de Grésivaudan, 111
  des Hauts de Badens, 45
  d'Allobrogie, 111
Vins de Savoie AC, 111
Vins des Sables, **130**, 133
Vins Doux Naturels (VDN) AC, 120, **131**
Vins gris, 31, 109
Vins jaunes, 109
Vinsobres, 119
Vintage, 77
  Cognacs, 309
  Madeiras, 219
  Port, 223
Vintners', Worshipful Company of, 261
Viognier grape, 116
Visan, 119
Viseu, 214
Viticulture, **3**
Vitis:
  berlandieri, 3, 269
  labrusca, 3, 179, 269, 310
  riparia, 3, 269
  rupestris, 3, 269
  vinifera, 3, 128, 269
Viura (Macabeo) grape, 196, 199
Vivarais, Côtes de, 119
Vodka, **331**, 337
  flavours, 337
  production, 331
  strength, 331
Vöslau, 250
Vojvodina, **252**
Volatile acidity, **33**
Volcanic soil*, 83, 157, 185–6, 215, 229, 251
Volnay, 104
Volstead Act, 271
Vosges, mts, 89, 312

Vosne-Romanée, 102
Vougeot, Clos-de-, 96
Vougeot, 102
Vouvray AC, 80, 85
Vranc grape, 252
VSOP, 308

Wachau, 250
Wachenheim, 150
Wälschrizling (Olasz-, Laski-, etc.) grape, 249–52, 255
Wagner, 160
Wash, 317
Wash still, 317
Washbacks, 317
Washington State, USA, 277
Watervale, 285
Weathering of wines, 131, 240
Weinbrand, 303
Weissherbst, 157, 158
Wermut, 239
West Germany, **135**
  climate, 135, **136**
  consumption habits, 136
  grape varieties, **137**, *138*
  labelling regulations, *142,* **143**
  QbA regions, *139*
  red wines, 146, 149–50, 158–9
  soil, **136**
  vinification, **140**
  viticulture, **137**
  wine production, 139
  wine regions, **141**
West Indies, 329
Wheat, 319, 325
Whiskey, **325**
  American, **326**
  Corn, 327
  Irish, **325**
  Japanese, 327
  Rye, **326**
Whisky, 315
  blending, 316, **322**
  distillation:
    grain, **319**
    malt, **317**
  distilleries, 322
  duty, 325
  legal definition, **324**
  Malt, districts, 318–19
  maturation, 318, **321**
  production, 322
  stills, 318
  storage, **323**
  strengths, **323**
'White lightning', 327
White Port, 223
White wines:
  fermentation, **31**
  fermentation temperature, 31, 60
  maturation, 34, 60
Wigwam training, 114
Williamette, River, 277
Willmes press, 22
Wind, katabatic, 113, 176
Wind machines, 273
Wind-breaks, 8
Wine:
  constituents of, **32**
  strengths of, 52, 349
Wine, spirit, and liqueur duty bands, **354**
Wine Institute, Israeli, 264
Wine Standards Board, 351
Wine and Spirit Association of Great Britain, 41
Wine and Spirit Education Trust, 261
Winery equipment, **23**
  cleaning, 24
  hygiene, **23**
  materials, 23
Wines, preparation for sale, **35**
Wines of Origin (Cape), 290
Wine-zones, EEC, 26, *27*
Winterhoek, range, 290
Wonnegau, Bereich, 149
Wood Ports, 223
Woodhouse, and Smith, 188
Woodruff, 242
Worcester, 293
Worms, 147
Wormwood, 239, 333
Wort, 317
Wrotham Pinot grape, 234
Württemberg, **158**
Würzburg, 156

Xarel-lo grape, 199
Xynisteri grape, 261
Xynomavro grape, 230

Yarra, 286
Yeasts, **20**
  baker's, 336
  cultured, 31
  decomposition of, 31
  strains, 21
  wild, 23, 54, 60, 77
  wine, 20
Yecla, 202
Yield, control of, 13, **45**, 76, 137, 181
Yolo, 271
Yonge Genever, 336
York Mountain AVA, 275
Yugoslavia, **252**
  production, 252
  regions and grape varieties, 252–4

Zacatecas, 280
Zadar, 254
Zell, Bereich, 155
Zibibbo grape, 189
Zierfandler grape, 250
Zikhron-Ya'aqov, 264
Zilavka Mostar, 254
Zinfandel grape, 275
Zitsa, 230
Zubrowska Vodka, Polish, 337
Zürich, canton, 248